DATE DUE

GAYLORD			PRINTED IN U.S.A.

1939–1989: Fifty Years Progress in Allergy

Chemical Immunology

(formerly 'Progress in Allergy')

Vol. 49

KARGER

Basel · München · Paris · London · New York · New Delhi · Bangkok · Singapore · Tokyo · Sydney

1939–1989: Fifty Years Progress in Allergy

A Tribute to Paul Kallós

Volume Editor
Byron H. Waksman, New York, N.Y.

47 figures and 16 tables, 1990

KARGER

Basel · München · Paris · London · New York · New Delhi · Bangkok · Singapore · Tokyo · Sydney

Chemical Immunology

Formerly published as 'Progress in Allergy'
Founded 1939 by *Paul Kallós*, Helsingborg

Bibliographic Indices
This publication is listed in bibliographic services, including Current Contents® and Index Medicus.

Drug Dosage
The authors and the publisher have exerted every effort to ensure that drug selection and dosage set forth in this text are in accord with current recommendations and practice at the time of publication. However, in view of ongoing research, changes in government regulations, and the constant flow of information relating to drug therapy and drug reactions, the reader is urged to check the package insert for each drug for any change in indications and dosage and for added warnings and precautions. This is particularly important when the recommended agent is a new and/or infrequently employed drug.

ISBN 3–8055–5076–6

Paul Kallós
(1902–1988)

Contents

Anaphylactic Phenomena

From Prausnitz-Küstner to Passive Cutaneous Anaphylaxis and Beyond

Histamine, Mast Cells and Basophils

New Insight into the Mechanisms of Histamine Release from Rat Peritoneal Mast Cells

Infectious Disease and Autoimmunity

Introduction

Waksman BH (ed): 1939–1989: Fifty Years Progress in Allergy.
Chem Immunol. Basel, Karger, 1990, vol 49, pp 1–4

In memoriam Paul Kallós (1902–1988)

Paul Kallós has left a lasting imprint on the history of immunology and allergology. Through his own experimental research, that was nourished by the world of ideas of Ehrlich and von Pirquet, fortified by his personal clinical experience, and through his role as a distinguished and creative editor, convener of scientists and founder of learned societies, he contributed richly to the conceptual and factual foundations of our discipline, he catalyzed novel ideas and research activities, encouraged and supported young scientific talents and, above all, promoted the rapid dissemination of important new biomedical knowledge.

The outer stations of his early private and scientific life reflect the impact of the radical and destructive changes that have transformed the old European order during the first half of this century. Born in Budapest at the peak of its glory in 1902, Kallós graduated from the University of Pressburg Medical School at Budapest and started his career as a promising physician and brilliant young scientist under the constant threat of persecution and eviction.

His clinical and experimental work at the Hungarian Queen Elizabeth Sanatorium led to his first publications on tuberculosis and earned him an invitation by Bessau to continue his research – together with Hans Fernbach – at the University of Leipzig Childrens Hospital. In 1930 he had to move to Nürnberg, where, in 1932, he married Liselotte Deffner, a beautiful young emancipated lady doctor and scientist, who became his inseparable life-time companion and professional alter ego.

Later that year, they were forced to move again, to Orselina and Davos in Switzerland, from where they were evicted in 1934 to find permanent shelter, residence, and eventually citizenship in Sweden.

The 3 years at the laboratories of the Academic Hospital in Uppsala, where they were welcomed by the Nobel laureate Bárány, led to Kallós' definite entry into the realm of clinical and experimental allergology. From 1937, there followed a 7-year research period at the Wenner-Gren Institute of Experimental Biology in Stockholm.

Finally, in 1944, Paul and Liselotte Kallós settled permanently in Helsingborg, dividing their time between clinical activity in a medical practice for allergic diseases of adults and children, experimental research in their own Wihlberg laboratory, which was supported by a variety of private and public grants, and, with increasing impact, editorial work in the rapidly expanding field of allergy and immunology.

Several segments of biomedical research have derived a particular benefit from the personal contributions of Kallós: The immunobiology of tuberculosis, allergy and asthma, and pseudo-allergic reactivity.

His research activities relating to predominantly immunological aspects of tuberculosis span the period of 1925 to 1940 and have been summarized in two major monographs [51, 80].[1]

Kallós emphasized the importance of host and mycobacterial genetic factors for the susceptibility to tuberculosis and the organ localization and the course of the disease. He recognized the relationship between specific anti-tuberculous allergy and immunity, and their mediation by macrophages. Tubercles were shown to display immunological functions, and the importance of macrophage activation in resistance to tuberculosis was elegantly demonstrated. Together with Nathan, he discovered the epicutaneous (eczematous) tuberculin reaction [11, 19–21] and was thus the first to describe delayed-type hypersensitivity to contact allergens. Moreover, he clearly recognized the haptenic nature of tuberculin [51].

Ever since 1934, it was the pathophysiology of allergic diseases sensu strictiori that has provided the focal interest of Kallós' professional life. His most important contribution in this domain is the development and thorough characterization of animal models of experimental asthma. The key paper by Kallós and Pagel [60] demonstrates the striking clinical, histological (eosinophil infiltration), radiological and immunological similarities between guinea pig bronchoconstriction elicited by aerosolized allergen inhalation and human allergic asthma. Moreover, in the same publication the authors also for the first time introduced histamine- and acetylcholine-

[1] References in the text refer to numbers in Kallós D.; Kallós L.: Paul Kallós (1902–1988): an almost complete bibliography, this vol.

aerosol models of guinea pig bronchoconstriction, and highlighted the important histological and pharmacological differences which set them apart from allergen-induced bronchoconstriction and human allergic asthma. Indeed, the authors concluded that the manifestation of human allergic asthma cannot be due to histamine release alone.

Kallós later introduced a number of other animal models of allergen-induced experimental asthma [88, 89, 94], and eventually demonstrated inhibition of guinea pig histamine bronchoconstriction, but not of experimental allergic asthma, by H1 antagonists [99].

In a series of other clinically relevant investigations, Kallós developed quantitative dermatopharmacological methods for the assessment of H1 antagonists in man [100, 106, 109, 112, 136, 160] and made numerous significant contributions towards the therapy of human asthma [51, 56, 66, 99, 104, 108, 109, 112, 120, 134, 135, 149, 157, 160].

Another area of Kallós' wide-ranging research interests was concerned with the nonallergic origin of migraine [123] and with novel pharmacological approaches to migraine therapy [117, 123, 145, 150, 159].

The evidently nonallergic nature of aspirin intolerance in what appeared to be genetically predisposed individuals [124, 163] led Kallós to coin the term pseudoallergic reactivity (PAR) [176]. Pseudoallergic reactions mimicking the signs and symptoms of type I–IV hypersensitivity, but lacking their immunologic specificity, are triggered in predisposed individuals by pharmacologic mechanisms inherent to the eliciting agents. In the meantime, the usefulness of the concept for drug development and monitoring of adverse reactions to drugs and chemicals has been widely accepted. A series of relatively recent papers by Kallós and his colleagues have centered on various aspects of pseudoallergic reactivity [171, 172, 178, 179, 182, 186, 189, 191–193].

However, the crowning achievement of Kallós' lifetime effort has been his extremely successful editorial and organizational activity.

Fortschritte der Allergielehre – Progress in Allergy [196] was founded in 1939 by Kallós to preserve the intellectual tradition of Pirquet's scientific school at a troubled time when his pupils had become dispersed all over the world. Under Kallós' stewardship, the series has earned international recognition for its outstanding reviews and developed into a high-impact medium that managed to distill the essence of scientific advance in applied immunology. Forty-two volumes had appeared by 1988.

The *International Archives of Allergy and Applied Immunology* [197] were founded in 1950 by Kallós, Löffler and Wittich, and blossomed into a respected journal for original full papers that had filled 87 volumes by 1988, again under Kallós meticulous supervision and untiring leadership. This publication was originally conceived as the official journal of the International Association for Allergology, but soon became linked to the Collegium Internationale Allergologicum (CIA), another child of Kallós, spawned in 1952, that has developed into a club of clinical and experimental allergologists of high international repute. Later, Kallós became the founding editor of two further successful review series, *Monographs in Allergy* [198; 25 vol. from 1966 to 1988] and *PAR: Pseudo-Allergic Reactions* [199; 4 vol. from 1980 to 1985].

The contributions and achievements of Kallós have earned him the respect, friendship and affection of a large international community of scientists, to some of whom he had become an almost mystical figure. Quite naturally, numerous honors and awards were bestowed on him: Honorary medical degrees of the Universities of Lund (1969) and Göteborg (1983), an honorary professorship in Medicine, the Pirquet Gold Medal (1950) and the Certificate of Merit of the American College of Allergists (1983), the Culture Award of the City of Helsingborg (1969), the Robert Koch Medal (1974), and the Karl Hansen Medal (1981) and a whole series of honorary memberships of illustrious scientific bodies, academies and societies.

Kallós has led a rich and fulfilled life. But it is almost bewildering to think of the enormous work load that he has carried until the last weeks of his allotted time, with exemplary mental and physical discipline, scholarly thorough, sober, but eagerly interested in anything novel and original, wise, but excited by the cutting edge of expanding knowledge. He was an avid reader and a tireless correspondent, constantly communicating with an international network of contributors and colleagues. But he was also witty, warm and loyal, and a loving husband, father, grandfather and friend. And as he always told those close to him, he could not have achieved what he did and would not have been where he got to, without the untiring all-embracing support of his wife and professional partner, Liselotte Kallós.

Peter Dukor

Waksman BH (ed): 1939–1989: Fifty Years Progress in Allergy.
Chem Immunol. Basel, Karger, 1990, vol 49, pp 5–6

Loss of a Sage

Michael Heidelberger

New York University Medical Center, New York, N.Y., USA

As I have described in greater detail elsewhere [1], Nina, who was our son Charles' mother, and I were in Uppsala in 1934 for about 6 weeks where I was studying the molecular weight of thyroglobulin in The Svedberg's ultracentrifuge with the help of Kai Pedersen. As I needed to lessen attacks of hay fever I was speedily introduced to Dr. Paul Kallos, who with his wife, Liselotte, was working in the Biochemistry Department and gave me injections of grass pollen extract. Both the Kalloses were interesting, friendly and kind, and we visited them down through the years whenever we were in the Scandinavian countries, enjoying their hospitality, conversation, and growing art collection. We also met at microbiological and immunological meetings at which Paul's questions and discussions were real contributions because of his wide knowledge and interests. Accordingly, it was with great sorrow that I received a letter from Liselotte telling of his death.

From among the many happy incidents of our long acquaintance I would like to relate one from those early days in Uppsala. I had been asked by the professor of microbiology to lecture to his students, which I was glad to do. On the way to the lecture hall he asked if I would give the talk in German, as few of his students had learned English. I said I would try, although I had never done so before. At the end of the lecture, Paul and Liselotte came to me in a state of great amusement: in referring to the law of mass action, Massenwirkungsgesetz, I had said 'Messenwirkungsgesetz', Messe, of course being a religious mass. Although I regretted my error, I couldn't help laughing, too, at the unwitting creation of a near pun in a serious lecture.

The loss of Paul is a serious blow to science as well as to Liselotte, to whom I hope this book will bring some degree of consolation.

Reference

1 Heidelberger M: The years at P and S. Ann Rev Biochem 1979;48:1–2/ Immunol Rev Reminiscences 1984;32:7–27.

Prof. Michael Heidelberger, New York University Medical Center,
550 First Avenue, New York, NY 10016 (USA)

Waksman BH (ed): 1939–1989: Fifty Years Progress in Allergy.
Chem Immunol. Basel, Karger, 1990, vol 49, pp 7–20

Paul Kallós (1902–1988): An Almost Complete Bibliography

Daniel Kallós, Liselotte Kallós

We regard this bibliography as 'almost complete'. Some of the omissions made are due to the fact that Paul Kallós did not keep a record of all his publications and that we did not have time enough to track all of his published work. Some of the very early publications – in Hungarian – have for example been impossible to verify at present. They are briefly mentioned in some of the early publications [2, 3, 4, 5, 6, 7, 8]. We have furthermore for reasons of space not included references to the numerous book reviews that Paul published chiefly in *International Archives of Allergy and Applied Immunology*. In the first 15 volumes of that journal he published approx. 160 reviews and he continued his output of reviews and book notices. The interested reader is referred to the indexes of that journal. We have, finally, omitted some of his shorter publications in Swedish where he tried to present his research in a popularized form. We have only included some examples of such work [84, 90, 101, 118, 122, 147, 161, 168].

The bibliography is organized chronologically by years. Full titles of journals are given, since many of them are not published any more. In a few instances full page referrals are not given since we only had access to reprints that did not always include page numbers referring to the printed paper [e.g. 27, 38, 49, 77, 78]. All titles of articles published in Swedish are given in English within brackets. In section II of this bibliography the book series and journals that Paul founded, edited (or coedited) are presented together with a complete list of coeditors.

The bibliography covers more than 60 years of scientific publication. It is our sincere hope that it will encourage researchers today to study the past and to understand – as Paul did – that Progress and not only in Allergy depends on our ability to incorporate new results of research into the existing body of knowledge and in that process reach a new understanding of both past and present.

I

1925

1 Kallós P: Az életvegytan compendiuma (biochémia). Budapest, Novák Rudolph és Társa Kiadasá, 67pp.

1927

2 Németh L, Kallós P: Über die Vitalfärbung der Erythrocyten. Protoplasma 3:11–16.

1928

3 Kentzler J, Kallós P: Bemerkung zu der Arbeit des Herrn Dr. T.D. Kahn: Potentielle Blutalkaleszena und deren Schwankungen bei Lungentuberkulose. Zeitschrift für Tuberkulose 51:478.

1929

4 Bajza E, Kallós P: Experimentelle Beiträge zur künstlichen Immunisierung gegen Tuberkulose. II. Mitteilung. Verwendung des künstlichen Primärherdes. Beiträge zur Klinik der Tuberkulose 71:617–624.
5 Kallós P: Tuberkulose und Konstitution. Zeitschrift für Konstitutionslehre 15: 35–42.
6 Kallós P: Experimentelle Beiträge zur künstlichen Immunisierung gegen Tuberkulose. I. Mitteilung. Theoretische Grundlagen. Beiträge zur Klinik der Tuberkulose 71:604–616.
7 Kentzler J, Kallós P: Bemerkung zu der Arbeit des Herrn Dr. H. Zain: Die Wasserstoffionenkonzentration im Blute tuberkulöser Kaninchen. Zeitschrift für Tuberkulose 54:234–235.

1930

8 Kallós P, Bajza E: Experimentelle Beiträge zur künstlichen Immunisierung gegen Tuberkulose. III. Mitteilung. Hämolysinversuche. Beiträge zur Klinik der Tuberkulose 73:323–324.

1931

9 Kallós P: Thyroxin und Tuberkulinallergie. Klinische Wochenschrift 10:1404–1405.
10 Kallós P: Zur Frage der Tuberkulosebazillämie bei Hauttuberkulösen. Münchener medizinische Wochenschrift 45:1901–1904.
11 Nathan E, Kallós P: Über eine epikutane Tuberkulinreaction bei Hauttuberkulose. Zugleich ein Beitrag zur immunbiologischen Sonderstellung der Haut. Klinische Wochenschrift 10:2392–2396.

1932

12 Kallós P: Über die Züchtung der Tuberkulosebazillen aus dem Blute nach Löwenstein. Zeitschrift für Tuberkulose 64:371–379.
13 Kallós P: Weitere Züchtungsversuche von Tuberkulosebazillen aus dem Blute nach Löwenstein bei Hauttuberkulosen. Dermatologische Wochenschrift 94: 861–863.
14 Kallós P: Beitrag zur Frage der morphologischen Veränderlichkeit des Kochschen Bacillus (Typus humanus) in der Kultur. Beiträge zur Klinik der Tuberkulose 79: 688–690.
15 Kallós P, Kentzler J: Tuberkulose und innere Sekretion. Beiträge zur Klinik der Tuberkulose 79:584–615.

16 Kallós P, Müller W: Über den Einfluss von Thyroxin bzw. Kalilauge auf die Tuberkulinhautreaktion. Klinische Wochenschrift 11:504–506.

17 Kallós P, Nathan E: Über die Einwirkung menschlicher Sera auf humane Tuberkulosebazillen. Zeitschrift für Immunitätsforschung 76:343–354.

18 Nathan E, Kallós P: Zur Frage des Wirkungsmechanismus der kochsalzarmen Diät. Archiv für Dermatologie und Syphilis 164:545–549.

19 Nathan E, Kallós P: Über eine epikutane Tuberkulinreaktion bei Hauttuberkulosen bzw. Tuberkuliden (zugleich ein Beitrag zur immunbiologischen Sonderstellung der Haut bei Hauttuberkulose). Dermatologische Zeitschrift 64:146–162.

20 Nathan E, Kallós P: Experimentelle Erzeugung epidermidaler Tuberkulinempfindlichkeit mittels Ektebin. Klinische Wochenschrift 11:1909–1912.

21 Nathan E, Kallós P: Über Sensibilisierung der Haut gegenüber epicutaner Tuberkulinapplikation mittels Ektebin (nebst Beiträgen zur Kenntnis der Tuberkulin-Epicutanreaktion). Archiv für Dermatologie und Syphilis 167:123–128.

1933

22 Bodmer H, Kallós P: Mast- und Entfettungskuren bei Lungentuberkulösen. Medizinische Klinik 14:1–7.

23 Bodmer H, Kallós P: Ein Fall von Corriganscher Lungenzirrhose mit seltener Ätiologie. Deutsche medizinische Wochenschrift 59:847–851.

24 Bodmer H, Kallós P: Über schwere Lungenschädigung (Lungenzirrhose) infolge Aspiration von Paraffinöl bei therapeutischer Anwendung. Archiv für Ohren-, Nasen- und Kehlkopfheilkunde 136:40–45.

25 Bodmer H, Kallós P: Über schwere Lungenschädigungen infolge Aspiration von Paraffinöl bei therapeutischer Anwendung. Schweizerische Medizinische Wochenschrift 63:618–621.

26 Kallós P: Über die Züchtung von Tuberkulosebazillen aus dem Blute nach Löwenstein. Schweizerische Medizinische Wochenschrift 63:624–637.

27 Kallós P: Beiträge zur Frage der morphologischen Veränderlichkeit des Kochschen Bazillus (Typus humanus) in der Kultur. Giornale di Batteriologia e Immunologia 10(3).

28 Kallós P: Über die Züchtung von Tuberkulosebazillen aus dem Blute nach Löwenstein. Giornale di Batteriologia e Immunologia 10(6):1–21.

29 Kallós P: Beeinflussung des Blutdepots in der Milz durch Thyroxin. Klinische Wochenschrift 12:352.

30 Kallós P, Hoffmann G: Einfache Methode zur Herstellung von Ultrafiltersäckchen aus Kollodium und ein neues Ultrafiltrationsgerät. Biochemische Zeitschrift 266:128–131.

31 Kallós P, Hoffmann G: Über Darstellung und chemische Eigenschaften eines aus der Nährbouillon von Tuberkulosekulturen isolierten biologisch wirksamen Körpers (β-Tuberkulin). Biochemische Zeitschrift 266:132–136.

32 Kallós P, Kallós-Deffner L: Über den Nachweis der Tuberkulosebazillämie in Kultur- und Tierversuch. Kritische Übersicht. Zentralblatt für die gesamte Tuberkuloseforschung 39:145–164.

33 Kallós P, Kallós-Deffner L: Über Allergie und Immunitätsverhältnisse bei Hauttuberkulose. Zentralblatt für Haut- und Geschlechtskrankheiten 46:273–289.

34 Kallós P, Nathan E: Über die Einwirkung von Sera Hauttuberkulöser auf humane

Tuberkulosebazillen in der Tiefenkultur nach Kirchner. Archiv für Dermatologie und Syphilis 167:333–343.
35 Kallós P, Nusselt H: Zur Kenntnis der Thromboangiitis obliterans (Bürger). Klinische Wochenschrift 12:425–427.
36 Nathan E, Kallós P: Serumkrankheit beim Menschen nach experimenteller Sensibilisierung mit Meerschweinchenserum (zugleich ein Beitrag zur Kenntnis der Idiosynkrasie). Archiv für Dermatologie und Syphilis 167:129–135.

1934
37 Kallós P, Kallós-Deffner L: Über den Nachweis der Tuberkulosebazillämie in Kultur- und Tierversuch. II. Kritische Übersicht. Zentralblatt für die gesamte Tuberkuloseforschung 41:145–161.
38 Kallós P, Kallós-Deffner L: Die Wirkung des Ultraviolettlichtes auf den Kohledratstoffwechsel. Strahlentherapie 50.
39 Kallós P, Nathan E: Über die Darstellung und biologischen Eigenschaften des wirksamen Prinzips des Tuberkulins (β-Tuberkulin). I. Mitteilung. Acta medica Scandinavica 83:130–164.
40 Kallós P, Nathan E: Über die Darstellung und biologischen Eigenschaften des wirksamen Prinzips des Tuberkulins (β-Tuberkulin). II. Mitteilung. Sensibilisierungsversuche mit β-Tuberkulin. Acta medica Scandinavica 83:165–168.
41 Kallós P, Nathan E: Über die Darstellung und biologischen Eigenschaften des wirksamen Prinzips des Tuberkulins (β-Tuberkulin). III. Mitteilung. Über Allergie und Immunitätsverhältnisse bei Tuberkulose. Acta medica Scandinavica 83:169–196.
42 Kallós P, Zoboli C: Über die Darstellung und biologischen Eigenschaften des wirksamen Prinzips des Tuberkulins (β-Tuberkulin). IV. Mitteilung. Weitere Sensibilisierungsversuche mit β-Tuberkulin. Acta medica Scandinavica 83:197–211.
43 Kallós P, Zoboli C: Die Beeinflussung der Atmung durch ein- und beidseitigen Oleothorax im Tierexperiment. Zeitschrift für die gesamte experimentelle Medizin 93:7–11.

1935
44 Kallós P: Über Tuberkulosebazillämie. Zeitschrift für ärztliche Fortbildung 32: 11–13.
45 Kallós P: Über Fehlerquellen bei Tuberkelbazillenzüchtungen, insbesondere bei Blutkulturen. Deutsche medizinische Wochenschrift 61:2098.
46 Kallós P: Über das Vorkommen von Eosinophilzellen im Sekret der Nase, der Nebenhöhlen und der Tonsillen. Acta oto-laryngologica Scandinavica 22:107–110.
47 Kallós P: Om forekomsten af eosinophilceller i sekret fra naesen, bihulerne og tonsillerne. [On the occurrence of eosinophil cells in the secretion from the nose, the sinus and the tonsils. In Danish.] Nordisk Medicinsk Tidskrift 9:964–965.
48 Kallós P: Über das Trockenbild des Tuberkulins. Beiträge zur Klinik der Tuberkulose 86:378–380.
49 Kallós P, Kallós-Deffner L: Die Wirkung des Ultraviolettlichts auf den Kohlehydratstoffwechsel. Bemerkungen zu den Ausführungen Rothmanns zu unserer gleichnamigen Arbeit. Strahlentherapie 52.
50 Kallós P, Kallós-Deffner L: Experimentelle Untersuchungen über Salvarsanallergie. Klinische Wochenschrift 14:1074–1076.

51 Kallós P, Kallós-Deffner L: Tuberkuloseallergie. Ergebnisse der Hygiene, Bakteriologie, Immunitätsforschung und experimentellen Therapie 17:76–146.
52 Kallós P, Kallós-Deffner L; Experimentelle Untersuchungen zur Calciumtherapie allergischer Zustände. I. Mitteilung. Klinische Wochenschrift 14:1247–1250.
53 Kallós P, Kallós-Deffner L: Die Bedeutung der chemischen Analyse des Tuberkuloseerregers und des Tuberkulins für die Tuberkuloseforschung. Zentralblatt für die gesamte Tuberkuloseforschung 42:1–25.
54 Kallós P, Kallós-Deffner L: Über die Ausscheidung von Tuberkulosebazillen durch gesunde Organe. Kritische Übersicht. Zentralblatt für die gesamte Tuberkuloseforschung 43:433–444.
55 Kallós P, Kallós-Deffner L: Über die Beeinflussung von Antigen (Pferdeserum) durch Ultraviolettbestrahlung. Klinische Wochenschrift 14:1392f.

1936

56 Kallós P: Den experimentella astmans betydelse för utforskandet av de s.k. idiosynkratiska sjukdomarna. [The importance of experimental asthma for research on so-called idiosyncratic diseases. In Swedish.] Nordisk Medicinsk Tidskrift 12:1268–1277.
57 Kallós P: Ein Fall von Urticaria. Deutsche medizinische Wochenschrift 6:181f.
58 Kallós P, Kallós-Deffner L: Über den Nachweis der Tuberkulosebazillämie in Kultur- und Tierversuch. III. Kritische Übersicht. Zentralblatt für die gesamte Tuberkuloseforschung 45:1–21.

1937

59 Kallós P, Kallós-Deffner L: Über die Beeinflussung von Antigen (Pferdeserum) durch Ultraviolettbestrahlung. Klinische Wochenschrift 16:313f.
60 Kallós P, Pagel W: Experimentelle Untersuchungen über Asthma bronchiale. Acta medica Scandinavica 91:292–305.

1938

61 Ewert B, Kallós P: Elektrokardiographische Untersuchungen im experimentell hervorgerufenen Asthmaanfall. Cardiologia 2:147–169.
62 Kallós P: Experimentelle Beiträge zur Membranhypothese der allergischen Reaktion. Schweizerische Zeitschrift für allgemeine Pathologie und Bakteriologie 1:192–200.
63 Kallós P: Studien über die Organwahl der Tuberkulose. I. Mitteilung. Studia Tuberculosea Pragensia 3:173–189.
64 Kallós P, Kallós-Deffner L: Experimentelle Beiträge zur Calciumtherapie allergischer Zustände. Acta medica Scandinavica 96:519–522.
65 Kalós P, Kallós-Deffner L: Experimentelle Beiträge zur Calciumtherapie allergischer Zustände. Archives Internationales de Pharmacodynamie et de Thérapie 59:253–268.
66 Kallós P, Kallós-Deffner L: Die experimentellen Grundlagen der Erkennung und Behandlung der allergischen Krankheiten. Ergebnisse der Hygiene, Bakteriologie, Immunitätsforschung und experimentellen Therapie 19:178–307.

1939

67 Kallós P: Über nutritive Allergie. Gastroenterologia 64:234–249.
68 Kallós P: Neuere Entwicklung der Bakteriologie und Immunbiologie der Tuberku-

lose. Schweizerische Zeitschrift für allgemeine Pathologie und Bakteriologie 2: 47–64.
69 Kallós P: Vorwort, in Kallós P (Hrsg): Fortschritte der Allergielehre. Basel, S Karger, p 3.
70 Kallós P: Pharmakotherapie der Migräne, in Kallós P (Hrsg): Fortschritte der Allergielehre. Basel, S Karger, pp 393–395.
71 Kallós P, Kallós-Deffner L: Die experimentellen Grundlagen der Allergielehre, in Kallós P (Hrsg): Fortschritte der Allergielehre. Basel, S Karger, pp 5–18.
72 Kallós P, Kallós-Deffner L: Die Therapie der allergischen Krankheiten: C. Spezifische Therapie, in Kallós P (Hrsg): Fortschritte der Allergielehre. Basel, S Karger, pp 382–392.

1940

73 Kallós P: Några synpunkter på immunbiologien med särskild hänsyn till tuberkulosen. [Some comments on immunobiology with particular regard to tuberculosis. In Swedish.] Nordisk Medicin 8:2025–2030.
74 Kallós P, Kallós-Deffner L: Karzinomstudien. I. Experimentelle Beiträge zur Frage der Unempfänglichkeit der Milz für Karzinom. Schweizerische Zeitschrift für allgemeine Pathologie und Bakteriologie 3(1):11–22.
75 Kallós P: Karzinomstudien. II. Stoffwechseluntersuchungen. Schweizerische Zeitschrift für allgemeine Pathologie und Bakteriologie 3(2):75–83.
76 Kallós P, Kallós-Deffner L: Karzinomstudien. III. Die akute 'Milzschwellung'. Schweizerische Zeitschrift für allgemeine Pathologie und Bakteriologie 3(3): 1–8.
77 Kallós P, Kallós-Deffner L: Karzinomstudien. IV. Weitere Beiträge zur Frage der Unempfänglichkeit der Milz für Karzinom. Schweizerische Zeitschrift für allgemeine Pathologie und Bakteriologie 3(3):160.
78 Kallós P, Kallós-Deffner L: Methodischer Beitrag zur Anwendung der Komplementbindungsreaktion zur serologischen Differenzierung von Hefearten. Arkiv för kemi, mineralogi och geologi 14B(5).
79 Kallós P, Kallós-Deffner L: Über die Empfindlichkeit weiblicher Kaninchen für das Impfkarzinom von Brown und Pierce. Arkiv för Zoologi 33B(5).

1941

80 Kallós P: Beiträge zur Immunbiologie der Tuberkulose. Stockholm, AB Hasse W Tullbergs Förlag, 120 pp. [Monographie aus dem 'Wenner-Gren Institut für experimentelle Biologie'].
81 Kallós P, Kallós-Deffner L: Weitere Beiträge über die biologischen Eigenschaften des Tuberkulins und über die Tuberkulin-Allgemeinreaktion. I. Untersuchungen über den anaphylaktischen Shock bei tuberkulösen Tieren. Acta medica Scandinavica 109:115–133.
82 Kallós P, Kallós-Deffner L: Weitere Beiträge über die biologischen Eigenschaften des Tuberkulins und über die Tuberkulin-Allgemeinreaktion. II. Untersuchungen über die Tuberkulin-Hautreaktion. Acta medica Scandinavica 109:574–600.
83 Kallós P, Kallós-Deffner L: Experimentelle Beiträge zur Calciumtherapie allergischer Zustände. II. Mitteilung. Die Entstehung und Verhütung der Pleuritis exsudativa. Archives Internationales de Pharmacodynamie et de Thérapie 65: 249–258.

1942

84 Kallós P: Om kräfta och andra svulstsjukdomar. [On cancer and other tumourous diseases. In Swedish.] Tidens Kalender 22:146–151.

85 Kallós P: Zur Frage der Organspezifizität von Nierenextrakten. Schweizerische Zeitschrift für Pathologie und Bakteriologie 5:119–121.

86 Kallós P: Inlägg vid förhandlingar vid Nordiska Tuberkulosläkarföreningens tolfte möte i Stockholm 1–2 november 1941. [Contribution to the 12th Meeting of the Nordic Society for Tuberculosis. In Swedish.] Nordisk Medicin 15:2576–2577.

87 Kallós P, Kallós-Deffner L: Weitere Beiträge über die biologischen Eigenschaften des Tuberkulins und über die Tuberkulin-Allgemeinreaktion. III. Untersuchungen über die Wirkung des Tuberkulins im Inhalationsversuch. Acta medica Scandinavica 110:493–498.

88 Kallós P, Kallós-Deffner L: Beiträge zum Verständnis des allergischen Bronchialasthmas (Versuche mit Forssman-Antiserum). Schweizerische Zeitschrift für Pathologie und Bakteriologie 5(1–2):97–118.

89 Kallós P, Kallós-Deffner L: Beiträge zum Verständnis des allergischen Bronchialasthmas. Versuche mit Forssman-Antiserum. Schweizerische medizinische Wochenschrift 72:361–365.

1943

90 Kallós, P: Farliga botemedel. [Dangerous drugs. In Swedish.] Tidens Kalender 23:137–142.

91 Kallós P, Kallós-Deffner L: Karzinomstudien. V. Über die Empfänglichkeit männlicher und weiblicher Kaninchen für das Impfkarzinom von Brown und Pierce und über die Lokalisation der Karzinommetastasen bei beiden Geschlechtern. Schweizerische Zeitschrift für Pathologie und Bakteriologie 6(3):180–191.

1944

92 Kallós P: Die sog. 'ablastische Funktion' der Milz. Schweizerische medizinische Wochenschrift 74:192–195.

93 Kallós P: Über die Beeinflussbarkeit des experimentellen Asthmas durch ein Adrenalinpraparat mit Depotwirkung. Acta medica Scandinavica 116:441–446.

94 Kallós P, Kallós-Deffner L: Modellversuche zum Verständnis des allergischen Bronchialasthmas (Versuche mit Ziegen- und Aalserum). Acta medica Scandinavica 116:409–440.

95 Kallós P, Kallós-Deffner L: Über Arzneimittelexantheme. Acta pathologica et microbiologica Scandinavica, Suppl 54:525–529.

1946

96 Kallós P: Immunbiologische Untersuchungen über das Impfkarzinom von Brown und Pierce. Acta medica Scandinavica, Suppl 170:239–251.

97 Kallós P: Über die Histadinbehandlung allergischer Krakheiten. Gastroenterologia 71:231–236.

98 Kallós P, Kallós-Deffner L: Über die Behandlung von urtikariellen Krankheiten mit K-Vitamin. Gastroenterologia 71:171–174.

1947

99 Kallós P: Om 'histaminantagonisternas' betydelse vid behandling av allergiska sjukdomar. [On the importance of histamine antagonists in the treatment of allergic diseases. In Swedish.] Nordisk Medicin 34:1015–1032

100 Kallós P, Deffner-Kallós L: En ny kvantitativ metod för bedömning av hudreaktioner. [A new quantitative method for the determination of skin reactions. In Swedish.] Nordisk Medicin 35:1878–1882.

1948

101 Kallós P: Allergiska sjukdomar. [Allergic diseases. In Swedish.] Tidens Kalender 29: 112–119.

102 Kallós P: Dystrophia unguium mediana canaliformis (Heller). Dermatologica 96: 432–433.

1949

103 Kallós P: Einleitung, in Kallós P (ed): Progress in Allergy – Fortschritte der Allergielehre. Basel, S Karger, vol 2, pp 1–10.

104 Kallós P: Asthma bronchiale. Schweizerische Rundschau für Medizin, No. 24.

105 Kallós P, Kallós-Deffner L: Die klinische Anwendung der Histaminantagonisten, in Kallós P (ed): Progress in Allergy – Fortschritte der Allergielehre. Basel, S Karger, vol 2, pp 329–352.

106 Kallós P, Kallós-Deffner L: Studien über die Antihistaminwirkung an der menschlichen Haut. Acta paediatrica 38:351–358.

1950

107 Kallós P: Some aspects of allergy. [Oration of the Recipient of the von Pirquet Gold Medal of the American College of Allergists.] Annals of Allergy 8:251–254.

108 Kallós P: Editorial. II. ACTH und cortisone. International Archives of Allergy and Applied Immunology 1:334–335.

109 Kallós P, Kallós-Deffner L: Experimentelle und klinische Untersuchungen über die Wirkung des Tephorins. International Archives of Allergy and Applied Immunology 1:189–216.

1951

110 Kallós P: Klinische Umfrage. Für welche Dermatosen sind Fokalinfektionen bedeutsam? Dermatologische Wochenschrift 123:102–104.

111 Kallós P: Editorial: Nahrungsmittelallergie. International Archives of Allergy and Applied Immunology 2:76–82.

112 Kallós P: Standardisierung der Allergene, in: Premier Congrès International, International Association of Allergists. Resumés des Rapports et Communications. Zürich, IAA, pp 109–112.

113 Kallós P, Kallós-Deffner L: Specific serum agglutination of corpuscles (collodium pellets or erythrocytes) coated with allergenic material. International Archives of Allergy and Applied Immunology 2:70–75.

114 Kallós P, Kallós-Deffner L: Betrachtungen über einige aktuelle Allergieprobleme. International Archives of Allergy and Applied Immunology 2:198–206.

1952

115 Kallós P: Introduction, in Kallós P (ed): Progress in Allergy – Fortschritte der Allergielehre. Basel, S Karger, vol 3, pp 1–20.

116 Kallós P: Editorial. Charles Richet. International Archives of Allergy and Applied Immunology 3:1–4.

117 Kallós P, Kallós-Deffner L: The use of ergot derivates in prophylaxis and therapy of migraine, in Kallós P (ed): Progress in Allergy – Fortschritte der Allergielehre. Basel, S Karger, vol 3, pp 485–499.

1953

118 Kallós P: Yrkesallergi – hälsofara för typografer. [Occupational allergy – a health hazard to typographers. In Swedish.] Svensk Typograftidning 443–445.

119 Kallós P: Some immunochemical aspects of allergy. International Archives of Allergy and Applied Immunology 4:291–306.

1954

120 Diamant M, Kallós P: Intrabronchiale Cortisonbehandlung von schweren Asthmafällen. International Archives of Allergy and Applied Immunology 5:283–288.

121 Kallós P: Introduction, in Kallós P (ed): Progress in Allergy. Basel, S. Karger, vol 4, pp 1–30.

1955

122 Kallós P: Allergiska yrkessjukdomar hos bagare. [Allergic occupational diseases in bakers. In Swedish.] Mål och Medel 34:72–73.

123 Kallós P, Kallós-Deffner L: Allergy and migraine. International Archives of Allergy and Applied Immunology 7:367–372.

1956

124 Kallós P: Violent reactions to food. Letters of the International Corresponding Society of Allergists 19:70–72.

125 Kallós P: Penicillin Allergy. Journal of American Medical Association 161:900.

126 Kallós P: Über einige allergische Vorgänge im Auge. Medizinische Probleme der Ophthalmologie 1:346f.

127 Kallós P, Kallós-Deffner L: Allergie und Antibiotika, in Bloch H et al (Hrsg): Antibiotica et Chemotherapia. Basel, S Karger, vol 3, 145–182.

1957

128 Kallós P, Kallós-Deffner L: Durch Arzneimittel bedingte Dyskrasien des Blutes. International Archives of Allergy and Applied Immunology 10:23–31.

129 Kallós P, Kallós-Deffner L: Untersuchungen über die rektale Zuführung wasserlöslicher Theophyllin-Derivate zur Behandlung des Bronchialasthmas. International Archives of Allergy and Applied Immunology 10:129–140.

130 Kallós P, Kallós-Deffner L: Effect of inoculation with *H. pertussis* vaccine on susceptibility of albino mice to 5-hydroxytryptamine (serotonin). International Archives of Allergy and Applied Immunology 11:237–245.

1958

131 Kallós P: Introduction, in Kallós P (ed): Progress in Allergy. Basel, S Karger, vol 5, pp ix-xxxix.

132 Kallós P, Deffner-Kallós L: Anervan mot migrän – terapi och profylax. en studie över ett nytt preparat. [Anervan against migraine – therapy and prophylaxis. A study of a new medication. In Swedish.] Svenska Läkartidningen 55:1782–1788.

133 Kallós P, Deffner-Kallós L: Medihaler ett nytt hjälpmedel för den symtomatiska astmabehandlingen. [Medihaler – a new device for use in the symptomatic treatment of bronchial asthma. In Swedish.] Svenska Läkartidningen 55:3367–3372.

1959

134 Kallós P: Advances in allergy. The Practitioner 183:317–327.

135 Kallós P, Kallós-Deffner L: A new device for use in the symptomatic treatment of bronchial asthma. International Archives of Allergy and Applied Immunology 15: 343–349.

136 Kallós P, Melander B: Experimental studies on the antianaphylactic and antihistaminic properties of Lergopenin. International Archives of Allergy and Applied Immunology 15:308–316.

1960

137 Diamant M, Kallós P, Rubensohn G: Gammaglobulinbehandling och infektionsskydd. [Gammaglobulin treatment and protection against infections. In Swedish.] Svenska Läkartidningen 57:3575–3588.

138 Diamant M, Kallós P, Rubensohn G: Gammaglobulin treatment and protection against infections. Acta oto-laryngologica Scandinavica 53:317–327.

139 Kallós P, Kallós-Deffner L: Untersuchungen über die Wirkung eines appetitreduzierenden Mittels (Tylinal) an Asthmapatienten mit Übergewicht. Nutritio et Dieta 2:229–238.

1961

140 Diamant M, Kallós P, Rubensohn G: Familial agammaglobulinemia. International Archives of Allergy and Applied Immunology 19:193–201.

141 Kallós P: Obituary. Professor Gunnar Dahlberg. International Archives of Allergy and Applied Immunology 19:127–128.

1962

142 Kallós P, Deffner-Kallós L: Behandling av obesitas med dietiska och aptitreducerande medel. [Treatment of obesitas with dietetic and appetite-reducing medication. In Swedish.] Svenska Läkartidningen 59: 3588–3593.

143 Kallós P, Waksman, B H: Introduction, in Kallós P, Waksman B H (ed): Progress in Allergy. Basel, S Karger, vol 6, pp 1–29.

1963

144 Kallós P: Gilt das Unterlassen eines Penicillin-Überempfindlichkeitstests bei Kindern im Krankenhaus als Kunstfehler? Kinderärztliche Praxis 31:262–263.

145 Kallós P: En jämförelse mellan olika former av migränterapi. [Treatment of migraine attacks: a comparative study. In Swedish.] Svenska Läkartidningen 60:3829–3835.

1964

146 Kallós P: Vorwort, in Miller J F A P, Dukor P: Die Biologie des Thymus nach dem heutigen Stande der Forschung. Basel, S Karger, PP v-vi.

147 Kallós P: De allergiska yrkessjukdomarna hos bagare. [Allergic occupational diseases among bakers. In Swedish.] Bröd 5:41–42.

148 Kallós P: Introduction, in Kallós P, Waksman B H (ed): Progress in Allergy. Basel, S Karger, vol 8, pp vi-xvi.

149 Kallós P, Kallós-Deffner L: Comparison of the protective effect of isoproterenol with isoproterenol-phenylephrine aerosols in asthmatics. International Archives of Allergy and Applied Immunology 24:17–26.

150 Kallós P, Kallós-Deffner L: Treatment of migraine attacks: a comparative study. Headache 4:250–254.

1966

151 Kallós P: Introduction, in Kallós P (ed): Fortschritte der Allergielehre – Progress in Allergy; 2nd ed. Basel, S Karger, vol 1, pp v-xxiii.

152 Kallós P, Kallós-Deffner L: Polyangitis in allergic conditions. Acta medica Scandinavica, Suppl 445:444–447.

1967

153 Kallós, P. Introduction, in Kallós P, Waksman B H (ed): Progress in Allergy. Basel, S Karger, vol 11, pp vi-xx.

1968

154 Kallós P: Die Immunsysteme beim Menschen. Verhandlungen der Deutschen Gesellschaft für innere Medizin 74:708–711.

1969

155 Burdzy, K, Kallós P (ed): Pathogenesis and etiology of demyelinating diseases. Basel, S Karger, 1969.

156 Kallós P: Introduction, in Kallós P, Waksman B H (ed): Progress in Allergy. Basel, S Karger, vol 13, pp xi-xxii.

157 Kallós P: Experimentell och klinisk undersökning av en ny histaminantagonist. [An experimental and clinical study of a new histamine antagonist. In Swedish.] Svenska Läkartidningen 66:57–61.

1970

158 Kallós P: Bestämning av antihistamineffekt hos människor. [The determination of antihistamine effects in humans. In Swedish.] Forskning och Praktik 2(5):8–59.

1971

159 Kallós P: Clinical and experimental evaluation of a new ergot derivative (ergostine). Headache 11:68–75.

160 Kallós P: Laboratory and clinical investigations of the antihistamine clemastine (Tavegyl). Clinical Trials Journal 8:23–26.

161 Kallós P: Föredrag vid Svenska Livsmedelarbetareförbundets 11:e kongress 13–18 juni 1971. [Lecture given at the Occasion of the 11th Congress of the Swedish Union of Food Industry Workers. In Swedish.], in Svenska Livsmedelsarbetareförbundet, Kongressprotokoll 1971. Stockholm, Svenska Livsmedelsarbetareförbundet, pp 182–187.

1972

162 Kallós P: Introduction, in Kallós P, Waksman B H, de Weck, A L (ed): Progress in Allergy. Basel, S Karger, vol 13, pp 1–8.

1974

163 Schlumberger H D, Löbbecke E A, Kallós P: Acetylsalicylic acid intolerance. Acta medica Scandinavica 196:451–458.

1975

164 Kallós P: Introduction, in Kallós P, Waksman B H, de Weck A L (ed): Progress in Allergy. Basel, S Karger, vol 19, pp xi-xx.

165 Kallós P: 'Sofortreaktionen' ausgelöst durch Arzneimittel, in Schwick H G (Hrsg): Immunologie und Gesellschaft. Marburg, Die Medizinische Verlagsgesellschaft, pp 149–153.

1976

166 Blands J, Diamant B, Kallós P, Kallós-Deffner L, Löwenstein H: Flour allergy in bakers. I. Identification of allergenic fractions in flour and comparison of diagnostic methods. International Archives of Allergy and Applied Immunology 52:392–406.

167 Kallós P: Beiträge zur Immunbiologie der Tuberkulose. Nach einem Vortrag in Bonn 9.12.1974 anlässlich der Verleihung der Robert-Koch-Medaille. Naturwissenschaften 63:185–189.

168 Kallós P: Föredrag vid Svenska Livsmedelsarbetareförbundets 12:e ordinarie kongress 30 maj-4 juni 1976 [Lecture given at the Occasion of the 12th Congress of the Swedish Union of Food Industry Workers. In Swedish.], in Svenska Livsmedelsarbetareförbundet, Kongressprotokoll 1976. Stockholm, Svenska Livsmedelsarbetareförbundet, pp 248–252.

1977

169 Diamant B, Kallós P: Spörgsmål og svar. Acetylsalicylsyre, asthma, urticaria. [Questions and answers: acetylsalicylic acid, asthma and urticaria. In Danish.] Ugeskrift for Laege 38:2273.

170 Kallós P: Introduction, in Kallós P, Waksman B H, de Weck A L (ed): Progress in Allergy. Basel, S Karger, vol 23, pp xi-xiv.

1978

171 Diamant B, Kallós P: Intolerans mot acetylsalicylsyra – ett icke immunologiskt fenomen. [Acetylsalicylic acid intolerance – a non-immunological phenomenon. In Swedish.] Nordisk Medicin 93:31–32.

172 Kallós P, Schlumberger H D: Letter to the editor. 'Immunogenic impurities' in acetylsalicylic acid. Journal of Pharmacy and Pharmacology 30:67–68.

173 Kallós P, Schlumberger H D: Allergie und allergische Krankheiten. Eine Einführung. Köln, Troponwerke, 200 pp. [Medizin von heute 29.]

174 Kallós P, Westphal O: Michael Heidelberger. International Archives of Allergy and Applied Immunology 57:97–100.

1979

175 Kallós P: Introduction, in Ishizaka K, Kallós P, Waksman B H, de Weck A L (ed): Progress in Allergy. Basel, S Karger, vol 26, pp xi–xvii.

1980

176 Dukor P, Kallós P, Schlumberger H D, West G B: Introduction, in Dukor P et al (ed): PAR – Pseudo-Allergic Reactions. Basel, S Karger, vol 1, pp ix–xiv.

177 Kallós P: 'I dare say the volume was a success...'. Karger Gazette, Suppl. 41:3.

178 Kallós P, Kallós L: Histamine and some other mediators of pseudo-allergic reactions, in Dukor P et al (ed): PAR – Pseudo-Allergic Reactions. Basel, S Karger, vol 1, pp 28–55.

179 Kallós P, Schlumberger, H D: The pathomechanism of acetylsalicylic acid intolerance. A hypothesis. Medical Hypothesis 6:487–490.

1981

180 Diamant B, Kallós P, Kallós L, Löwenstein, H: Isolering av mjölallergen som orsakar bagarastma. [Isolation of flour allergen that causes asthma in bakers. In Swedish.] Arbetarskyddsfondens rapporter, nr 331.

181 Kallós P: Standardisierung von Allergenextrakten. Geschichtliches und Aktuelles. Köln, Troponwerke, 31pp.

1982

182 Gerhardt W, Kallós P, Lundquist, A: Plasma carboxypeptidase activity and intolerance to acetylsalicylic acid. Unimpaired CP-N activity in ASA-intolerant individuals. Intenational Archives of Allergy and Applied Immunology 69:206–209.

183 Kallós, P: Gegenwäriger Stand der Allergielehre, in Allergie-Kolloquien Hollister-Stier 1980. Köln, Troponwerke pp 11–17.

184 Kallós, P: Introduction, in Kallós P (ed): Recent trends in allergen and complement research. Progress in Allergy. Basel, S Karger, vol 30, pp ix–xiv.

185 Kallós P: Introduction, in Kallós P (ed): Immunity and concomitant immunity in infectious diseases. Progress in Allergy. Basel, S Karger, vol 31, pp ix–xv.

186 Kallós P, Kallós, L: Pseudo-allergic reactions due to disodium cromoglycate, in Dukor P et al (ed): PAR – Pseudo-Allergic Reactions. Basel, S Karger, vol 3, pp 122–132.

1983

187 Hanson L-Å, Kallós P: Introduction, in Kallós P, Hanson, L-Å, Westphal O (ed): Host-parasite relationships in gram-negative infections. Progress in Allergy. Basel, S Karger, vol 33, pp 1–8.

188 Kallós P, Schlumberger H D: Allergie und allergische Krankheiten. Eine Einführung. 2., überarbeitete Auflage. Köln, Troponwerke, 209 pp. [Medizin von heute 29.]

189 Kallós P, West G B: Pseudo-allergic reactions in man, in Turner P, Shand D G (ed): Recent Advance in Clinical Pharmacology. Edinburgh, Churchill Livingstone, vol 3, pp 235–252.

1984

190 Kallós P, Kallós L: Experimental asthma in guinea pigs revisited. International Archives of Allergy and Applied Immunology 73:77–85.

1985

191 Kallós P, Schlumberger H D, West G B: Introduction, in Dukor P et al (ed): PAR – Pseudo-Allergic Reactions. Basel, S Karger, vol 4, pp 1–12.

1987

192 Kallós P, Schlumberger H D: Editorial. Pseudo-Allergic Reactions – PAR. International Archives of Allergy and Applied Immunology 82:1–3.

1988

193 Kallós P, West G B: Letter to the editor. Categories of drug allergy. The Lancet 1988 (II) 399.

1989

194 Hansson L-Å, Kallós, P: Allergy – Immunology. Karger Gazette, in press.
195 Kallós P, Schlumberger H D: Allergie und allergische Krankheiten. Eine Einführung. 3., überarbeitete Auflage. Köln, Troponwerke, 1989.

II

196 Fortschritte der Allergielehre – Progress in Allergy. Vol 1 (1939) – Vol 45 (1988).

Founded in 1939 by P Kallós (1939–1988). Continued by B H Waksman (1962ff), A L de Weck (1971–1984), K Ishizaka (1979ff) and P J Lachmann (1986ff). Starting with volume 28 (1981) the editors functioned as 'Series Editors' with (a) special invited editor(s) for each of the succeeding volumes.

197 International Archives of Allergy and Applied Immunology. Vol 1 (1950) – Vol 87 (1988).

Founded in 1950 by P Kallós (1950–1988), W Löffler (1950–1961) and F W Wittich (1950–1954). Continued by E A Brown (1952–1954), D Harley (1952–1968), W Kaufman (1955–1967), R R A Coombs (1955–1984), F Hahn (1961–1972), H C Goodman (1963–1975), G B West (1963ff), F Milgrom (1965ff), P Miescher (1970–1972), Z Trnka (1972–1988), L M Lichtenstein (1976–1982), L Å Hansson (1981ff), R A Reisfeld (1984–1986), K Rother (1984ff), K Kano (1986ff).

198 Monographs in Allergy. Vol 1 (1966) – Vol 26 (1989).

Founded in 1966 by P Kallós (1966–1988), H C Goodman (1966–1969) and T Inderbitzin (1966–1977). Continued by M Hasek (1968–1975), P A Miescher (1969–1974), B H Waksman (1969ff), Z Trnka (1977–1988), A L de Weck (1977–1983), P Dukor (1978ff), L Å Hansson (1987ff), F Shakib (1987ff). Starting with volume 10 (1977) the editors functioned as 'Series Editors' with (a) special invited editor(s) for each of the succeeding volumes.

199 PAR – Pseudo-Allergic Reactions. Vol 1 (1980) – Vol 4 (1985).

Founded in 1980 by P Dukor, P Kallós, H D Schlumberger and G B West.

Professor Daniel Kallós, Pedagogiska institutionen, Umeå universitet, S-901 87 Umeå (Sweden)

Dr. Liselotte Kallós, Villatomtsvägen 12, S-252 34 Helsingborg (Sweden)

Waksman BH (ed): 1939–1989: Fifty Years Progress in Allergy.
Chem Immunol. Basel, Karger, 1990, vol 49, pp 21–34

The First Set of Antigens Confronted by the Emerging Immune System

Susumu Ohno

Beckman Research Institute of the City of Hope, Duarte, Calif., USA

Self-nonself discrimination mechanisms of various sorts have been in existence since time immemorial, for they are seen in certain invertebrates and plants as well as in certain unicellular eukaryotes. Yet, the line of reasoning that sought the origin of the vertebrate adaptive immune system in one of these mechanisms appears to have shot its bolt. As an alternative, I propose that the immune system started with a few B cell clones whose IgM-like antibodies were directed against pathogen-specific small molecules such as phosphocholine and equally pathogen-specific polysaccharide side chains. In order to deal with the ever-changing proteins of viral pathogens, B cells eventually succumbed to the tempation of Prometheus and began to generate numerous antibodies directed against any and all antigens of the universe. The marriage of convenience between B cells and T cells, the latter being specialized in self and nonself discrimination, was consumated at this stage for the necessary suppression of anti-self B cell clones. This late marriage, however, is not functioning very well, and the adaptive immune system is slightly out of control; hence, numerous autoimmune diseases and unpleasant side effects such as allergy.

Today's adaptive immune system operating in mammals and birds appears hopelessly complicated. There are antibody-producing B cells whose clonal expansion depends upon antigen-specific help offered by helper T cells. Yet, B cell antibody and helper T cell receptor apparently prefer to recognize different epitopes on the same antigen [1], and that the recognition by helper T cells of a specific antigen fragment is class II MHC restricted, simply because macrophages present that specific antigenic fragment after the intracellular digestion of antigen in association with class II MHC antigen. In addition, the tolerance of self requires the suppression of B cell clones producing antibodies directed against multitudes of self antigens. It would appear that the symmetry of the system dictates the presence of suppressor T cells that can prevent clonal expansion of anti-self B cells in the antigen-specific manner. Yet, the very existence of suppressor T cells came to be doubted in recent years. Nevertheless, there are at least cytotoxic T cells which, in theory, can carry out the role that should have been assigned to suppressor T cells. But alas! Antigen recognition by cytotoxic T cells is, for the most part, class I MHC restricted. While the expression of class II MHC antigen is largely restricted to antigen-presenting macrophages and antibody-producing B cells, class I MHC antigens are ubiquitously expressed by all somatic cell types. If class I MHC antigen associated with a specific antigen fragment, as does class II MHC antigen [2], the presentation involves the intracellular enzymatic cleavage of antigens most likely occurring inside lysosomes. What is then preventing the presentation of fragments of intracellular proteins to the extracellular world in association with class I MHC antigens? Normally one thinks of the induced tolerance to involve only circulating antigens (e.g. serum albumin and insulin) and plasma membrane antigens, e.g. insulin receptor.

These represent only a small fraction of the proteins encodable by the mammalian genome. Although, only a few percent of the 3×10^9 base pairs of DNA representing the mammalian genome are coding sequences, the total numbers may approach 100,000 [3]. If all these proteins mostly sequestered inside the cell are presented to the extracellular world as digested fragments in association with class I MHC antigen, the tolerance to self indeed becomes a formidable problem. At any rate, such anti-self cytotoxic T cell clones are expected to be eliminated during the self versus nonself education in thymus. It follows then, were cytotoxic T cells to play the role which should have been assigned to suppressor T cells, they should be of a subset whose antigen recognition is class II MHC restricted instead of class I MHC. It is granted that such a subset of cytotoxic T cells would be quite effective in eliminating anti-self B cell clones, while remaining harmless to most other somatic cell types of the self which do not express class II MHC antigen, but whence they can receive the antigen-specific stimulation, for they are liable to kill off relevant antigen-presenting macrophages as well. Thus, the adaptive immune system as it operates in mammals and birds appears to have become unmanageably complicated, or rather it remains not quite manageable in spite of these complications; witness multitudes of autoimmune diseases individuals are apt to suffer.

Redundancy as the Hallmark of Our Unmanageably Complicated Adaptive Immune System

The hallmark of any unmanageable complicated system is redundancy and there is plenty of it in the adaptive immune system. For example, the mammalian genome appears to contain a few hundred V_H genes encoding antigen-binding variable regions of Ig heavy chains, although one ought to expect that one-third to one-quarter of them have by now become degenerate pseudogenes. Nevertheless, a majority of these V_H genes appear readily dispensable. Heavy chain variable regions encoded by a great majority (90% or more) of the V_H genes of the domestic rabbit are marked by allotypes; a_1, a_2 and a_3. The transfer of embryos to the surrogate mother (e.g. a_1/a_1) immunized against the specific allotype (e.g. a_2/a_2) of embryos yield rabbits who are incapable of expressing a majority of the V_H genes. Yet, such suppressed rabbits apparently cope well with multitudes of pathogens present in the environment [4]. Human chromosome 14 carries 9 different Ig constant region genes for IgM, IgD, IgG-3, IgG-1, IgA-1, IgG-2, IgG-4, IgE and IgA-2. Yet, individuals homozygous for three types of extensive deletions of this region remain quite healthy. The first individual found as homozygous deficient for type I deletion was a 75-year-old Tunisian woman who was incapable of producing IgG-1, IgA-1, IgG-2 and IgG-4. In spite of this apparent handicap, she was a mother of 4 daughters and 1 son [5]. Of the four subclasses of IgG, IgG-1 is the major component of serum immunoglobulins, but the one which readily passes the placenta and protects fetuses is IgG-3. Similarly, of the two subclasses of IgA, IgA-1 outnumbers IgA-2 9:1 in serum. Yet, it is IgA-2 which is preferentially secreted into milk (colostrum) to protect newborns. It would thus appear that even for women, IgG-1, IgA-1, IgG-2 and IgG-4 are redundant and quite dispensable. For the well-being of men, even fewer subclasses and classes of immunoglobulins might suffice. More relevant is type III deletion subsequently identified by the extensive survey of European populations. An apparently healthy individual homozygous deficient for type III deletion was incapable of producing IgA-1, IgG-2, IgG-4 and IgE [6]. This begs a question: What is the *raison d'être* of IgE. Perhaps one can dispense with the misery of allergy by eliminating $C_{\varepsilon H}$ locus from our genome.

Why the Unmanageable Complexity?

A priori, there ought to be an inverse correlation between the appropriate error rate of nucleic acid replication and the number of gene loci an organism possesses. The generation time of an organism possessing but a small number of gene loci is very short which permits extremely rapid population expansion. Hence, even if a substantial fraction of individuals sustain deleterious mutations at each generation, their loss is of little consequence. Such an organism, therefore, can afford to be adventurous and utilize a high mutation rate for the successive emergence of new mutants better adapted than before to changing environment. Indeed, reverse transcriptase of retroviruses is extremely error prone; the rate being of the order of 10^{-3}/base pairs/year [7]. Bacteria endowed with a few thousand gene loci can no longer afford extremely error prone DNA polymerase, for the greater the number of gene loci, the higher the possibility that one of them would sustain a deleterous mutation. The error rate of bacterial DNA polymerase is of the order of 10^{-9}/base pairs/generation. The invention of various DNA repair mechanisms no doubt contributed to this relatively high fidelity in bacterial DNA replication. Nevertheless, when the above rate expressed as per generation is translated to per year rate, it is likely to become at least 10^{-7}/base pairs/year, since 100 generations per year is a considerable underestimate if anything, for bacteria.

As for mammals, the total number of gene loci in our genome approaches 100,000 as already noted. The unbridled error rate, as estimated on defunct pseudogenes, is 3×10^{-9}/base pairs/year for mammals. At this high fidelity of DNA replication, even if all of the mutations sustained are permitted to accumulate, it will take one million years for proteins to undergo 1% amino acid sequence change. Herein lies the cause for the unmanageable complexity of the adaptive system.

A difference in the rate of evolutionary changes between retroviruses and mammals is one million-fold, while that between bacteria and mammals is still a formidable 100-fold. In order for the mammalian adaptive immune system to half successfully cope with these ever-changing viral and bacterial pathogens, it has to generate an innumerable variety of antigen-combining sites. Multitudes of V_L and V_H genes of the order of 10^2 each and multitudes of D_H and five or so each of J_L and J_H no doubt contribute to the diversity of antibodies the mammalian adaptive immune system is capable of generating. However, more important is the extremely high error rate of DNA replication that variable regions (more precisely $V_L + J_L$ and $V_H + D_H + J_H$) of rearranged immunoglobulin genes undergo during the antigen-stimulated clonal expansion of B cells. This error rate is thought to be of the order of 10^{-3}/base pairs/cell generation [8]. It follows then that after completing productive rearrangements, antigen-combining regions of immunoglobulins undergo evolutionary changes even faster than proteins of retroviruses. It would be recalled that the 10^{-3} error rate for retroviruses was per year, while the same rate for rearranged variable regions is per cell generation. Diverse antigen-combining sites of immunoglobulins generated at this rate can indeed cope not only with all antigens that exist in the present world but also with antigens that might come into being in the future. The very fact that specific antibodies can almost invariably be generated against all sorts of odd chemicals of human concoction used as haptens attests to the adaptive immune system's capability to anticipate the future. By anticipating the future, however, the core of the adaptive immune system was placed above and beyond the reach of Darwinian evolution by natural selection. Think of an epidemic caused by a new lethal mutant of the influenza virus. Conventional wisdom of neo-Darwinists would have predicted a very profound alteration in composition of V_L, V_H and D_H genes of a population after a recovery from such a devastating epidemic, for survivors are expected to be those chosen few who managed to generate sufficiently effective antibodies directed against that mutant antigen. But alas! By anticipating the future, a great majority of individuals were already endowed with the capacity to generate specific antimutant antibodies. If such an epidemic excercised Darwinian selection, the subject was likely to be polymorphic class II MHC antigen loci.

In one of the several versions of creation in Greek mythology, the task of populating the just formed earth with living creatures was assigned by Zeus to the brothers, Prometheus and Epimetheus. Unwittingly, Prometheus endowed with foresight fell asleep, thus, most living creatures were made by the hands of Epimetheus. As he was endowed only with hindsight, he nearly exhausted a preciously small number of available attributes on other animals, e.g. giving strength to the elephant, beauty to the horse and majesty to the lion. Accordingly, when time came to create man, there were so few attributes left that man was made naked and weak. At this point, Prometheus awoke and took pity on man and gave him intelligence which would culminate in eventual mastery of fire as an instrument to forge many tools. The profoundness of this version lies in the nature of Zeus's judgement passed on these two brothers. After the completion of the task, Zeus severely punished not Epimetheus, but Prometheus, for the audacity of his foresight.

By acquiring the ability to generate specific antibodies directed against any and all antigens of the past, present and future, the adaptive immune system too, is forced to pay dearly for the audacity of foresight. Since mammalian DNA replication is of such a high fidelity as already noted, self as such is relatively immutable. Were the generation of antibody diversity to exclusively rely upon rearrangements of V_L and J_L as well as V_H, D_H and J_H genes in the germ line, it would have been relatively easy to selectively eliminate most of those genes that can generate autoreactive antibodies. But, since the diversity is also generated by extremely high somatic mutations affecting variable regions after the successful rearrangement, mammals and birds have to live with the ever-present danger of developing autoimmune disease. The very elaborate and complex T cell system developed for suppression of autoreactive B cell clones is not as effective as its complexity suggests. Furthermore, most of the antibodies generated are bound to be directed against antigens of the past as well as against those that might come into being in the future. Thus, they are of no use to the well-being of individuals in their lifetime. It follows that the reasonable production of specific antibodies demanded by each occasion has to depend upon the antigen-dependent clonal expansion of relevant B cells. For this antigen-dependent clonal expansion, another equally elaborate and complex T cell system had to evolve, but its efficacy came to depend upon polymorphic class II MHC antigens.

What B Cells See versus What T Cells See

Circumsporozoite antigens of parasitic malaria protozoa are rather interesting proteins, their central domain is comprised of near exact repeats of specific oligopeptides, whereas the amino- and carboxyl-terminals are more ordinary. When animals were immunized against whole sporozoites purified from mosquito salivary glands, they generated antibodies preferentially directed against oligopeptide repeats in the central domain of circumsporozoite antigens [9]. In the case of human parasite, *Plasmodium falciparum*, this central domain is tetrapeptidic repeats: $(\text{Asn-Ala-Asn-Pro})_n$. However, when mice were immunized with a synthetic $(\text{Asn-Ala-Asn-PRO})_3$ oligopeptide, only mice of $H\text{-}2^b$ MHC haplotype (e.g. C57BL/6 strain) responded, while those of other haplotypes were nonresponders. Thus, it became evident that only when $(\text{Asn-Ala-Asn-Pro})_3$ peptide fragment is presented together with class II MHC antigens of $H\text{-}2^b$ haplotype by antigen-presenting macrophages, were there T cells equipped with specific receptors to recognize this complex. Class II MHC antigens of other H-2 haplotypes were either incapable of associating with (Asn-Ala-Asn-Pro)$_3$ or there were no T cell receptors capable of recognizing such complexes. Without the antigen-specific T cell help, clonal expansion of anti-(Asn-Ala-Asn-Pro)$_3$ B cell clones did not take place. The synthetic 21-residue-long peptide representing a part of the carboxyl-terminal domain of *P. falciparum* circumsporozoite antigen, on the other hand, was capable of eliciting the in vitro helper response from human T cells of diverse HLA haplotypes [1]. Similarly, when (Ans-Ala-Asn-Pro)$_3$ was conjugated with the aforementioned 21-residue-long fragment (Asp-Ile-Glu-Lys-Lys-Ile-Ala-Lys-Met-Glu-Lys-Ala-Ser-Ser-Val-Phe-Asn-Val-Val-Asn-Ser) and the resulting conjugate was used as an immunogene, mice of all 7 different strains with divergent H-2 haplotypes became responders [1]. It would appear that either class II MHC antigens have the uniform affinity to associate with α-helical fragments, particularly if they are amphipathic (hydrophobic residues clustered on one side of the cylinder, while charged residues being clustered on the opposite side), or T cell receptors have a definite preference for the above-noted type of complexes. Herein lies an inherent awkwardness in the cooperation between antibody-producing B cells and helper T cells whose assigned task is the antigen-specific expansion of particular B cell clones. Since many proteins lack α-helical structure, let alone the amphipathic kind, a sizeable fraction of a population is expected to be nonresponders against each of these protein antigens. These individuals are nonresponders, not because they are incapable of generating specific antibodies directed against that protein antigen, but because their class II MHC alleles are incapable of eliciting helper T cell response to that antigen.

As to equally important suppression of autoreactive B cell clones, the now-doubted existence of antigen-specific suppressor T cells casts an ominous shadow. If this task has been assigned to cytotoxic T cells, it creates more problems than solutions, as already noted in the introduction. First of all, they would almost have to be an unusual subclass of cytotoxic T cells whose antigen recognition is class II MHC-restricted instead of class I. If that is the case, one expects the same hole, as discussed in connection with the helper T cell repertoire, to exist in the antigen-specific suppression of autoreactive B cell clones. Exposed parts of many of the plasma membrane proteins of the self are made mostly of β-pleated sheet segments lacking α-helix.

The above consideration on inherent awkwardness in the necessary cooperation between free-wheeling B cells and MHC-restricted T cells led me to believe that MHC antigens, therefore T cells, were not the original components of the primordial immune system, but rather they were late recruits borrowed from certain other preexisting biological systems. Only when the repertoire of B cell antibodies became so enormous, that specific antibodies could be generated against any and all antigens of the universe, did this recruitment become necessary to give the immune system an adaptive character. Let me expand on this line of reasoning.

Late Recruitment of the Self-versus-Nonself Discrimination Mechanism to the Immune System

There is little doubt as to the antiquity of the self-versus-nonself discrimination mechanism. In the life of unicellular eukaryotes, the sporadic insertion of sexual reproduction into otherwise vegetative propagation would already have necessitated the discrimination of those who are still engaged in vegetative growth from those that were either prepared to go into meiosis or are already undergoing meiosis, for conjugation should have taken place only among the latter. In higher plants, self-fertilization is often avoided via polymorphic proteins of pollen and stigma. In corals, the formation of mosaic colonies is avoided by this self-versus-nonself discrimination mechanism, and the same is practiced by certain members of the most primitive of *Chordata,* colony-forming tunicates.

Even during ontogeny of each multicellular individual, organogenesis begins with sequestering of like cells from the rest. This recognition among like cells is dependent upon the class of plasma membrane proteins variously known as CAM (cell adhesion molecule) proteins or cadherins, e.g. one neuronal cell recognizes other neuronal cells via N-CAM protein to organize the brain and spinal cord from the neural tube [10, 11]. There is little doubt that the so-called β_2-microglobulin-like domain which is a building block of nearly all the components of the adaptive immune system arose in these CAM proteins [12]. Shown in figure 1 are the 9th-to-118th residues of CsA CAM of the slime mold *(Dictyostelium discoideum)* arranged in a β_2-microglobulin-like fold. It should be noted that two amino acid substitutions, one replacing 51st Cys (codon T G T or T G C) with Trp (codon T G G) and the other replacing 93rd Phe (T T T or T T C) with Cys, would have changed it to the classical β_2-microgobulin-like domain. The life-style of slime mold

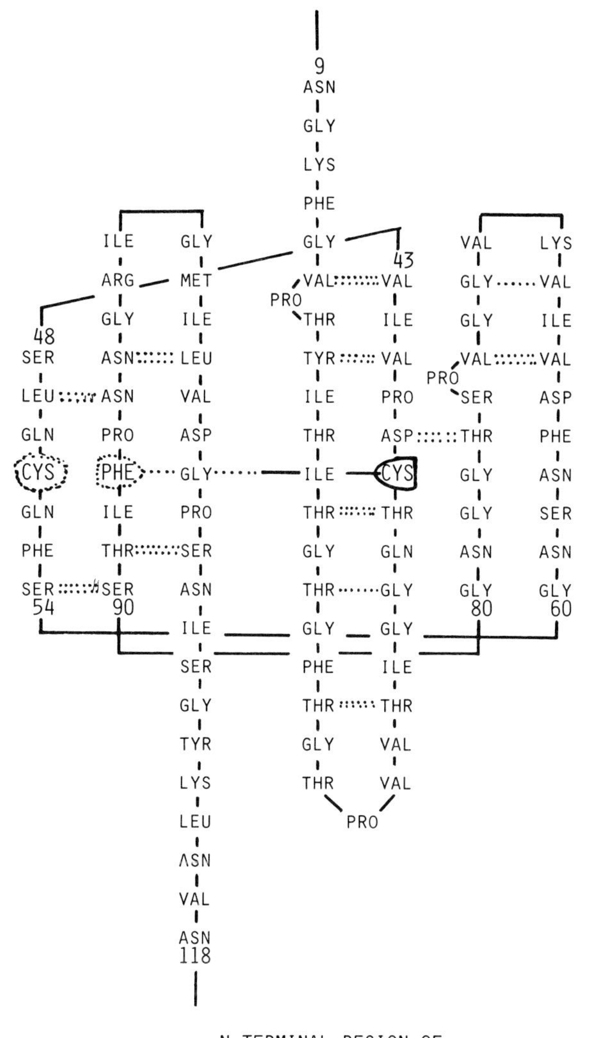

N-TERMINAL REGION OF

CsA CAM OF THE SLIME MOLD *DICTYOSTELIUM DISCOIDEUM*

Fig. 1. The amino-terminal region of Csa CAM protein of the slime mold [13] is arranged in the β_2-microglobulin-like fold [12]. 9th to 118th positions are rich in Gly, Val Ile, Phe and other hydrophobic residues. Yet, the region also contains 8 each of Ser and Thr residues. These are hallmarks of polypeptide chains that form a series of β-pleated sheet strands. There are also a pair of cysteines separated by 14 residues. The distance between two cysteines is too short. Yet, if 51st Cys is replaced by Trp by a single base substitution and 87th Phe by Cys also be another single base substitution, the disulfide bridge formed between 38th Cys and 87th Cys would make this segment a very typical β_2-microglobulin-like domain.

borders between unicellular and multicellular. When the environment is favorable, they live as ameboid unicellular organisms. When the environment becomes harsher, they aggregate to form a miniature mushroom-like fruiting body, thus engaging in the primordial organogenesis. This aggregation is the task assigned to CsA CAM [13].

What were the tasks assigned to the ancestor of class I and class II MHC antigens before the emergence of the adaptive immune system? The task in ontogeny assigned to various CAM proteins dictates that they should remain monomorphic and conserved. Very extensive polymorphism at both class I and class II MHC loci in nearly all mammalian and avian species may indicate that they were redundant members of the CAM family not performing any critical function. Certainly, MHC differences do not interfere with organogenesis, since two mouse blastocysts of different H-2 haplotypes can be fused to form one very healthy mouse. It may be this very redundancy which permitted later recruitment of class I and class II MHC antigens to the adaptive immune system, and their extensive polymorphisms are a mere consequence of this original redundancy.

The Original Immune System Might Have Started with a Few Very Useful Antibodies

The defense against infectious pathogens of invertebrates primarily depends upon phagocytic activities of macrophage-like and granulocyte-like wandering cells. Assisting the above are lectin-like molecules that aggregate bacteria as well as small proteins with broad bactericidal activities. I can think of only one role which might have been played by MHC antigens in this rather indiscriminate defense mechanism against infectious pathogens. They might have served as cell-surface markers which enabled phagocytes to distinguish healthy cells from infected dying cells. If the half-life of MHC antigens was reasonably short, dying cells might have expressed less MHC antigens compared to healthy cells. Conversely, MHC antigens of cells massively infected by viruses might have been less conspicuous because of their association with fragments of viral proteins. If the latter was the case, the ultimate ancestor of T cell receptors might be sought on invertebrate macrophages.

On this background, how was the antigen-specific antibody response beneficially imposed? Here, it should be pointed out that the production of a very small variety of specific antibodies, if directed against particular antigens, is by itself extremely beneficial to the possessor. One antibody immediately comes to mind; antiphosphocholine antibody. While phosphocholine is an integral component of all plasma membranes, in the case of vertebrates, phosphocholine is sequestered inward being not exposed to the surface while it protrudes outward in the bacterial plasma membrane and cell wall. Therefore, an antiphosphocholine antibody is an almost ubiquitous bactericidal element while remaining harmless to the host. Indeed, the immune response of BALB/c mice to phosphocholine is almost monoclonal, most of the antiphosphocholine antibodies bearing T15 idiotype. This T15 idiotype was attributed to one V_H gene (S107) in the genome [14]. Interestingly, in the primitive shark *(Heterodontus francisci)*, each V_H gene is accompanied by its own J_H and $C_{\mu H}$ and presumably also D_H genes [15]. This cassette arrangement of immunoglobulin genes in the primitive shark may indeed indicate that the immune response originally started with few universally beneficial antibodies such as that directed against phosphocholine. Aside from phosphocholine, there might be few other small molecules that were and still are present on the outer surface of infectious pathogens but not on that of vertebrate cells. Antibodies against these small molecules would have been immediately beneficial, for there is no risk that they would turn against the self. Antibodies are endowed with the inherent capacity to discriminate small molecules with little structural differences when used as haptens, e.g. dinitrophenyl versus trinitrophenyl. This inherent capacity might hark back to the original set of a small number of primordial antibodies. If one follows the above line of reasoning, antibodies that were next in line of succession would have been those directed against sugar residues of glycoproteins and glycolipids. The kinds of polysaccharide side chains an organism is capable of generating are solely determined by gene loci in the genome that encodes the fixed variety of sugar tranferases. It follows then that there must necessarily be antigenic determinants unique to polysaccharide side chains of infectious pathogens, therefore not to be found on the host cells. Monoclonal antibodies secreted by multiple myelomas are quite often directed against polysaccharides, e.g. β-1, 6-*d*-galactan, β-N-acetyl-*d*-glucosamine [16]. It looks as though a considerable portion of the antibody repertoire of mammals even today is directed against polysaccharide side chains. The early possession of such antipolysaccharide antibodies posed no threat to the host. At this date, no one seems to be certain as to whether or not clonal expansion of antipolysaccharide B cells requires T cell help. Nevertheless, a certain polysaccharide (LPS) is a polyclonal B cell mitogen, and probably because of this, the immune response to certain polysaccharides are T cell independent [17]. Such T cell independence may attest to the antiquity of some of the antipolysaccharide antibodies. The immune system might indeed have started with a small variety of B cells that produced antibodies directed against pathogen-specific small molecules (e.g. phosphocholine) as well as polysaccharide side chains. In this line of reasoning, B cells evolved before T cells. On the contrary, if one assumes the self-versus-nonself discrimination mechanism as the ancestor of the adaptive immune system, T cells must necessarily have preceded the emergence of B cells.

The above-noted antipolysaccharide strategy does not function well against viruses, for polysaccharide side chains of viral proteins are assembled not by the viral genome but by sugar transferases of the host. Thus, the emerging immune system sooner or later must have begun to produce antipeptide antibodies, and, by so doing, encountered an enormous problem. On the one hand, keeping up with a nearly one million-fold higher mutation rate of pathogenic viruses, and on the other hand avoiding self destruction by anti-self antibodies. In order to keep up with viral mutation rates, the somatic mutation rate of the order of 10^{-3}/base pairs/cell generation affects variable regions of productively rearranged Ig light and heavy chain loci, as already noted. With this audacity, the adaptive immune system gained the capacity to cope not only with antigens of the past and the present but also with antigens that might come into being in the future. The price for this audacity was constant generation of anti-self antibodies that have to be suppressed by one means or the other. It would be recalled that a single somatic mutation that changed 35th Glu to Ala affecting the aforementioned S107 V_H of BALB/c mice changed a very useful antiphosphocholine antibody of T15 idiotype to self-destructive anti-DNA antibody [18].

The Generally Poor Strategy of Suppressing Anti-Self B Cell Clones

It is the current dogma that T cells undergo self-nonself education in the thymus. For this strategy to work, thymic epithelia ultimately derived from one or two of the ancient gill arches in phylogenic as well as ontogenic sense have to be endowed with the truely mystical property of being able to express every one of the proteins encodable by upward of 100,000 genes in the genome. Be as it may, these proteins are bound to be presented as digested fragments in association with either class I or class II MHC antigens. This invariably creates various holes in the self-nonself education system, for each individual is bound to be bestowed with the nonresponder status with regard to a number of self proteins. These holes are because of extensive allelic polymorphism at the MHC loci. The problem is multiplied by the apparent absence of suppressor T cells, and cytotoxic T cells disguised as suppressors are inherently incapable of playing the role very well for the reasons already noted.

As of today, it would thus appear that the adaptive immune system has no mechanism of actively suppressing antiself B cell clones. One might argue that the apparent suppression is due to the removal of antiself helper T cells in thymus. Yet, antiself B cell cones must invariably contain those directed against polysaccharide side chains of the self. If polysaccharides in general are T cell-independent antigens, as they appear to be, the thymic removal of helper T cells is of no consequence in these instances.

Predictions

From the foregoing discussions, I conclude that the cooperation between T and B cells as the two operate within the framework of the modern adaptive immune system is awkward and haphazard at best. Such awkwardness implies, at least to my mind, the marriage of convenience consumated between B and T cells at a relatively late stage in development of the adaptive immune system. Therefore, I propose that the immune system, as it operates in vertebrates, started with a small number of B cell clones whose IgM-like antibodies were directed against pathogen-specific small molecules such as phosphocholine and equally pathogen-specific polysaccharide side chains. The origin of T cells engaged in discrimination should be sought elsewhere, either in the rejection mechanism of the colony-forming tunicates or in macrophages and other phagocytes of invertebrates who had to distinguish diseased, dying cells from healthy cells of the self. The marriage of convenience between B and T cells was consumated when B cells succumbed to the temptation of Prometheus and began to generate numerous antibodies directed against any and all antigens of the universe.

Genes for antibodies of elasmobranches have already been analyzed in detail by Litman's group [15]. What remain are hagfish and lampreys of the cycoltomata. I predict that the immune system of these jawless fish would be comprised solely of a small number of B cell clones whose antibodies are directed against pathogen-specific small molecules and polysaccharide side chains.

References

1 Sinigaglia F, Guttinger M, Kilgus J, et al: A malaria T-cell epitope recognized in association with most mouse and human MHC class II molecules. Nature 1988; 336:778.
2 Bjorkman P, Saper M, Samraoui B, et al: The foreign antigen binding site and T cell recognition regions of class I histocompatibility antigens. Nature 1987;329:512.
3 Ohno S: The total number of genes in the mammalian genome. Trends Genet 1986;2:8.
4 Vice J, Gilman-Sachs A, Hunt W, et al: Allotype suppression in a^2a^2 homozygous rabbits fostered in uteri of a^2-immunized a^1a^1 homozygous mothers and injected at birth with anti-a^2 antiserum. J Immunol 1970,104:550.
5 Lefranc G, Chaabani H, van Loghem E, et al: Simultaneous absence of the human IgG1, IgG2, IgG4 and IgA1 subclasses immunological and immunogenetical considerations. Eur J Immunol 1983;13:240.
6 Migone N, Oliviero S, De Lange G, et al: Multiple gene deletions within the human immunoglobulin heavy-chain cluster. Proc Natn Acad Sci USA 1984;81:5811.
7 Gojobori T, Yokoyama S: Rates of evolution of the retroviral oncogene of Moloney murine sarcoma virus and of its cellular homologues. Proc Natn Acad Sci USA 1985; 82:4198.
8 Wable M, Burrows P, von Gabin A, et al: Hypermutation at the immunoglobulin heavy chain locus in a pre-B-cell. Proc Natn Acad Sci USA 1985;82:479.
9 Zavala F, Cochrane A, Nardin E, et al: Circumsporozoite proteins of malaria parasites contains a single immunodominant region with two or more identical epitopes. J Exp Med 1983;157:1947.
10 Hemperly J, Murray B, Edelman G, et al: Sequence of a cDNA clone encoding the polysialic acid-rich and cytoplasmic domains of the neural cell adhesion molecule N-CAM. Proc Natn Acad Sci USA 1986;83:3037.
11 Takeichi M: Cadherins. A molecular family essential for selective cell-cell adhesion and animal morphogenesis. Trends Genet 1987;3:213.
12 Ohno S: The ancestor of the adaptive immune system was the CAM system for organogenesis. Exp Clin Immunogenet 1987;4:181.
13 Noegel A, Gerisch G, Stadler J, et al: Complete sequence and transcript regulation of a cell adhesion protein from aggregating *Dictyostelium* cells. EMBO J 1986;5:1473.
14 Crews S, Griffin J, Huang H, et al: A single V_H gene segment encodes the immune response to phosphocholine: somatic mutation is correlated with the class of the antibody. Cell 1981;25:59.
15 Litman G, Hinds K, Berger L, et al: Structure and organization of immunoglobulin V_H genes in *Heterodontus*, a phylogenetically primitive shark. Dev Comp Immunol 1985;9:749.
16 Kabat E, Wu T, Bilofsky H, et al: Sequences of Proteins of Immunologial Interest; in US Department of Health and Human Services. Bethesda, National Institute of Health, 1983.
17 Coutinho A, Miller A: Thymus-independent B cell induction and paralysis. Adv Immunol 1975;21:114.
18 Diamond B, Scharff M: Somatic mutation of T-5 heavy chain gives rise to an antibody with autoantibody specificity. Proc Natn Acad Sci USA 1984;81:5841.

Susumu Ohno, MD, Beckman Research Institute of the City of Hope,
1450 East Duarte Road, Duarte, CA 91010–0269 (USA)

Waksman BH (ed): 1939–1989: Fifty Years Progress in Allergy.
Chem Immunol. Basel, Karger, 1990, vol 49, pp 35–50

Origin and Function of *Mhc* Polymorphism

Jan Klein[a,b], *Masanori Kasahara*[b], *Jutta Gutknecht*[a], *Felipe Figueroa*[a,1]

[a]Max-Planck-Institut für Biologie, Abteilung Immungenetik, Tübingen, BRD;
[b]Department of Microbiology and Immunology, University of Miami School of Medicine, Miami, Fla., USA

Prologue at Pierre à la Fontaine Gaillon

At a dinner table, the conversation turned to the future of *Mhc* studies. 'Sequencing of *Mhc* alleles is boring', a noted molecular biologist asserted emphatically. 'Only someone with no inspiration or with nothing better to do would embark on such a pedestrian project. Regulation of gene expression, transgenic mice, site-directed mutagenesis, exon-shuffling ... that's where the action and the excitement is.'

It is amazing how we scientists differ in our perception of what is important in our fields of study. We would argue just the opposite of what the molecular biologist pontificated. Regulation can be studied on any one of the thousands of genetic systems; it is not the expressed domain of *Mhc* research. And as for transgenic mice, site-directed mutagenesis, exon shuffling, etc., they are mere tools, methods, rather than topics of a study. The true outstanding domains of *Mhc* research are, in our opinion, polymorphism, function, and evolution. Polymorphism because it is without parallel, function because discrimination between self and nonself is one of the most basic characteristics of life, and evolution because its study might be the only way of making sense of the system. Gene sequencing, as unostentatious as it may be, is essential to all three of these key topics of *Mhc* studies. Sequencing of all *HLA* alleles, for example, should become a joint project in which many laboratories ought to participate in a planned, noncompetitive fashion. Although each individual sequence may not offer any new insights, the

[1] We thank Ms. Lynne Yakes for editorial assistance. The experimental work described in this communication was supported in part by grant No. RO1 AI23667 from the National Institutes of Health, Bethesda, Md.

impact of the overall project should be considerable. In the end, it might bring us closer to answering the ultimate question: What is the *Mhc* for?

In paying tribute to the *Progress of Allergy* on the occasion of its golden anniversary, and in particular to Prof. Paul Kallos, the founder of the series, we have therefore chosen the *Mhc* polymorphism and its function as the topic of our contribution to this anniversary volume. We are saddened by the news that Prof. Kallos has not lived to see the publication of this volume which he obviously took such a pleasure to plan. In his memory, we have prepared our contribution as he requested it – with a touch of philosophy, concentrating on a few selected topics, rather than reviewing the entire subject.

Order from Chaos?

The true nature of scientific inquiry is to seek order out of chaos (only most recently has chaos itself become the subject of study). A group of things is taken by a scientist as a challenge to organize it into a system in which the relationship of one thing to all other things in the group becomes known. Biologists often summarize their inquiries into who is related to whom by drawing trees in which close relatives are twigs on the same branch and more distant relatives are on different branches. The phylogenetic tree is then presumed to reflect the evolution (phylogeny) of the system, evolution in which less related things are presumed to have separated from a common ancestor a longer time ago than more related things [1].

The major histocompatibility complex *(Mhc)* alleles, of which there may be hundreds at each functional locus [2], constitute such groups begging to be lifted from chaos and organized into an ordered system. Curiously enough, immunologists have resisted the temptation to draw phylogenetic trees of *Mhc* alleles and even loci, arguing that such an effort would be futile because the genes have been reshuffled so much that true relationships among them have been obscured [3, 4, 38]. Refusing to accept this argument, we have maintained all along that the reshuffling idea has been exaggerated and that in reality is should be possible to draw realistic phylogenetic trees of *Mhc* alleles once enough sequence information has become available [5]. In the human *Mhc*, the *HLA* complex, a modest amount of sequence information has been generated for alleles at the *HLA-A, HLA-B, DP, DQ,* and *DR* loci. We have therefore collated this information and used it to construct phylogenetic trees of alleles at these loci. Of the many tree-construction methods available, we

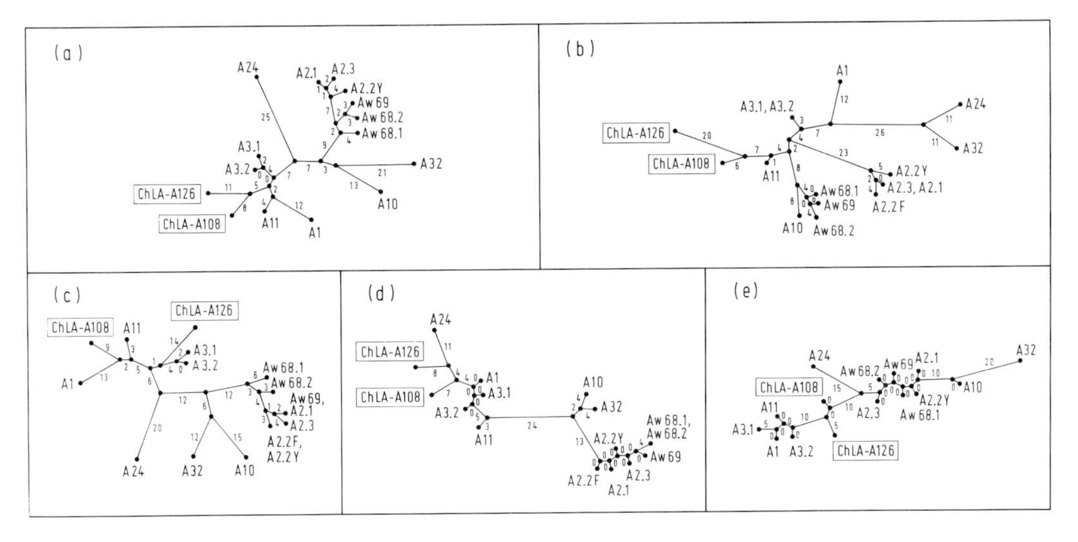

Fig. 1. Phylogenetic trees of human *HLA-A* and chimpanzee *ChLA-A* (boxed) alleles (coding sequences). *a* Entire gene. *b* α1 domain-encoding exon. *c* α2 domain-encoding exon. *d* α3 domain-encoding exon. *e* Exons encoding for the transmembrane and cytoplasmic regions.

have selected that of Saitou and Nei [6], which we believe suits our purpose best. We have constructed trees separately for the entire genes and for their main exons. Had there been exon shuffling between alleles during their evolution, the trees constructed separately for the individual exons should have revealed it. Trees showing the relationships among *HLA-A* alleles are depicted in figures 1 and 2. Discussion of the main conclusions that can be drawn from these trees follows. Trees depicting the relationships among the class II *HLA-DR* alleles are described elsewhere [64].

Evolutionary Relationships among HLA-A *Alleles*

To date, 14 *HLA-A* alleles have been sequenced [7–18]. These, as well as two chimpanzee *ChLA-A* alleles [18], were compared with one another, genetic distances between them were calculated, and the values obtained were used to construct unrooted phylogenetic trees. As figure 1 shows, the tree constructed for the protein-coding sequence has four main branches: one branch bears the *A02, Aw68,* and *Aw69* alleles; another consists of the *A10*

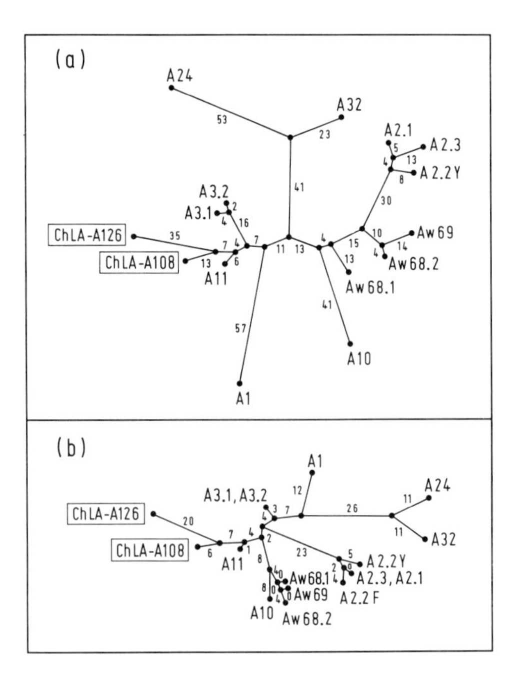

Fig. 2. Phylogenetic trees of human *HLA-A* and chimpanzee *ChLA-A* (boxed) alleles (coding sequences). *a* Nucleotides coding for the antigen-binding sites. *b* Nucleotides coding for the rest of the molecule.

and *A32* alleles; the third branch bears the *A24* allele; and the fourth branch leads to the *A01, A11, A03,* as well as the two *ChLA-A* alleles. The *A10* and *A32* alleles are so far apart that they can be regarded as each representing a main branch, equivalent to the remaining main branches. In turn, the *A10* and *A32* branches fork off from a common stem that leads also to the (*A02, Aw68, Aw69*)-bearing branch. Some of the main branches then display second- or third-order ramifications, bearing the so-called subtype alleles (e.g. *A02.1, A02.2Y,* and *A02.3*). All in all, the product of these comparisons is a respectable phylogenetic tree, comparable to that obtained for other loci, except that it represents a tree of *alleles* rather than of homologous genes in different animal species.

Virtually the same branching pattern is obtained when one draws separate trees for the exons that encode the α2, α3, and transmembrane (TM) plus cytoplasmic (CY) domains of the HLA-A molecule. This consistency of trees representing the individual exons argues against homogenization of the

HLA-A gene. Homogenization through gene conversion or some other similar mechanism would be expected to act over distances shorter than an exon and should therefore lead to an independent diversification of the individual exons. Had the bulk of the *Mhc* polymorphism been generated by gene conversion as has been argued [3, 4, 19, 38, and others], one would expect different exons to show different affinities among the alleles, and if such were the case, the trees for the individual exons would not match. The observed consistency of trees constructed for the different exons must therefore mean that the *HLA-A* alleles have diversified largely by random mutations and that the branching pattern of the phylogenetic trees reflects true kinships among the alleles. Kinships, in turn, must reflect the evolutionary history of the alleles.

The exon coding for the α1 domain also fits into the divergence pattern of the other exons, with the exception of the *A24* and *A32* alleles. Here, the two alleles form a branch of their own which is distinct from the other main branches. This change of branching pattern is most easily explained by intragenic recombination during which the *A24* and *A32* alleles acquired exon 2 from another, already highly diversified gene, while retaining the original α2, α3, TM + CY exons. The first domain of the A24 and A32 allomorphs is peculiar in that it contains, at positions 79 through 83, the amino acid sequence RIALR, which is absent in all other HLA-A allomorphs thus far sequenced but is present in some HLA-B allomorphs, for example B27.2 or Bw58 [16, 20]. The sequence is probably responsible for the presence of the Bw4 epitope on these molecules, which the A24 and A32 allomorphs share with some B allomorphs. All other A and some B allomorphs lack this sequence (and the Bw4 epitope). The sharing of polymorphic sequences (and of public epitopes) by some allomorphs encoded in two different loci has been explained by gene conversion between the corresponding loci: a short stretch from the first domain-encoding exon of an *HLA-B* gene has been transferred to an *HLA-A* gene [16]. However, a more likely explanation is that an unequal intragenic recombination exchanged the entire second exon between two different loci. The exchange need not have taken place between the *HLA-A* and *HLA-B* loci, because the *pHLA-12.4* pseudogene [21] and the *HLA-E* gene [22] also encode the RIALR sequence. Supporting the crossing-over hypothesis is the observation that the homology of the *A24* flanking region changes abruptly 5′ of the gene: While downstream from the second exon the gene and its flanking region are homologous to other *HLA-A* alleles, upstream the restriction map is very different from that of other *A* alleles [15]. It is conceivable therefore that sometime in the distant past, one of the

HLA-A alleles recombined with *pHLA-12.4, HLA-E,* one of the *HLA-B* alleles, or a gene at some other locus that codes for the RIALR sequence. The recombinant allele could have then become the ancestor of the *A24* allele, and the *A32* allele could have arisen by further recombination of *A24* with another (*A10*-like?) allele. This hypothesis can be tested by restriction mapping and sequencing of the flanking regions in selected *HLA-A* and *HLA-B* alleles. Whatever the explanation of the interlocus sharing of a sequence, the observation does not detract from the conclusion that meaningful phylogenetic trees can be constructed for *HLA* alleles. Intragenic recombination is well documented for *Mhc* loci [17, 23, and others] and as long as its possible occurrence during the evolution of alleles is taken into account, the problems in the construction of phylogenetic *Mhc* trees should be minimal.

The function of the Mhc molecules is to present antigens to T cell receptors on thymus-derived lymphocytes [2]. This presentation is believed to be effected by a specialized region of the Mhc molecule, the antigen-binding site, to which some 50 amino acids from different parts of the α1 and α2 domains contribute [24, 25]. The antigen-binding site represents the most variable segments of the class I Mhc molecule. In figure 2 we have constructed separate evolutionary trees for the segments that code for the antigen-binding sites of the individual HLA-A allomorphs, and for the regions outside the site (the rest of the molecule). The tree of the binding site resembles that of exon 2, which is not surprising in lieu of the fact that this exon contributes a lion's share to the site and that the positions constituting the site are the most variable in the entire gene. The tree of the remainder of the gene is similar to the trees constructed for the entire gene or the α2-, α3-, TM + CY-encoding exons. The significance of this observation will become clear later, when we discuss the *trans*-species origin of *Mhc* alleles.

A remarkable feature of the *HLA-A* phylogenetic tree is that its main branches correspond very closely to the so-called cross-reactive groups (CREGS) defined by serological analysis [26]. Alleles *A02, Aw68,* and *Aw69* are members of one such group; alleles *A01, A03,* and A11 are members of another group; *A10* and *A32* form a third group; and *A23-A24* form yet another group. This observation confirms, on the one hand, the power of the serological analysis and, on the other hand, allows one to make some predictions about alleles that have not been sequenced as yet but that have been analyzed serologically. One such prediction concerns the total number of main branches in the phylogenetic tree of the *HLA-A* genes. Not counting the subtypes, there are 21 officially recognized *HLA-A* alleles [27]. Of these, only nine have been sequenced (not counting the subtypes). However, virtually all the major serologically cross-reactive groups are represented by at least one member among the sequenced alleles; the only groups not represented are *Aw36* and *Aw43*. We predict therefore that not many more main branches will be added to the *HLA-A* tree through the sequencing of additional alleles. This prediction should not be interpreted as a discouragement for further sequencing of *HLA-A* alleles. On the contrary, since the main branches are more like trunks than branches, much more sequence information will be needed to understand in greater detail the evolutionary inter-relationships among the *HLA-A* alleles.

How Old Are Mhc *Alleles?*

Although the phylogenetic trees in figure 1 are mainly for human *HLA* alleles, they include also two chimpanzee *ChLA* alleles. Normally, genes from a distant group should provide a root to an otherwise unrooted tree. The human alleles should therefore form a single group apart from the chimpanzee alleles and so root down the tree. This, however, clearly has not happened in the trees of *Mhc* alleles. Here the chimpanzee alleles, rather than providing a root, are twigs on one of the main branches. This fact means that if the labels were removed from the figures, there would be no way of telling which are human and which chimpanzee alleles. The only sensible interpretation of this observation is that the main branches of the tree separated before humans and chimpanzees diverged from a common ancestor, and that many of the *Mhc* alleles are older than the species.

This conclusion comes as a surprise to those who have advocated rapid evolution of *Mhc* alleles by some sort of a homogenization process [3, 4, 19]. In contrast to these investigators, we have been arguing for some time [5, 28, 29, 30] that there really is no evidence for a rapid evolution of *Mhc* alleles [31], that the reported cases of gene conversion can either be explained by other mechanisms or are irrelevant to the generation of the *Mhc* polymorphism [32], and that if *Mhc* genes do not evolve faster than other genes, the only explanation for the origin of the *Mhc* polymorphism is *trans*-species evolution of alleles [5, 28, 31, 63]. We first proposed the *trans*-species hypothesis of *Mhc* polymorphism in 1980 [5] and have since been steadily accumulating evidence supporting it [18, 23, 28, 29, 33]. Only recently has support for the hypothesis been coming also from other laboratories [34, 35].

The evidence for *trans*-species evolution of *Mhc* alleles is by no means restricted to the primate *Mhc*. In fact, one of the strongest arguments in favor of the hypothesis has come from the study of rodent *Mhc* [23]. The *H-2Aβ* alleles, which are the mouse homologs of the human class II *HLA-DQB* alleles, fall into two groups: those that have two codons deleted at positions 65 and 67 ($A\beta^{\Delta}$ alleles), and those with an undeleted stretch of codons ($A\beta^{v}$ alleles). About half of the sequenced alleles are of the $A\beta^{\Delta}$ type, while the other half is of the $A\beta^{v}$ type, so that the two types constitute a true polymorphism in the species *Mus domesticus* [36]. We could show that the same two types of allele are also present not only in other species of the genus *Mus* (e.g. *Mus musculus*), but also in the rat, *Rattus norvegicus* [23]. In the rat, the prevalent allele is of the $A\beta^{\Delta}$ type, but the $A\beta^{v}$ type is also present [Figueroa et al., unpubl. data]. All this means that the two groups of allele (two branches of a phylogenetic tree) separated before the separation of the rat and the mouse from a common ancestor, an event that probably occurred more than 20 million years ago [37]. Some of the *Mhc* alleles must therefore be at least this old. Their separation predated not only speciation, but also the emergence of different genera (such as the *Mus* and *Rattus* or *Homo* and *Pan*).

One point deserves emphasis at this juncture. As we have seen, much of the variability that differentiates alleles occurs at the positions constituting the antigen-binding site. Could it therefore be that these positions diversify rapidly, while the rest of the gene evolves at the same pace as may other genes, and that the *impression* of *trans*-species evolution of alleles somehow arises from this fact, while in reality all the polymorphism is generated after speciation? Figure 1 clearly demonstrates that such is not the case. *All* regions of the gene show evidence of *trans*-species evolution, regardless of whether they are or are not part of the functional site. The antigen-binding site itself gives in fact the strongest evidence for *trans*-species evolution. It is therefore the entire gene that evolves *trans*-specifically, although of course exchanges of exons between main branches of alleles may occur so that some alleles may have regions that are younger (or older) than the rest of the gene.

The above considerations have led us to conclude that some alleles may have been separated for *at least* 20 million years. But how long have they really been separated? A clear-cut answer to this question will become available only when many *Mhc* alleles from different primates or different rodents have been sequenced; the available data allow one only to make approximate estimates of the age of the main allelic branches. The assumptions on which these estimates are based are that all sequence divergence is caused by point mutations; that transitions and transversions occur with the same frequency; that the molecular clock runs at a constant rate during primate evolution; and that the evolutionary rate is given by the number of silent substitutions [39]. Based on these assumptions, one can calculate the time *(T)* of divergence of the individual alleles from the formula $T = K_s^c/2Vs$, where K_s^c is the average number of silent substitutions per site corrected for multiple events, and Vs is the mean evolutionary rate [39]. For the chimpanzee-human comparisons, Li et al. [40] estimated Vs to be equal to 1.2×10^{-9} per site per year; for Old World monkey-human comparisons, Sakoyama et al. [41] came up with a value of Vs of 1.56×10^{-9} per site per year. Using these two values, we calculate that *HLA-A01* and *HLA-A02*, two of the most distant *HLA-A* alleles thus far sequenced (excluding *A24* and *A32* which, as was discussed earlier, are probably 'contaminated' by *B*-locus sequences), separated some 28–38 million years (myr) ago. Similarly, we estimate that *HLA-DRB1*07* and *HLA-DRB1*10*, the two most unrelated *DRB1* alleles of those sequenced thus far, separated from a common ancestor some 24–32 myr ago. These estimates place the origin of the main branches on the phylogenetic tree of class I and class II *Mhc* alleles all the way back to the time of radiation of the major mammalian groups. Some of the allelic branches might in fact be almost as old as the loci themselves. In other words, the ancestral genes might have duplicated to establish the individual loci and almost at the same time the newly generated functional loci began to diversify to establish the major allelic branches.

How Many Mhc *Alleles Are Older than the Species?*

Two points should be made before we attempt to answer this question. The first point is that *Homo sapiens* has of course not evolved from the chimpanzee but from *Homo erectus*, which has been extinct for some 200,000 years now [42]. Therefore, technically speaking, when we count alleles that predated speciation, we should not go all the way back to the common ancestor of *Homo* and *Pan* but to *H. erectus*. The second point is that speciation is not instantaneous but a process that never ends. The species *H. sapiens* is still evolving and has been evolving since it first emerged. This fact puts some ambiguity into the counting of prespeciation *Mhc* alleles.

If we take 200,000 years as the age of *H. sapiens* and consider the evolutionary distances discussed in the preceding section, we reach the conclusion that all the branches on the allelic trees except the ones leading to some of the minor subtypes separated before speciation. Even some of the subtypes may actually have predated the emergence of *H. sapiens*. The actual counts may therefore be a minimum of 20 *HLA-A*, 40 *HLA-B*, and 15 *HLA-DRB1* alleles being inherited by *H. sapiens* from *H. erectus*. If we go back in time and ask how many alleles present in the human population today existed in precursory forms in the last common ancestor of humans and chimpanzees, the estimates are some 10 *HLA-A*, 20 *HLA-B*, and 7 *HLA-DRB1* alleles. The extant alleles were, of course, not present in the ancestral species in their present forms. Since the emergence of a new species, the alleles have continued to evolve so that what was present in the ancestors were the forerunners of the present-day alleles.

Equally obvious should be the notion that different species may have retained different sets of allele from their common ancestors. It is actually somewhat surprising that thus far all chimpanzee *Mhc* alleles that have been sequenced have their counterparts in human *Mhc* alleles. Eventually, alleles should be found that form major branches in one species but are not represented in another species. Therefore, whenever alleles are found to be more closely related to certain alleles in another species than they are to other alleles in the first species, they provide evidence for *trans*-species evolution of *Mhc* polymorphism. However, whenever alleles are found that do not show this characteristic, they do not constitute evidence *against* the hypothesis.

Selection for Mhc *Polymorphism?*

The polymorphism sets the functional *Mhc* loci apart from all other known loci. Because they are so special, the *Mhc* loci must have been molded by special forces not acting, or at least not to the same degree, on other loci. Intuitively, one would like one of the special forces to be selection. Surprisingly, however, it has, until recently, proved difficult to come up with convincing evidence in support of the selection hypothesis. To be sure, there is no question that *Mhc* loci are subject to negative (purifying) selection which eliminates functionally unfit variants. But this sort of selection probably acts on most functional genes, so there is nothing special about it. The question is: Are the *Mhc* loci also subject to positive selection which, for example, provides an advantage to individuals heterozygous at *Mhc* loci or to a population with certain frequency of *Mhc* alleles? Certain observations seem, in fact, to go against the idea of positive selection: Functional *Mhc* loci in certain species are apparently not polymorphic or at least not highly polymorphic [43–46, 62]; some populations of species with generally high *Mhc* polymorphism are virtually monomorphic at their *Mhc* loci [47]; and it has proved to be exceedingly difficult to identify agents that could be driving the positive selection [31]. While these observations remain unexplained, recently evidence for positive selection *has* been reported. One of the most convincing arguments for positive selection has been provided by Hughes and Nei [47, 48] who analyzed the rates of nucleotide substitutions at synonymous and nonsynonymous positions in the codons (the former do not and the latter do lead to amino acid substitutions in the encoded protein). While such analyses had been done before [49], the novelty in the Hughes-Nei approach was to compare the nucleotides coding for the antigen-binding site with the nucleotides constituting the rest of the gene. Under these conditions a clear-cut indication of positive selection was obtained: the rates of nonsynonymous substitutions in the antigen-binding site were higher than would be expected if the substitutions were neutral. (In the rest of the molecule, the nonsynonymous substitutions were lower than expected, indicating that negative selection is acting on the corresponding portions of the gene.) While this sort of analysis and the data it has produced are not completely immune to certain criticisms, on the whole they strongly support the notion of positive selection acting on the antigen-binding site of the functional *Mhc* loci. The most likely form the positive selection takes is heterozygous advantage (overdominant selection), in which heterozygous individuals have a better chance of survival than homozygotes. Slight excess of heterozygotes over the theoretically expected number has actually been reported for human populations by Klitz et al. [50] and for mouse populations by Potts and Wakeland [51]. The conservation of allelic lineages over millions of years is also most easily explained by assuming that positive selection has been acting on the functional *Mhc* genes.

Driving Forces of Mhc *Polymorphism*

If *Mhc* polymorphism is generated and maintained by positive selection, what forces exert the selection pressure? Here again, the intuitive answer seems obvious but the evidence in support of it is scarce. Since the function of *Mhc* molecules is to present to T lymphocytes foreign antigens on the surface of antigen-presenting cells and since this presentation is part of the immune response against pathogens, it appears logical to conclude that the *Mhc* polymorphism has something to do with the effectiveness of this defensive reaction. Simplistically, it has been argued that the polymorphism is a way of making sure that the critical antigens are presented [52, 53]. Since the presentation, for reasons that are still being debated, suffers from 'blind spots' in the sense that certain combinations of foreign antigen and self *Mhc* molecule do not stimulate an immune response [54], a heterozygote expressing two different Mhc molecules has an obvious advantage over homozygotes expressing only one kind of Mhc molecules. The problem with this explanation is that it has proved to be very difficult to marshal evidence for it. Although susceptibility or resistance to many pathogens has been shown to be genetically controlled [55], with the exception of a few pathogens, it could not be demonstrated that the controlling loci reside in the *Mhc.* At best, *Mhc* could be only peripherally implicated in the susceptibility control [55] but seemingly not enough to account for the tremendous selection pressure that seems to be driving the *Mhc* polymorphism.

Alternative selection forces have therefore been sought and proposed. One idea is that *Mhc* is involved in mate selection operating through pheromone scenting [56]. That mice can identify each other when they seem to differ only at their *Mhc* loci has been reported [57] and the function of Mhc molecules in marking individuality has been proposed [58]. There are, however, two serious problems with using these observations as an explanation for the *Mhc* polymorphism. First, species with high *Mhc* polymorphism have often poorly developed olfactory senses and clearly do not use pheromone scenting for mate recognition and selection. Humans and other higher primates are a good example of this [59]. Second, the selection is clearly exerted on the antigen-binding site, a site that is known to function in immune response and not in mate recognition. Although one could argue that the antigen-binding site is also used for pheromone binding, one would have to then postulate that the pheromones are peptides (which they are not) because the *Mhc* molecules can interact only with other proteins and not with lipids, sugars, and other natural substances. Moreover, one would also have to reconcile pheromone and antigen recognition, two very clearly unrelated functions. On the whole, the mate selection hypothesis therefore seems to us a highly improbable explanation of the *Mhc* polymorphism.

Another often considered possibility is that the *Mhc* polymorphism has something to do with maternal-fetal interactions. One idea, for example, is that a histoincompatible fetus stimulates the mother's immune response which, in turn, makes the fetus more vigorous [60]. A large amount of literature exists on this sort of interaction [61], but its critical review has failed to convince us that any experimental support of the hypothesis has thus far been provided. Moreover, there is a very simple but extremely powerful argument against the hypothesis: in birds there is no maternal-fetal interaction, yet birds have polymorphic *Mhc.*

All in all, the pathogen resistance hypothesis is still the best explanation of what the *Mhc* polymorphism might be for. It may, however, be that we have not as yet identified the agents against which the polymorphism is protecting us. They may be of a different class than has been studied thus far or they may be agents that generally do not cause disease because the *Mhc* protects us well against them. The pathogens that do cause diseases may be the wrong ones to test for susceptibility association with *Mhc:* they may cause diseases precisely because resistance to them, only marginally if at all, depends on the *Mhc.* And here, we believe, lies the future of *Mhc* studies: in the elucidation of what the *Mhc* polymorphism really is for and what the true *Mhc* function might be.

References

1 Nei M: Molecular Evolutionary Genetics. New York, Columbia University Press, 1987.
2 Klein J: Natural History of the Major Histocompatibility Complex. New York, Wiley 1986.
3 Brégégère F. A directional process of gene conversion is expected to yield dynamic polymorphism associated with stability of alternative alleles in class I histocompatibility antigens gene family. Biochimie 1983;65:229–237.
4 Ohta T: Some models of gene conversion for treating the evolution of multigene families. Genetics 1984;106:517–528.
5 Klein J: Generation of diversity at MHC loci: implications for T cell receptor repertoires; in Fougereau, Dausset (eds): Immunology 80, Progress in Immunology IV. London, Academic Press, 1980.
6 Saitou N, Nei M: The neighbor-joining method. A new method for reconstructing phylogenetic trees. Mol Biol Evol 1987;4:406–425.
7 Parham P, Lomen CE, Lawlor DA, et al: Nature of polymorphism in HLA-A, -B, and -C molecules. Proc Natn Acad Sci USA 1989;85:4005–4009.
8 Koller BB, Orr HT: Cloning and complete sequence of an *HLA-A2* gene: analysis of two *HLA-A* alleles at the nucleotide level. Proc Natn Acad Sci USA 1985;134:2727–2733.
9 Mattson DH, Handy DE, Bradley DA, et al: DNA sequence of the genes that encode the CTL-defined HLA-A2 variants M7 and DK1. Immunogenetics 1987;26:190–192.

10 Holmes N, Ennis P, Wan AM, et al: Multiple genetic mechanisms have contributed to the generation of the HLA-A2/A28 family of class I Mhc molecules. J Immunol 1987;139:936–941.
11 Strachan T, Sodoyer R, Dmotte M, et al: Complete nucleotide sequence of a functional class I HLA gene, *HLA-A3:* implications for the evolution of *HLA* genes. EMBO J 1984;3:887–894.
12 Cowan EP, Jordan BR, Coligan JE: Molecular cloning and DNA sequence analysis of genes encoding cytotoxic T lymphocyte-defined HLA-A3 subtypes. The E1 subtype. J Immunol 1985;135:2835–2841.
13 Cianetti L, Testa U, Scotto L, et al: Three new class I *HLA* alleles: structure of mRNAs and alternative mechanisms of processing. Immunogenetics 1989;29:80–91.
14 Cowan EP, Jelachich ML, Biddison WE, et al: DNA sequence of HLA-A11. Remarkable homology with HLA-A3 allows identification of residues involved in epitopes recognized by antibodies and T cells. Immunogenetics 1987;25:241–250.
15 N'Guyen C, Sodoyer R, Trucy J, et al: The *HLA-Aw24* gene. Sequencing, surroundings and comparison with the *HLA-A2* and *HLA-A3* genes. Immunogenetics 1985; 21:479–489.
16 Wan AM, Ennis P, Parham P, et al: The primary structure of HLA-A32 suggests a region involved in formation of the Bw4/Bw6 epitopes. J Immunol 1986;137:3671–3674.
17 Holmes N, Parham P: Exon shuffling in vivo can generate novel HLA class I molecules. EMBO J 1984;4:2849–2854.
18 Meyer WE, Jonker M, Klein D, et al: Nucleotide sequences of chimpanzee MHC class I alleles. Evidence for *trans*-species mode of evolution. EMBO J 1988;7:2765–2774.
19 Kourilsky P: Genetic exchanges between partially homologous nucleotide sequences: possible implications for multigene families. Biochimie 1983;65:85–93.
20 de Waal LP, van der Meer C, van der Horst AR, et al: A new public antigen shared by all HLA-A locus products except HLA-A23, -A24, -A32, and -A25 is probably influenced by the amino acid residue at position 79 in the α1 domain. Immunogenetics 1988;28:211–213.
21 Malissen M, Malissen B, Jordan BR: Exon/intron organization and complete nucleotide sequence of an *HLA* gene. J Immunol 1982;79:893–897.
22 Koller BH, Geraghty DE, Shimizu Y, et al: *HLA-E*. A novel class I gene expressed in resting T lymphocytes. J Immunol 1988;141:897–904.
23 Figueroa F, Günther E, Klein J: MHC polymorphism pre-dating speciation. Nature 1988;335:265–267.
24 Bjorkman PJ, Saper MA, Samraoui B, et al: Structure of the human class I histocompatibility antigen, HLA-A2. Nature 1987;329:506–512.
25 Bjorkman PJ, Saper MA, Samraoui B, et al: The foreign antigen binding site and T cell recognition regions of class I histocompatibility antigens. Nature 1987;329: 512–518.
26 Rodey GE, Fuller TC: Public epitopes and the antigenic structure of the HLA molecules. CRC Crit Rev Immunol 1987;7:229–267.
27 Committee: Nomenclature for factors of the *HLA* system, 1987. Immunogenetics 1988;28:391–398.
28 Arden B, Klein J: Biochemical comparison of major histocompatibility complex molecules from different subspecies of *Mus musculus*. Evidence for *trans*-specific evolution of alleles. Proc Natn Acad Sci USA 1982;79:2342–2346.

29 Arden B, Wakeland EK, Klein J: Structural comparisons of serologically indistinguishable *H-2K*-encoded antigens from inbred and wild mice. J Immunol 1980;125: 2424–2428.
30 Klein J: Evolution and function of the major histocompatibility complex; in Parham, Strominger, Histocompatibility Antigens: Structure and Function. London, Chapman & Hall, 1982, pp 223–239.
31 Klein J, Figueroa F: Evolution of the major histocompatibility complex. CRC Crit Rev Immunol 1986;6:295–386.
32 Klein J: Gene conversion in *Mhc* genes. Transplantation 1984;38:327–329.
33 Figueroa F, Golubic M, Nižetić D, et al: Evolution of mouse major histocompatibility complex genes borne by *t* chromosomes. Proc Natn Acad Sci USA 1985;82: 2819–2823.
34 McConnell TJ, Talbot WS, McIndoe RA, et al: The origin of MHC class II gene polymorphism within the genus *Mus*. Nature 1988;332:651–654.
35 Lawlor DA, Ward FE, Ennis PD, et al: *HLA-A* and *B* polymorphisms predate the divergence of humans and chimpanzees. Nature 1988;335:268–271.
36 Mengle-Gaw L, McDevitt HO: Allelic variation in the murine Iaβ chain genes; in Cantor, Chess, Sercarz (eds): Regulation of the Immune System. New York, Liss, 1984, pp 29–46.
37 Sarich WM: Rodent macromolecular systematics; in Luckett, Hartenberger, Evolutionary Relationships among Rodents. New York, Plenum Press, 1985, pp 423–452.
38 Pease LR: Diversity in H-2 genes encoding antigen-presenting molecules is generated by interactions between members of the major histocompatibility complex gene family. Transplantation 1985;39:227–231.
39 Kimura M, Ohta T: On the stochastic model for estimation of mutational distance between homologous proteins. J Mol Evol 1972;2:87–90.
40 Li W-H, Tanimura M, Sharp PM: An evaluation of the molecular clock hypothesis using mammalian DNA sequences. J Mol Evol 1987;25:330–342.
41 Sakoyama Y, Hong K-J, Bynn SM, et al: Nucleotide sequences of immunoglobulin ε genes of chimpanzee and orang utan: DNA molecular clock and hominoid evolution. Proc Natn Acad Sci USA 1987;84:1080–1084.
42 Lewin R: Human Evolution. An illustrated introduction. New York, Columbia University Press, 1984.
43 Darden AG, Streilein JW: Syrian hamsters express two monomorphic class I major histocompatibility complex molecules. Immunogenetics 1984;20:603–622.
44 Trowsdale J, Groves V, Arnason A: Limited MHC polymorphism in whales. Immunogenetics 1989;29:19–24.
45 Watkins DI, Hodi FS, Letvin NL: A primate species with limited major histocompatibility complex class I polymorphism. Proc Natn Acad Sci USA 1988;85:7714–7718.
46 Figueroa F, Tichy H, Berry RJ, et al: The polymorphism in island populations of mice. Curr Top Microbiol Immunol 1986;127:100–105.
47 Hughes AL, Nei M: Pattern of nucleotide substitution at major histocompatibility complex class I loci reveals overdominant selection. Nature 1988;335:167–170.
48 Hughes AL, Nei M: Nucleotide substitution at major histocompatibility complex class II loci. Evidence for overdominant selection. Proc Natn Acad Sci USA 1989; 86:958–962.
49 Gustaffson K, Wiman K, Emmoth E, et al: Mutations and selection in the generation of class II histocompatibility antigen polymorphism. EMBO J 1984;3:1655–1661.

50 Klitz W, Thomson G, Baur MP: The nature of selection in the HLA region based on population data from the ninth workshop; in Albert, Baur, Mayer (eds): Histocompatibility Testing 1984. Berlin, Springer, 1984, pp 330–332.
51 Potts WK, Wakeland EK: The maintenance of MHC polymorphism. Disease, social competition, and reproductive selection. Immunol Today 1989; in press.
52 Doherty PC, Zinkernagel RM: A biological role for the major histocompatibility antigens. Lancet 1975;i:1406–1409.
53 Klein J: An attempt at an interpretation of the mouse *H-2* complex. Contrib Top Immunobiol 1976;5:297–335.
54 Klein J: What causes immunological nonresponsiveness? Immunol Rev 1984;81: 177–202.
55 Rosenstreich DL, Weinblatt AC, O'Brien AL: Genetic control of resistance to infections in mice. CRC Crit Rev Immunol 1982;3:263–330.
56 Flaherty L: Major histocompatibility complex polymorphism. A nonimmune theory of selection. Human Immunol 1988;21:3–13.
57 Yamaguchi M, Yamazaki K, Beauchamp GK et al: Distinctive urinary odors governed by the major histocompatibility locus of the mouse. Proc Natn Acad Sci USA 1981;78:5817–5820.
58 Thomas L: Symbiosis as an immunologic problem; in Neter, Milgrom (eds): The Immune System and Infectious Diseases. Basel, Karger, 1975, pp 2–11.
59 Passingham RE: The Human Primate. Oxford, Freeman, 1982.
60 Palm J: Maternal-fetal histocompatibility in rats. An escape from adversity. Cancer Res 1974;34:2061–2065.
61 Wegmann TG, Gill TJ III (eds): Immunology of Reproduction. New York, Oxford University Press, 1983.
62 O'Brien SJ, Roelke ME, Marker L, et al: Genetic basis for species vulnerability in the cheetah. Science 1985;227:1428–1434.
63 Klein J: Origin of major histocompatibility complex polymorphism. The trans-species hypothesis. Hum Immunol 1987;19:155–162.
64 Fan W, Kasahara M, Gutknecht J, et al: Shared class II polymorphisms between human and chimpanzees. Hum Immunol 1989; in press.

Prof. Dr. Jan Klein, Max-Planck-Institut für Biologie,
Abteilung Immungenetik, Correnstrasse 42, D–7400 Tübingen (FRG)

Waksman BH (ed): 1939–1989: Fifty Years Progress in Allergy.
Chem Immunol. Basel, Karger, 1990, vol 49, pp 51–68

The Thymus and Its Role in Immunity

J.F.A.P. Miller

The Walter and Eliza Hall Institute of Medical Research,
Post Office Royal Melbourne Hospital, Melbourne, Vic., Australia

'The outstanding feature of the development of immunology in the last 10 years has been the recognition of the function of the lymphocyte and of the importance of the thymus in the immune process' [1].

From Leukaemogenesis to Immunogenesis

In the late 1950s, the competence of the lymphocyte was being debated [2], and many did not consider the thymus capable of taking part in immune processes. It had even been stated: 'We interpret these observations as evidence that the thymus gland does not participate in the control of the immune response' [3]. The thymus was known to be involved in the development of lymphocytic leukaemia in mice and thymectomy at 6–8 weeks of age prevented leukaemogenesis in high-leukaemic strain mice as well as in low-leukaemic strains given ionizing irradiation, chemical carcinogens or cell-free filtrates [4]. Thymus implantation 6 months after thymectomy (performed at 1 month of age) restored the potential for leukaemia development in mice inoculated at birth with filtrates containing the Gross virus [5]. The virus was thus considered to have remained latent and could in fact be recovered from the non-leukaemic tissues of thymectomized mice [6]. To determine whether virus could multiply *outside* thymus tissue, mice were thymectomized *before* the filtrates were inoculated, i.e. immediately after birth. The experiment met with some difficulties: 'Subsequent mortality was very high in mice that had been thymectomized at 1 day of age, more than 50% of these dying between 2 and 4 months (whether they had been inoculated with virus or not), suggesting that the thymus at birth may be essential to life' [7]. This was totally unlike the situation in mice thymectomized in adult life which had never shown untoward effects or curtailment of lifespan [8]. It was clear that mice without a thymus from the time of birth were susceptible to infections because, when kept in 'clean' conditions, the incidence of wasting and death was less. Examination of their tissues revealed 'atrophy' of the lymphoid system and blood counts showed a marked diminution in lymphocyte levels and a reversal of the lymphocyte polymorph ratio [9]. Foreign skin grafts survived, even those incompatible at the major histocompatibility complex (the H-2 locus, abbreviated MHC) and even skin from other species such as rats [9–11]. These experiments led to the conclusion that 'during embryogenesis the thymus would produce the originators of immunologically competent cells many of which would have migrated to other sites at about the time of birth. This would suggest that lymphocytes leaving the thymus are *specially selected* cells' [9]. These results, or their interpretation, were not universally accepted, probably because most considered the thymus as having no role to play in immune processes. In many experiments, thymus cells had failed to transfer immunity or break tolerance [12]. Medawar [13] suspected 'that we shall come to regard the presence of lymphocytes in the thymus as an evolutionary accident of no very great significance'. Nevertheless, several groups in 1962 published the results of their own experiments indicating the importance of the thymus in establishing immune competence [14, 15] and within a few years no one would have even thought of questioning the immunological role of the thymus. In the 1970s the athymic nu/nu mouse strain became available [16] and showed the typical phenomena associated with neonatally thymectomized mice of several other strains.

Two Universes of Lymphocytes

A different line of investigation led Glick et al. [17] to conclude that the bursa of Fabricius, a lymphoid organ of birds somewhat analogous to the thymus, was essential in early life for normal antibody-forming capacity in chickens. Szenberg and Warner [18] were the first to show a division of labour among chicken lymphocytes: early thymectomy impaired cellular immunity and early bursectomy humoral immunity. In contrast to other vertebrates which do not have a bursa, lymphoid differentiation in birds would thus seem to occur in 2 separate and distinct sites giving rise to 2 populations subserving different functions. Since, in the mouse, neonatal thymectomy not only impaired cellular immunity but also antibody production to some antigens [19, 20], the mammalian thymus was believed to fulfill the function of *both* the thymus and bursa of birds. Yet an explanation was required for the observation that neonatal thymectomy in mice was associated with a marked reduction of lymphocytes normally found in those areas of the lymphoid tissues associated with cellular immunity (e.g. paracortical areas of lymph nodes and periarteriolar lymphocyte sheaths in spleen), and not so much in those areas where antibody formation normally took place (e.g. follicles and germinal centres) [21]. An important observation was made in 1966 by Claman et al. [22]: Irradiated mice given mixtures of cells from marrow and thymus together with antigen produced far more antibody than when given either cell population alone. It was not possible in this system to define the precise role of the thymus and marrow cells. This was solved in a systematic study of the role of various cell types in the reconstitution of immune functions in immuno-incompetent animals [23–27]. Marrow cells had no effects in either irradiated or thymectomized mice, but surprisingly thymus cells were as effective as recirculating lymphocytes (obtained by thoracic duct cannulation) in restoring antibody-forming capacity when given simultaneously with antigen, though only in neonatally thymectomized hosts, not in irradiated mice. The latter required marrow cells to be given as well and responded better with thoracic duct cells than with thymus cells. If, however, the thymus cells had previously been exposed to antigen in a first irradiated host, their ability to restore responsiveness to a second irradiated host given marrow cells was considerably enhanced for that specific antigen. This introduced the novel concept of 'thymus cell education' and implicated some interaction between educated thymus cells and marrow cells [26]. With genetically marked cells susceptible to destruction by specific antibodies it was 'demonstrated that the precursors of the hemolysin-forming cells are derived not from either thymus or thoracic duct lymphocytes but from bone marrow' [23]. This was the first unequivocal proof that thymus-derived cells did not become antibody-forming cells but were required in many antibody responses to help potential antibody formers produce antibody.

The establishment of cell interactions in antibody responses explained the 'carrier effect', namely the fact that successful immunization to a hapten required the latter to be coupled to a carrier molecule. Proof was obtained by groups led by Mitchison [28] and Rajewsky [29] for the cooperation of 2 distinct and separate classes of lymphocytes, one recognizing the carrier and the other the hapten, cooperation between these involving a carrier-hapten 'antigen bridge'. It was soon shown that the carrier-reactive cells were thymus derived and hapten-sensitive cells marrow derived. The former were designated 'helper' cells since they did not produce the antibody but enabled the latter to do so [30]. For simplicity, thymus-derived cells were called T cells and bone marrow-derived cells (or bursa-derived cells) B cells. The requirement for lymphocytes to cooperate in some antibody responses led to a reappraisal of numerous immunological phenomena including tolerance, autoimmunity, genetically determined unresponsive states, 'original antigenic sin', and immunogenicity.

Markers of Lymphocyte Subsets

By the late 1960s, both the immunological importance of the thymus and the existence of 2 major subsets of lymphocytes had been clearly established. T and B cells could not, however, be easily distinguished by morphological criteria alone. Hence, a major step forward was made when various cell-surface components became identifiable by the use of specific antibodies: the Thy-1 antigen on T cells [31]; high density of surface immunoglobulin on B cells [32]; Ly-2 [33], now called CD8, on a subset of T cells with cytotoxic [34] and suppressor activities [35, 36]; L3T4, now called CD4, on another subset of T cells with helper functions [37]. The CD8 molecule is a disulfide-bonded glycoprotein of M_r 70 kD (kilodaltons) and consists of two chains, α (M_r 34–38 kD) and β (M_r 30 kD). It is a member of the immunoglobulin supergene family with one extracytoplasmic domain closely resembling an immunoglobulin light chain V region. The CD4 molecule is a glycoprotein (M_r 52 kD) of 435 amino acids with little homology to CD8. Both CD4 and CD8 are believed to assist T cell activation by stabilizing low affinity interactions between the T cell antigen receptor (TCR) on the T cell membrane and the MHC-peptide complex on the antigen-presenting cell (APC).

A plethora of other T cell subsets, such as various types of helper and suppressor cells, contrasuppressors and inducers have also been postulated and in some cases identified. Suppressor T cells probably play a crucial role in regulating a great variety of cellular and humoral immune responses, e.g. in antigenic competition, in the induction and maintenance of some forms of immune tolerance and in the control of allergic and autoimmune reactions. Unravelling the complex interactions occurring between the various lymphocyte subsets and other haemopoietic cells, notably macrophages, is a goal of cellular immunologists today. For it is clear that the immune response does not unfold following a simple interaction of antigen and lymphocyte: it results from the activation of an elaborate network of interacting cells [38]. In addition, each response is a highly amplified reaction and, as with other amplification systems, each step must be subjected to appropriate control mechanisms. Cells involved in this control must be identified if we are to learn how to manipulate the immune response effectively.

The Major Histocompatibility Complex Controls Antigen Recognition by T Lymphocytes

Burnet's [39] clonal selection theory not only accounted for antigen recognition and the distinction between self and non-self but predicted that only a minority of lymphocytes would bind a given antigenic determinant. This was clearly demonstrated for B cells: $< 0.1\%$ bound radiolabelled antigen specifically, the binding being inhibited by anti-immunoglobulin reagents [40]. Hence, the immunoglobulin detected on the B cell surface acted as specific receptors for antigen. The situation with T lymphocytes is much more complex as T cells do not bind naked antigenic determinants. Using the graft-vs.-host reaction in chicken embryos, Simonsen [41] had estimated the frequency of alloreactive lymphocytes (later shown to be T cells) to be some 100–1,000 times as high as the frequency of cells reacting to other antigens. A similar conclusion was reached using in vitro cytotoxic T cell assays [42]. To account for such a high frequency of alloreactive T cells, Jerne [43] proposed that the repertoire of T cell reactivities depended on a set of germline V genes coding for receptors complementary to the MHC alleles of the species. Within the thymus, potentially alloreactive T cells, accounting for a relatively large proportion of the T cell pool, would not be influenced. By contrast, T cells with anti-self-H-2 reactivities would proliferate in response to H-2 antigens in the thymus. This could not be allowed to continue as such cells would kill self-H-2-bearing cells. Random somatic mutations in the genes coding for self-H-2 receptors would accumulate and thus decrease the strength of anti-self-H-2 binding (negative selection). Hence, only T cells without anti-self-H-2 reactivities would mature and would have their receptors directed to non-H-2 antigens. Each specific set would thus constitute a much smaller proportion of the total T cell pool than alloreactive T cells. This particular way in which T cell reactivities are selected would allow virtually all T cells to retain some 'memory' of their anti-self-H-2 past and to be directed to antigens which contained some measure of H-2ness. This negative selection theory thus accounted for the high frequency of alloreactive T cells and actually predicted the phenomenon of 'MHC restriction'.

Zinkernagel and Doherty [44], while studying resistance of mice to virus infections, documented a requirement for H-2 matching between cytotoxic T cells and their targets. They termed the phenomenon 'H-2 restriction' and mapped the genes imposing it in cytotoxic T cells to the K and D regions of H-2. Further experiments by Zinkernagel et al. [45], using thymectomized and thymus-grafted irradiated mice reconstituted with T cell-depleted marrow cells, gave data implying that H-2 restriction had been imposed at the level of differentiation of pre-T cells within the thymus, not as a result of priming. Once differentiated, mature T cells could then perceive antigen only if presented by antigen-presenting cells (APC) in association with the appropriate MHC gene product (called restriction element) identical to that present on thymus stromal cells; i.e. the T cells and the APC (or T cell target) need not have identical restriction elements. A model was formulated proposing that T cells had either one receptor directed towards both antigen and the MHC component involved in restriction, or two distinct receptors, one for antigen, the other for the MHC component. The $CD8^+$ T cells were found to be restricted by class I MHC molecules and the $CD4^+$ T cells by class II [37, 46]. In what manner these CD antigens govern the restriction of T cells by MHC molecules is the subject of intense investigations [47]. One possibility is that they increase the avidity of binding of T cells to their targets, CD8 having an affinity for class I and CD4 for class II, but the molecular details of such affinities have yet to be worked out.

That antigen recognition by T cells was indeed governed by MHC genes had been apparent since the discovery of MHC-linked immune responsiveness (Ir) genes [48, 49]. The immune response to a large number of antigenic determinants was controlled by genes mapping to the I region of the H-2 complex. These genes governed the activities of T cells (e.g. delayed-type hypersensitivity, in vitro-induced antigen proliferation) or T cell-assisted functions like antibody production. After the discovery of MHC restriction, it was possible to envisage identity of Ir gene products and I region gene-coded restriction elements [50] and also localization of the Ir genes controlling the activities of K and D restricted cytotoxic T cells to the K and D regions of the MHC [45]. In general, therefore, one could conclude that class II gene products govern the immune response of $CD4^+$ T cells and class I products influence that of $CD8^+$ T cells. The exact mechanism of this control has been debated and opposing views have been formulated to explain nonresponsiveness. One view seeks to explain nonresponsiveness in terms of a gap in the T cell repertoire [51]. Another view requires antigenic determinant and MHC gene product to associate to allow T cell activation [52, 53]. T cells recognize the 'associative' antigen and are thus MHC-restricted. A 'nonpermissive' interaction between an antigenic determinant and a particular MHC component leads to nonresponsiveness. The recent elucidation of the three-dimensional structure of an MHC molecule, which shows a large groove between α-helices forming a binding site for processed antigens [54], should provide a molecular explanation for MHC restriction and Ir gene effects.

A Plethora of Lymphokines

In the 1970s, evidence accumulated for the release by sensitized T lymphocytes of nonantigen-specific factors (now termed cytokines or lymphokines) which activate macrophages to kill some intracellular bacteria [55]. Similar mechanisms are likely to be involved in resistance to a variety of pathogens and hence in protection. Lymphokines have a variety of target cells. In fact, complex interactions involving cells of the immune system, their surface components and the various cytokines or lymphokines, occur during immune responses and are beginning to be unravelled. For example, the cytokine interleukin 1 (IL-1) is released from some antigen-presenting cells and causes T cells to expose on their surface a receptor able to bind a second lymphokine, interleukin 2 (IL-2). IL-2, which is produced by T cells, themselves, can now bind to the IL-2 receptor and stimulate T cell proliferation. When B cells present antigen in association with class II MHC molecules to T cells, a conjugate is formed and a variety of T cell lymphokines (e.g. interferon-γ (IFNγ) and IL-4) are involved in activating the B cell to respond. Their major effect is to promote the proliferation of activated B cells and to implement the switch from an IgM-producing B cell to an IgG-secreting one. At least 7 interleukins and other factors such as IFN-γ and tumor necrosis factor α and β have been cloned by genetic engineering techniques. T lymphocytes produce some 10 distinct lymphokines after activation: these include IL-2 to IL-6, IFNγ, granulocyte-macrophage colony-stimulating factor, and lymphotoxin [56–59]. These lymphokines are glycoproteins and are difficult to detect in serum, even during immune responses, probably because they are produced locally at the site of cell interactions and are rapidly consumed by their targets. They play a major role in immunoregulation having diverse and at times opposing effects on the growth, differentiation and activation of hemopoietic and lymphoid cells. There is evidence of coordinated regulation of their synthesis but much more work remains to be done to unravel the molecular events involved.

The T Cell Receptor for Antigen

The major discovery of the 1980s has been the elucidation of the molecular nature of the T cell antigen recognition receptor (TCR) and identification of the genes which code for it [60]. Use of monoclonal antibodies [61] and T cell clones [62] were instrumental in isolating the TCR [63, 64]. It is a disulfide-bonded heterodimeric glycoprotein composed of an α- and a β-chain. Its coding sequences were cloned independently in mouse [65] and man [66] by differential hybridization techniques followed by identification of rearranging genes. It is closely associated on the T cell membrane with the molecule CD3, a complex of 4 polypeptides, γ, δ, ε and the disulfide-bonded homodimer ς (M_r of these ranging from 15 to 25 kD) [60]. CD3 is thought to act as a signal transducer for the TCR enabling the T cell to sense the occupation of its TCR by antigen.

Two other genetic loci coding for a γ- [67] and a δ-chain [68] were identified but the function of T cells bearing such a heterodimer has yet to be clarified. These γδ T cells may conceivably be specialized for epithelial surveillance since they have mostly been found associated with epithelial cells in the epidermis and intestine [69].

Evidence soon became available to show that the TCR α-β heterodimer was entirely responsible for both antigen specificity and MHC restriction [64], thus vindicating the hypothesis of the one TCR model directed to both antigen and associated MHC component. There is, however, no correlation between α and β V region gene usage and MHC or antigen specificity [60]: any given V region may form part of a TCR restricted by class I or class II molecules [71].

Intrathymic Events

Relatively little is known of the intrathymic events which eventually produce the mature, non-self-reactive, T lymphocyte. The thymus generates vast numbers of lymphocytes each day (10^8 in mice) but exports only a minor fraction to the periphery (about 10^6 per day) [72]. Early thymocytes, constituting less than 3% of the cells in the thymus of a mature mouse, are referred to as double-negative cells because they bear neither the CD4 nor the CD8 molecules, which characterize mature T cells [73, 74]. Within the thymus, double-negative cells proliferate, rearrange and express TCR genes, and produce in a few days the predominant cell type: the cortisone-sensitive, double-positive cells ($CD4^+$, $CD8^+$) located in the cortex. More than 50% of these cells bear surface TCR, at a much lower concentration than peripheral T cells. The mature thymic lymphocytes bear either CD4 or CD8. They are derived from double negative cells [73, 75], presumably after first proceeding through an intermediate double-positive stage [76, 77], and are almost indistinguishable from peripheral T cells, being cortisone resistant and expressing high levels of TCR. They appear to be localized in the medulla.

The high rate of lymphopoiesis in the thymus is paralleled by a massive rate of cell death, the vast majority of the double-positive cells dying within the thymus [78]. This is attributed to a variety of factors among which are aberrant TCR gene rearrangement [79], failure to pass the positive selection test, and negative selection (see below). Since a high proportion of immature thymus lymphocytes express low levels of TCR [80, 81], aberrant TCR gene rearrangement may not be a major factor accounting for the large numbers of dying cells.

The T cell repertoire is shaped by the intrathymic microenvironment confronting the developing T lymphocytes. As a result, the mature T cells become MHC restricted. Most of the evidence favours a positive selection process whereby T cells that have some degree of binding avidity for polymorphic regions of MHC molecules encountered on thymic epithelial cells are selected for survival. The binding to epithelial cells is presumed to have some protective effect. T cells which fail to bind would then undergo apoptosis. The failure to pass the positive selection test is unlikely to account for the larger proportion of the cells dying within the thymus, because the germline repertoire of the early T cells appears to be skewed towards recognition of MHC molecules [82]. For example, some TCR Vβ genes seem to have germline specificity for MHC class II molecules and are expressed in high frequency in immature T cells [83, 84]. Furthermore, if early T cells survive because they recognize molecules on epithelial cells, conserving genes coding for TCR molecules with specificity for non-MHC determinants would not seem to be useful.

After, or before, positive selection by thymic epithelial cells, T cells are subjected to negative selection which deletes clones of cells with high avidity for self-MHC molecules. There is in fact convincing evidence for the imposition of self-tolerance by means of the clonal deletion of self-reactive T cells operating within the thymus. Experiments using monoclonal antibodies reacting with murine TCR binding the class II MHC molecule, I-E, clearly show the existence of cells with such TCR in I-E$^+$ mice in the immature population of thymus lymphocytes, but not in the mature intrathymic or peripheral pool [84]. Likewise, in transgenic mice expressing in many of their thymus cells a TCR specific for the male (H-Y) antigen in association with the class I H-2D^b MHC molecule, H-Y autospecific T cells are deleted in male though not in female mice [77]. The most widely accepted model of deletion of self-reactive T cells attributes a censorship role to cells of the dendritic or macrophage lineages, which are rich in class I and II molecules and are situated predominantly at the corticomedullary junction [85]. A simpler way in which deletion might occur, at least for class I reactive lymphocytes, would be via the postulated 'intracellular censorship' [86]: as the TCR is assembled in a self-reactive cell it should complex with its ligand (either self-class I or self-class I associated with self-peptide) and the accumulation of complexes in the endoplasmic reticulum would be incompatible with cell survival. This would easily account for the massive loss of immature CD4$^+$8$^+$ cortical-type cells in the anti-H-Y TCR transgenic mice [77].

Whatever the mechanism may be by which self-reactive thymus lymphocytes are deleted, the mature T cell population is clearly made up of cells unresponsive to self-MHC molecules but able to react to MHC alloantigens and to antigens associated with self-MHC.

Immunological Tolerance and Autoimmunity

Recent work on antigen presentation to T cells distinguishes two pathways: whereas endogenous molecules synthesized within cells associate with class I, exogenous antigens taken up by antigen-presenting cells usually associate with class II molecules [87]. If this model is generally correct, two implications follow: first, tolerance artificially induced by the systemic introduction of antigen may not mimic the natural way by which class I-restricted T cells are rendered tolerant to self-components; second, T cells capable of recognizing self-components not encountered in the thymus but unique to other tissues, will not be subjected to clonal deletion in the thymus. How, then, can T cells become tolerant to such unique extrathymic antigens? In some cases there need not be a requirement for T cell tolerance. Thus, some autoantigens may be anatomically secluded from immunocompetent cells; others may occur on cells which do not usually express MHC class I or II molecules and hence are ignored by T cells; others still may be exposed on cells which do express these MHC molecules but lack costimulator properties required for activation of T cells [88]. Some autoantigens could be shed in the circulation and reach thymic macrophages or dendritic cells which would then impart tolerance [89, 90]. This does not, however, seem to be the case for some antigens in the circulation: thus, in the case of transgenic mice producing soluble MHC class I antigens, the evidence is against circulating molecules inducing tolerance to the membrane-bound forms [91]. Likewise, in transgenic mice expressing an allogeneic class I molecule only in the β-cells of the pancreas [92] there is no evidence for clonal elimination of T cells reactive to this molecule [93]. Thus, peripheral T cells isolated from these transgenic mice at an early age did not react to the transgene MHC class I molecules in an in vitro cytotoxic T lymphocyte assay, but gave a response when isolated from old transgenics which had lost their β-cells and hence the allogeneic molecules. Thymic lymphocytes from young transgenic mice, on the other hand, were not tolerant in vitro suggesting that tolerance did not result from shed allo-antigen circulating back to the thymus and imposing clonal deletion on potentially reactive cells. Possibly, functional silencing of peripheral T cells, or 'T cell anergy', occurred in the absence of a costimulator signal which the β-cells could not provide. Some evidence in favour of this notion comes from recent experiments in which fetal islets expressing transgenic I-E molecules did not provoke an immune reaction when they were transplanted into normal I-E negative adult mice but were able to tolerize T cell lines in vitro [94]. Additional confirmation for the existence of a peripheral mechanism of tolerance induction in T cells has come from the study of other transgenic mice expressing allogeneic class I MHC molecules in hepatocytes [95], although in these a peripheral suppressor effect may play a role in the tolerogenic process. The details of such a mechanism have yet to be worked out.

Predisposing factors in autoimmune disease include several sets of genes, notably MHC genes. Some are not normally expressed on endocrine tissues but have been found on target cells in human autoimmune thyroid disease and several other disorders, such as type I diabetes [96]. Since the chief mediator of aberrant MHC gene expression is IFN-γ, it has been postulated that a local virus infection leads to interferon production, aberrant MHC gene expression and activation of T cells able to react to autoantigens because these are now presented in the context of self-MHC molecules [97]. This hypothesis may have to be revised in view of the recent work on transgenic mice, mentioned above, showing very clearly that MHC class I gene expression in the insulin-producing β-cells of the pancreas is per se sufficient to induce β-cell destruction at an early age and in the absence of T cells [92]. This suggests a direct, nonimmune, role for class I molecules in the impairment of β-cell function and raises questions as to what mechanisms may be involved. Apart from having a major role in antigen recognition by T lymphocytes, MHC class I molecules have been shown to influence cellular events in a variety of nonimmune ways [98]. Particularly relevant are the reports of an association of certain MHC molecules with peptide hormone receptors as demonstrated by antibody coprecipitation and other techniques [99]. Expression of MHC class I molecules in the β-cells might therefore be associated with the binding of peptides or the complexing of receptors essential for cell survival. Whether this effect is unique to pancreatic β-cells, whether it operates only in some subsets of autoimmune diseases, or whether it represents a general phenomenon associated with all these diseases, remains to be determined.

Conclusions

A remarkable amount of new knowledge has been gained in the last 3 decades on the role of the thymus and T cells in immunity. This has improved our understanding of resistance to infections, rejection of foreign transplants, MHC-linked diseases, self-tolerance and diseases of immunological aberrations including autoimmune diseases. Novel ways of manipulating the immune response can now be envisaged and explored. The pathogenesis of the acquired immunodeficiency syndrome (AIDS) would never have been unravelled 2 or 3 decades ago, before the discoveries of the functions of the lymphocyte and of the thymus in immune processes. Yet much work still remains to be done for, 'although the feeling has been expressed that immunology was born from the demand for protection against infectious disease, and that essentially what was needed was provided, it may be worthwhile pointing out that a recent World Health Organization Survey estimated that approximately 200 million people are infected with schistosomes, 300 million with filariae and other helminths, 250 million with malaria, and 1,000 million with a variety of intestinal parasites. Furthermore, of the immunological methods available, none can satisfactorily halt the spread of gonococcal and syphilitic infections' [100]. The situation, as it is now 15 years after this WHO survey, has not much improved and AIDS must be added to the list of communicable diseases. Thus, we have yet to work out how to immunize against a large variety of infectious and parasitic diseases. Effective adjuvants safe for human use are not yet available. We do not know exactly what constitutes good T and B cell epitopes for protective immunization nor how these should be put together in a synthetic vaccine. Hopefully, the immunochemical dissection of dominant T and B cell epitopes coupled with further studies of the cell biology and molecular biology of antigen processing may soon allow the engineering of safe and reliable synthetic vaccines.

References

1 Burnet FM: Chairman's opening remarks; in Wolstenholme GEW, Porter R (eds): The Thymus: Experimental and Clinical Studies. Ciba Fdn Symp. London, Churchill, 1966, pp 1–2.
2 Wolstenholme GEW, O'Connor M (eds): Cellular Aspects of Immunity. Ciba Fdn Symp. London, Churchill, 1960.
3 MacLean LD, Zak SJ, Varco RL, et al: The role of the thymus in antibody production: an experimental study of the immune response in thymectomized rabbits. Transplant Bull 1957;41:21–22.
4 Miller JFAP: Aetiology and pathogenesis of mouse leukaemia. Adv Cancer Res 1961;6:291–368.
5 Miller JFAP: Fate of subcutaneous thymus grafts in thymectomized mice inoculated with leukaemic filtrates. Nature 1959;184:1809–1810.
6 Miller JFAP: Recovery of leukaemogenic agent from non-leukaemic tissues of thymectomized mice. Nature 1960;187:703.
7 Miller JFAP: Analysis of the thymus influence in leukaemogenesis. Nature 1961; 191:248–249.
8 Miller JFAP: Role of the thymus in virus-induced leukaemia; in Wolstenholme GEW, O'Connor M (eds): Tumour Viruses of Murine Origin. Ciba Fdn Symp. London, Churchill, 1962, pp. 262–288.
9 Miller JFAP: Immunological function of the thymus. Lancet 1961;ii:748–749.
10 Miller JFAP: Role of the thymus in transplantation tolerance and immunity; in Wolstenholme GEW, Cameron MP (eds): Transplantation. Ciba Fdn Symp. London, Churchill, 1962, pp. 383–403.
11 Miller JFAP: Role of the thymus in transplantation immunity. Ann NY Acad Sci 1962;99:340–354.
12 Billingham RE, Brent L: Quantitative studies on tissue transplantation immunity. IV. Induction of tolerance in newborn mice and studies on the phenomenon of runt disease. Phil Trans R Soc Lond 1959;242B;439–477.
13 Medawar PB: In Discussion following Miller JFAP, Osoba D: The role of the thymus in the origin of immunological competence; in Wolstenholme GEW, Knight J (eds):

The Immunologically Competent Cell. Ciba Fdn Study Group. London, Churchill, 1963, pp. 70.

14 Arnason BG, Jankovic BD, Waksman BH: Effect of thymectomy on 'delayed' hypersensitive reactions. Nature 1962;194:99–100.

15 Martinez C, Kersey J, Papermaster BW, et al: Skin homograft survival in thymectomized mice. Proc Soc Exp Biol Med 1962;109:193–196.

16 Rygaard J: Thymus and Self: Immunology of the Mouse Mutant Nude. London, Wiley, 1973.

17 Glick B, Chang TS, Japp RG: The bursa of Fabricius and antibody production. Poultry Sci 1956;35:224–225.

18 Szenberg A, Warner NL: Dissociation of immunological responsiveness in fowls with a hormonally arrested development of lymphoid tissue. Nature 1962;194: 146–147.

19 Miller JFAP: Effect of neonatal thymectomy on the immunological responsiveness of the mouse. Proc R Soc Lond 1962;156B:410–428.

20 Miller JFAP: Tolerance in the thymectomized animal; in La Tolérance Acquise et la Tolérance Naturelle à l'égard de Substances Antigéniques Définies. Paris, CNRS Colloque, 1963, pp 47–75.

21 Parrott DMV, de Sousa MAP, East J: Thymus-dependent areas in the lymphoid organs of neonatally thymectomized mice. J Exp Med 1966;123:191–204.

22 Claman HN, Chaperon EA, Triplett RF: Thymus-marrow cell combinations – synergism in antibody production. Proc Soc Exp Biol Med 1966;122:1167–1171.

23 Miller JFAP, Mitchell GF: The thymus and the precursors of antigen-reactive cells. Nature 1967;216:659–663.

24 Miller JFAP, Mitchell GF: Cell to cell interaction in the immune response. I. Hemolysin-forming cells in neonatally thymectomized mice reconstituted with thymus or thoracic duct lymphocytes. J Exp Med 1968;128:801–820.

25 Mitchell GF, Miller JFAP: Cell to cell interaction in the immune respose. II. The source of hemolysin-forming cells in irradiated mice given bone marrow and thymus or thoracic duct lymphocytes. J Exp Med 1968;128:821–837.

26 Mitchell GF, Miller JFAP: Immunological activity of thymus and thoracic duct lymphocytes. Proc Natn Acad Sci USA 1968;59:296–303.

27 Miller JFAP, Mitchell GF: Thymus and antigen-reactive cells. Transplant Rev 1969; 1:3–42.

28 Mitchison NA: The carrier effect in the secondary response to hapten protein conjugates. II. Cellular cooperation. Eur J Immunol 1971;1:18–27.

29 Rajewsky K: The carrier effect and cellular cooperation in the induction of antibodies. Proc R Soc Lond 1971; 176B: 385–392.

30 Miller JFAP: Lymphocyte interactions in antibody responses. Int Rev Cytol 1972; 33:77–130.

31 Reif AE, Allen JMV: The AKR thymic antigen and its distribution in leukemias and nervous tissues. J Exp Med 1964;120:413–433.

32 Raff MC: Surface antigenic markers for distinguishing T and B lymphocytes in mice. Transplant Rev 1971;6:52–80.

33 Boyse EA, Old LJ: The immunogenetics of differentiation in the mouse. Harvey Lect 1975;71:23–53.

34 Cerottini J-C, Brunner KT: Cell-mediated cytotoxicity, allograft rejection and tumor immunity. Adv Immunol 1974;18:67–132.

35 McCullagh PJ: The abrogation of immunological tolerance by means of allogeneic confrontation. Transplant Rev 1972;12:180–197.
36 Gershon RK: T-cell control of antibody production. Contemp Top Immunobiol 1974;3:1–40.
37 Dialynas DP, Wilde DB, Marrack P, et al: Characterization of the murine antigenic determinant, designated L3T4a, recogized by monocloncal antibody GK1.5: expression of L3T4a by functional T cell clones appears to correlate primarily with class II MHC antigen-reactivity. Immunol Rev 1983;74:29–56.
38 Jerne NK: Towards a network theory of the immune system. Ann Immunol Inst Pasteur 1974;125C:373–389.
39 Burnet FM: The Clonal Selection Theory of Acquired Immunity. London, Cambridge University Press, 1959.
40 Ada GL: Antigen binding cells in tolerance and immunity. Transplant Rev 1970; 5:105–129.
41 Simonsen M: The clonal selection hypothesis evaluated by grafted cells reacting against their host. Cold Spring Harbor Symp Quant Biol 1967;32:517–523.
42 Skinner MA, Marbrook J: An estimation of the frequency of precursor cells which generate cytotoxic lymphocytes. J Exp Med 1976;143:1562–1567.
43 Jerne NK: The somatic generation of immune recognition. Eur J Immunol 1971;1: 1–9.
44 Zinkernagel RM, Doherty DC: Immunological surveillance against self components by sensitized T lymphocytes in lymphocytic choriomeningitis. Nature 1974;251: 547–548.
45 Zinkernagel RM, Callahan GN, Althage A, et al: On the thymus in the differentiation of 'H-2 self-recognition' by T cells. Evidence for dual recognition? J Exp Med 1978; 147:882–896.
46 MacDonald HR, Glasebrook AL, Bron C, et al: Clonal heterogeneity in the functional requirement for Lyt2/3 molecules on cytolytic T lymphocytes (CTL). Possible implications for the affinity of CTL antigen receptors. Immunol Rev 1982;68:89–115.
47 Owens T, Fazekas de St Groth B, Miller JFAP: Co-aggregation of the T cell receptor with CD4 and other T cell surface molecules enhances T cell activation. Proc Natn Acad Sci USA 1987;84:9209–9213.
48 Benacerraf B, Katz DH: The histocompatibility-linked immune response genes. Adv Cancer Res 1975;21:121–173.
49 McDevitt HO, Delovitch TL, Press JL, et al: Genetic and functional analysis of the Ia antigens. Their possible role in regulating the immune response. Transplant Rev 1976;30:197–235.
50 Meruelo D, Nimelstein SH, Jones PP, et al: Increased synthesis and expression of H-2 antigens on thymocytes as a result of radiation leukemia virus. A possible mechanism for H-2 linked control of virus-induced neoplasia. J Exp Med 1978;147: 470–487.
51 Nagy ZA, Klein J: Macrophage or T cell – that is the question. Immunol Today 1981; 2:228–229.
52 Benacerraf B: A hypothesis to relate the specificity of T lymphocytes and the activity of I-region-specific Ir genes in macrophages and B lymphocytes. J Immunol 1978; 120:1809–1812.
53 Rosenthal AS: Determinant selection and macrophage function in genetic control of the immune respose. Immunol Rev 1978;40:136–152.

54 Bjorkman PJ, Saper MA, Samrraoui B, et al: The foreign antigen binding site and T cell recognition regions of class I histocompatibility antigens. Nature 1987;329:512–518.
55 Mackaness GB: Delayed hypersensitivity and the mechanism of cellular resistance to infection. Progr Immunol 1971;1:413–424.
56 Gillis S, Ferm MM, Ou W, et al: T cell growth factor: parameters of production and a quantitative microassay for activity. J Immunol 1978;120:2027–2032.
57 Yung Y-P, Eger R, Tertian G, et al: Long-term in vitro culture of murine mast cells. II. Purification of a mast cell growth factor and its dissociation from TCGF. J Immunol 1981;127:794–799.
58 Klaus GGB: Unravelling the control of B cells. Nature 1986;324:16–17.
59 Metcalf D: The granulocyte-macrophage colony-stimulating factors. Science 1985; 229:16–22.
60 Davis MM, Bjorkman PJ: T-cell antigen receptor genes and T-cell recognition. Nature 1988;334:395–402.
61 Milstein C: Monoclonal antibodies; in: Accomplishments in Cancer Research. Philadelphia, Lippincot, 1981, pp 86–94.
62 Samelson LE, Schwartz RH: The use of antisera and monoclonal antibodies to identify the antigen-specific T cell receptor from pigeon cytochrome c-specific T cell hybrids. Immunol Rev 1983;76:59–78.
63 Allison JP, McIntyre BW, Block D: Tumor-specific antigen of murine T-lymphoma defined with monoclonal antibody. J Immunol 1982;129:2293–2300.
64 Haskins K, Kubo R, White J, et al: The major histocompatibility complex restricted antigen receptor on T cells. I. Isolation with a monoclonal antibody. J exp Med 1983; 157:1149–1169.
65 Hedrick SM, Cohen DI, Nielsen EA, et al: Isolation of cDNA clones encoding T cell-specific membrane-associated proteins. Nature 1984;308:149–153.
66 Yanagi Y, Yoshikai Y, Leggett K, et al: A human T cell-specific cDNA clone encodes a protein having extensive homology to immunoglobulin chains. Nature 1984;308: 145–149.
67 Saito H, Kranz DM, Takagaki Y, et al: A third rearranged and expressed gene in a clone of cytotoxic T lymphocytes. Nature 1984;312:36–40.
68 Chien Y-H, Iwashima M, Wettstein DA, et al: T-cell receptor δ gene rearrangements in early thymocytes. Nature 1986;330:722–727.
69 Janeway CA: Frontiers of the immune system. Nature 1988;333:804–806.
70 Goverman J, Minard K, Shastri N, et al: Rearranged β T cell receptor genes in a helper T cell clone specific for lysozyme. No correlation between Vβ and MHC restriction. Cell 1985;40:859–867.
71 Rupp F, Acha-Orbea H, Hengartner H, et al: Identical V_β T cell receptor genes in alloreactive cytotoxic and antigen plus I-A specific helper T cells. Nature 1985;315: 425–427.
72 Scollary R, Butcher E, Weissman I: Thymus migration: quantitative studies on the rate of migration of cells from the thymus to the periphery in mice. Eur J Immunol 1980;10:210–218.
73 Fowlkes BJ, Edison L, Mathieson BJ, et al: Early T lymphocytes. Differentiation in vivo of adult intrathymic precursor cells. J Exp Med 1985;162:802–822.
74 Scollay R, Bartlett P, Shortman K: T cell development in the adult murine thymus: changes in the expression of the surface antigens Ly2, L3T4 and B2A2 during development from early precursor cells to emigrants. Immunol Rev 1984;82:79–103.

75 Kingston R, Jenkinson EJ, Owen JJT: A single stem cell can recognize an embryonic thymus, producing phenotypically distinct T cell populations. Nature 1985;317: 811–813.
76 Smith L: CD4+ murine T cells develop from CD8+ precursors in vivo. Nature 1987; 326:798–800.
77 Kisielow P, Blüthmann H, Staerz UD, et al: Tolerance in T-cell-receptor transgenic mice involves deletion of nonmature CD4+8+ thymocytes. Nature 1988;333:742–746.
78 McPhee D, Pye J, Shortman K: The differentiation of T lymphocytes. V. Evidence for intrathymic death of most thymocytes. Thymus 1979;1:151–162.
79 Kronenberg M, Siu G, Hood LE, et al: The molecular genetics of the T cell antigen receptor and T cell antigen recognition. Ann Rev Immunol 1986;4:529–591.
80 Cristanti A, Colantoi A, Snodgrass R, et al: Expression of T cell receptors by thymocytes: in situ staining and biochemical analysis. EMBO J 1986;5:2837–2843.
81 Roehm N, Herron L, Cambier J, et al: The major histocompatibility complex-restricted antigen receptor on T cells. Distribution on thymus and peripheral T cells. Cell 1984;38:577–584.
82 Blackman, M, Yagüe J, Kubo R, et al: The T cell receptor may be biased in favor of MHC recognition. Cell 196;47:349–357.
83 Pullen AM, Marrack P, Kappler JW: The T cell repertoire is heavily influenced by tolerance to polymorphic self-antigens. Nature 1988,335:796–801.
84 Kappler JW, Roehm M, Marrack P: T cell tolerance by clonal elimination in the thymus. Cell 1987;49:273–280.
85 Sprent J, Webb SR: Function and specificity of T cell subsets in the mouse. Adv Immunol 1987;41:39–133.
86 Miller JFAP, Watson JD: Intracellular recognition events eliminate self-reactive T cells. Scand J Immunol 1988;28:389–395.
87 Braciale TJ, Morrison LA, Sweetser MT, et al: Antigen presentation pathways to class I and class II MHC-restricted T lymphocytes. Immunol Rev 1987;98:95–114.
88 Lafferty KJ, Andrus L, Prowse SJ: Role of lymphokine and antigen in the control of specific T cell responses. Immunol Rev 1980;51:279–314.
89 Lorenz RG, Allen PM: Direct evidence for functional self-protein/Ia-molecule complexes in vivo. Proc Natn Acad Sci USA 1988;85:5220–5223.
90 Boguniewicz M, Sunshine GH, Borel Y: Role of the thymus in natural tolerance to an autologous protein antigen. J Exp Med 1989;169:285–290.
91 Arnold B, Dill O, Küblbeck G, et al: Alloreactive immune responses of transgenic mice expressing a foreign transplantation antigen in a soluble form. Proc Natn Acad Sci USA 1988;85.2269–2273.
92 Allison J, Campbell IL, Morahan G, et al: Diabetes in transgenic mice resulting from over-expression of class I histocompatibility molecules in pancreatic β-cells. Nature 1988;333:529–533.
93 Morahan G, Allison J, Miller JFAP: Tolerance of class I histocompatibility antigens expressed extra-thymically. Nature 1989;339:622–624.
94 Markmann J, Lo D, Naji A, et al: Antigen presenting function of class II MHC expressing pancreatic beta cells. Nature 1988;336:476–479.
95 Morahan G, Brennan F, Bhathal et al: Expression in transgenic mice of class I histocompatibility antigens controlled by the metallothionein promoter. Proc Natn Acad Sci USA 1989;86:3782–3786.
96 Rose NR, Mackay IR: The Autoimmune Diseases. New York, Academic Press, 1985.

97 Bottazzo GF, Pujol-Borrell R, Hanafusa T, et al: Role of aberrant HLA-DR expression and antigen presentation in induction of endocrine autoimmune disease. Lancet 1983;ii:1115–1119.
98 Edidin M: MHC antigens and non-immune functions. Immunol Today 1983;4: 269–270.
99 Simonsen M, Skjødt K, Crone M, et al: Compound receptors in the cell membrane: ruminations from the borderland of immunology and physiology. Prog Allergy. Basel, Karger, 1985, vol 36, pp 151–176.
100 Miller JFAP: Experimental thymology has come of age. Thymus 1979;1:3–25.

J.F.A.P. Miller, MD, The Walter and Eliza Hall Institute of Medical Research, Post Office Royal Melbourne Hospital, Melbourne, Vic. 3050 (Australia)

Waksman BH (ed): 1939–1989: Fifty Years Progress in Allergy.
Chem Immunol. Basel, Karger, 1990, vol 49, pp 69–81

The T Cell Receptors

Philippa Marrack, John W. Kappler

Department of Medicine, Howard Hughes Medical Institute at
National Jewish Center for Immunology and Respiratory Medicine,
Denver, Colo., USA

Like all other biological systems, the immune response has proved to be more sophisticated, more subtle and, in a way, simpler than could possibly have been imagined. No aspect of the subject suggests this point of view more strikingly than that of T cells, their receptors and their functions. Immunologists were surprised to find that there was more than one kind of small lymphocyte and scientists were equally surprised to discover that these lymphocytes bore more than one kind of system for antigen recognition on their surfaces and yet, in retrospect, the matter seems perfectly straightforward and incontestable. Given the ease with which we can accept and account for these complexities after their discovery, it is perhaps difficult to understand how we all failed to appreciate and predict the properties of such systems *before* they were (in some cases accidentally) found.

During the last 30 years, our understanding of T cells has progressed from the recognition that such cells exist, with a role in the immune response apart from that of B cells, to a deep understanding of the mechanisms such cells use to detect antigen, and of the functions they perform.

Progress on the nature of the receptors of the major population of T cells in all species examined so far was initially slow. Several factors led to slow progress, including the fact that it was widely assumed that T cells would, like B cells, use immunoglobulin molecules to bind antigen. Equally significant was the fact that immunologists initially had no idea of the complexity of the ligand for αβ T cell receptors until after the pioneering work of several groups which demonstrated the connection, for T cells, between antigen recognition and association with major histocompatibility complex (MHC) proteins [1–9]. Even after the discovery of MHC-associated antigen recognition, arguments over whether T cells bore one or two receptors for antigen and MHC clouded the issue of receptor structure for some years.

Receptors on Most T Cells Recognize Antigen Complexed to MHC

In the 1970s, a number of investigators showed that T cells do not bind free antigen but rather, a complex ligand made up of antigen and MHC [1–9]. Several different types of experiment led to this conclusion: most involved the fact that T cells could respond only to cell-bound antigen. Moreover, the T cells concerned would respond to antigen only if the available antigen-presenting cells bore an appropriate allele of the right MHC molecule. Many experiments have indicated that the antigen recognized is not, in fact, native but rather a proteolytic product of the intact protein [10–13]. For example, it has long been known that T cells, unlike antibodies, do not distinguish between native and denatured protein [14]. Moreover, it was realized some time ago that presentation of antigen to T cells involved some cell-dependent 'processing' event. T cells do not recognize antigen added to aldehyde-fixed cells, although they will respond to cells which have been preincubated with antigen and then fixed [12]. The effect of antigen preincubation is abolished if reagents which inhibit lysosomal activity are present during this period. Finally, of course, several groups were able to demonstrate that the receptor on helper T cells, specific for antigen plus class II MHC molecules, could be satisfied by class II on fixed cells, or by isolated class II molecules, provided the appropriate, predigested antigenic peptide was added [12, 15, 16].

Later experiments showed that receptors on cytotoxic T cells were likewise specific for antigen peptides, rather than the intact antigenic protein. These receptors recognize antigenic peptide plus class I MHC molecules [17].

The solution of the crystal structure of a human class I MHC molecule has provided an excellent explanation for the mysterious ability of a single MHC molecule to bind many different antigenic peptides and present the peptides to T cells [18]. It seems that MHC molecules bear on their upper surfaces a long and well-defined groove, in which material not part of the MHC protein itself is found. This site is thought to be the place where antigen-derived peptides may bind. Since the groove is long, and flanked by many MHC amino acids, many different residues are available as potential contact points with antigenic peptides. This may account for the fact that a single MHC molecule can bind so many different peptides [16].

Recent research has uncovered an interesting dichotomy between class I and class II association with antigen. Peptides derived from exogenous proteins appear to associate freely with class II molecules. In cell biological terms this implies that the partially digested products of endocytosed or phagocytosed proteins must reach a compartment of the cell which also contains newly synthesized or recycling class II molecules. This compartment is thought to be endosomal, or prelysosomal. In contrast, it is very difficult to demonstrate an association between peptides from exogenous antigens, and class I molecules. Class I associates preferentially with peptides derived from proteins synthesized by its host cell [19]. In this case, the cell compartment in which association occurs is mysterious. The locale appears to be cytoplasmic, but class I molecules are traditional cell-surface-bound membrane proteins with no apparent access to the cell cytoplasm, so some doubt remains.

At first sight the specificity of αβ T cells for MHC-associated antigen seems to be a strange modus operandi, an evolutionary hangover, perhaps, from organisms in which receptor-like molecules and their MHC-like ligands were used to discriminate self from nonself in invertebrate colonies, or to control mating. Teleologically, however, there is a rationale for the phenomenon. It serves to focus the attention of T cells on locations at which they can properly perform their function. Class I-associated antigen concentrates cytotoxic T cells onto virus-infected cells, for example, and recognition of antigenic fragments in association with class II MHC molecules, of limited cell distribution, focuses helper T cells onto B cells and macrophages, the appropriate targets for helper T cell-derived lymphokines and signals. Secondly, although antibody molecules are extremely diverse, and have the ability to bind at an almost infinite number of structures, their ability to recognize antigen during an immune response is limited by the fact that they, in general, bind and act on the native structure of the antigen. An invading organism can therefore avoid antibody responses by small changes in its external surface, a strategy adopted by flu viruses among others. The fact that αβ T cell receptors react with processed products of the antigen means that such receptors screen the entire structure of an invading organism. To avoid attacks by T cells, the organism must be able to mutate all its proteins, not only those of its surface, and this may usually be too demanding a task.

A system which recognizes the combination of antigen and MHC is of course not without its drawbacks. There may be no association between peptides from some antigens and particular MHC products. In this case, no responding T cells will be generated. Moreover, the system is perilously close to recognition of self MHC without antigen, so T cells must be carefully screened for tolerance to self.

The Structure of αβ T Cell Receptors

Several investigations in the early 1980s showed immunologists that T cells bear a single receptor, able to bind antigen and MHC complexed in a way that was not, at that time, understood. The experiments which led to this finding depended on construction of, or isolation of, a single T cell with specificity for more than one combination of antigen and MHC. The demonstration that such cells were in every case able to recognize antigen X plus MHC^a or antigen Y plus MHC^b, but not the mixed combinations, X plus MHC^b or Y plus MHC^a, convinced most immunologists that a single receptor with dual specificity was involved [20–22]. The problem then was to discover such receptors, proteins presumably present in low amounts on the surfaces of cells which were difficult to obtain in large numbers.

Finally, two approaches were successful. Both depended for their success on the fact that they made the minimum number of assumptions about T cell receptors. The first method depended on the production of monoclonal antibodies against cloned T cell lines or hybridomas. The assumption in this work was that the antigen/MHC receptors on T cells would be surface proteins, and would vary in sequence (and therefore in antigenicity) from one T cell line to another. In the early 1980s, therefore, several groups set out to isolate anti-T cell receptor antibodies by screening monoclonal antibodies, made from animals immunized with T cell clones, for ability to react with the immunizing clone, but not other T cells.

This was a successful approach. Over the course of 6 months many groups described such antibodies, and the structure of the T cell surface proteins with which they reacted [23–28]. T cell receptors were thus identified as clonally variable heterodimeric disulphide-bonded glycoproteins with molecular weights of about 90 kdaltons.

A molecular biological approach proved equally rewarding. In this case it was assumed that T cell receptor genes would be expressed only in T cells, and that they would, like those of immunoglobulins, be rearranging genes. Again, these assumptions proved correct and screens of cDNA libraries for T cell-specific clones yielded the products of first β and then α T cell receptor genes [29–33].

Table 1. Variable elements used to construct T cell receptors

Mouse					Man				
a-chain		β-chain			α-chain		β-chain		
Vα	Jα	Vβ	Dβ	Jβ	Vα	Jα	Vβ	Dβ	Jβ
50	50	21	6	12	50	50	50	6	12
2,500		1,512			2,500		3,600		
3,780,000[1]					9,000,000[1]				

[1]These numbers do not include a 'fudge factor' of several orders of magnitude or more which should be included to account for N region-derived diversity.

Analysis of these genes showed that functional T cell αβ receptor genes were constructed very much like the genes for immunoglobulin molecules, by rearrangements involving several variable components and by the introduction of non-germline encoded bases at the joining points of these genes [34]. The number of variable components which can contribute to these genes for mouse and man is listed in table 1. The rules by which rearrangements are governed have been well described elsewhere [35–44]. At this point it is sufficient to say that receptor gene rearrangement occurs in T cell precursors after they have migrated to the thymus and that β-genes rearrange prior to those of α [45]. Since β-expression is allelically excluded, i.e. each T cell expresses only one functional β-protein, it is likely that expression of a functional β-chain suppresses further rearrangements at β-loci. It is also possible that β-protein induces α-gene rearrangements in the same cell.

The Structure of the αβ/Antigen/MHC Complex

In most cases, if animals are immunized with a particular antigenic peptide/MHC combination, the repertoire of T cells which are so primed is somewhat limited. This is well illustrated by the work of a couple of groups on responses of mice to pigeon cytochrome c plus IE^k. Most of the responding T cells use a particular Vβ, Vβ3 and one of two Vαs, as part of their receptors. Even the Dβ, Jβ, Jα and N regions are of limited diversity [46]. Similarly, in studies on responses to chicken ovalbumin 327–339 plus IA^d, we have found that a very small collection of T cell receptors can bind such a combination, with more than one-third of the response driven by T cells bearing a particular VαJα, VβDβJβ combination. Because responses to such peptides are so limited, it seems that all the variable components of the αβ receptor must contribute to antigen/MHC binding, with contact residues provided by Vα, Jα and Vβ, Dβ, Jβ as well as the N regions. In addition, because only a few receptor combinations are satisfied by any given antigen/MHC ligand, in unprimed animals the frequency of T cells able to respond to this ligand is low, probably less than $1/10^4$ or $1/10^5$.

We and others have recently discovered some interesting exceptions to this rule. It appears that certain antigens are able to combine with MHC and thus form a ligand which can interact with T cells via a single variable component of the receptor, in the cases we have studied, Vβ. For example, the product of the M1s-1^a gene, on mouse chromosome 1, associates with class II of the k, d or b haplotypes, and forms a ligand which can bind virtually all T cells bearing Vβ6 or Vβ8.1 [47, 48]. This reaction is almost unaffected by the other variable components of the receptors involved. We have called antigens of this type 'super antigens' to distinguish them from conventional peptides and to emphasize the fact that such antigens are very powerful T cell stimulants, because their reaction with T cells depends on only 1 of the many variable components of T cell receptors. A Vβ-specific superantigen/MHC combination will stimulate at least 5% of all mouse and 2% of all human T cell receptors (were such antigens to be expressed by human cells), a sizable percentage which is usually greater than that involved in response to allogeneic MHC, for example.

Superantigens are not without biological significance. Their constitutive expression in an animal leads to clonal elimination of all T cells bearing their target Vβ. T cells with receptors which include Vβ6 or Vβ8.1 are absent from animals which express M1s-1^a and a permissive MHC type [47, 48]. Conversely, introduction of a superantigen into a naive animal causes an enormous T cell response which can lead to shock and death, a phenomenon seen in man infected with superantigen-producing bacteria such as certain strains of *Staphylococcus aureus* [50]. We have in fact suggested that the genes for superantigens may exist in mice to prevent the consequences of bacterial superantigen-induced shock. M1s-1^a, for example, deletes Vβ6 and Vβ8.1-bearing T cells, and may reduce the toxic effects in mice of staphylococcal enterotoxin B, which includes among the T cells it stimulates, those bearing Vβ8.1.

We do not know the structure of the superantigen/MHC complex, nor do we know how this complex engages the T cell receptor through Vβ. It is possible that superantigens lie in the antigen-binding groove of MHC, and there engage Vβ. Alternatively, or as well, superantigens may bind to other sites on the MHC molecule and engage Vβ at a site which is not in contact with conventional antigen/MHC complexes.

The Repertoire of Mature αβ T Cells

It is clear that several phenomena control the diversity of receptors on mature T cells. These phenomena include the antigenic experience of the animal and positive and negative selection. While selection by antigen will primarily affect the numbers of T cells with any given specificity, restriction and tolerance affect the absolute specificities expressed, since they permit only a certain subset of thymocytes to mature, and allow the loss of, or cause the elimination of, cells which may be useless or harmful to the animal.

Although a number of mechanisms has been suggested to account for tolerance in T cells, there is at present solid evidence in vivo for only one of these. It has thus been shown that tolerance to many self antigens involves clonal deletion of T cells at a particular stage of their development, probably at some point in their existence as $CD4^+$, $CD8^+$, receptor-bearing immature cells. Evidence from both normal and transgenic mice supports this point of view [47–53]. Tolerance of this type can, however, only be established for antigens which are expressed in the thymus. T cells are also nonresponsive to antigens or MHC molecules expressed extrathymically in transgenic or normal animals. In this case, a different type of mechanism must account for nonresponsiveness, and evidence has been gathered for clonal anergy [54, 55]. It has, for example, been shown that challenge of T cells with antigen on fixed cells can lead to an unresponsive state, and similar events seem to occur in some circumstances in vivo.

Bevan and Zinkernagel et al. were the first to show that MHC molecules on thymus stroma select particular thymocytes for maturation and this idea has been supported by recent experiments in transgenic mice [51, 56, 57]. The result of this selection is clear: peripheral T cells in an animal bear receptors able to react with ligands which include self rather than allogeneic MHC molecules. The cause of positive selection is less clear. Broadly speaking, two hypotheses can be presented.One suggests that thymocytes are allowed to mature if the receptors they bear have a low affinity for self MHC, an affinity which is too low to allow peripheral reaction in the absence of foreign peptides [58]. Alternatively, it has been suggested that MHC molecules on the thymus cells which control positive selection, thymus cortical epithelial cells, may differ from those elsewhere [59, 60]. For example, they may be complexed with a different spectrum of self peptides than MHC on bone marrow-derived cells. There is some evidence that thymus cortical class II molecules differ from those elsewhere and, of course, it is almost inevitable that MHC in different locations will be bound by different self peptides. There is also evidence in favor of one form of the low-affinity hypothesis and since the two types of hypotheses are not mutually exclusive, both may apply.

γδ Receptors

Four years ago, immunologists were surprised to discover that a third pair of clonally variable genes existed in mouse and man. These genes and their protein products are called γδ, and like αβ and immunoglobulin genes, functional τδ-genes are formed by rearrangement of variable components [61–64].

At the time of writing, the molecular nature of τδ-receptors is well known. It is also known that γδ-cells arise in the thymus and that, like αβ-receptors, signal transduction by γδ involves the proteins of the CD3 complex. Almost everything else about γ- and δ-cells is a mystery, although some clues as to their function and specificity are known. γ- and δ-chains do not seem to be randomly distributed with respect to each other in different T cells, and particular Vγ, Vδ pairs are preferred [65–71]. This phenomenon is not noticeable for α and β T cell receptor polypeptides or immunoglobulin light and heavy chains. Cells bearing particular τδ pairs arise at particular times of development, at least in mice. τδ-Bearing cells are found as a low percentage of mouse lymph node, spleen and peripheral blood T cells, but they are the majority of T cells associated with epithelia. γ/δ-Bearing T cell clones specific for MHC proteins have been isolated, so some γδ-cells may have specificities similar to those of αβ-bearing T cells [72–75], but other $γδ^+$-cells appear to react with products of mycobacteria and/or heat shock proteins [76–79]. It is thought that γδ-cells exist as a first line of defense, in exposed surfaces of the body, and that there they respond to heat shock proteins or abnormal class I MHC molecules expressed by damaged or shocked cells.

Conclusions

Not surprisingly, the immune system has, in its complexities, easily outstripped the imagination of the scientists who set out to understand it. It seems that higher vertebrates have at least 3 ways of detecting an invading organism. These are antibodies, which bind the intact invader and contribute to its elimination, αβ-bearing T cells, which recognize and kill infected cells, or help immune responses of antigen-binding B cells, and γδ-bearing cells, which may respond to damaged cells of the host in the absence of antigen. These 3 systems seem to combine all imaginable ways of recognizing attack by microorganisms, although surprises are always possible and, in fact, experience has taught us that they are likely to occur.

Acknowledgment

The authors would like to thank Kelly Crumrine for her helpful secretarial assistance.

References

1 Katz DH, Hamaoka T, Dorf ME, et al: Cell interactions between histocompatible T and B lymphocytes. IV. Involvement of the immune response (Ir gene in the control of lymphocyte interactions in responses controlled by the gene). J Exp Med 1973; 138:734–750.
2 Shevach EM, Rosenthal AS: Function of macrophages in antigen recognition by guinea pig T lymphocytes. II. Role of the macrophage in regulation of genetic control of the immune response. J Exp Med 1973;138:1213–1225.
3 Zinkernagel R, Doherty P: H-2-compatibility requirement for T cell-mediated lysis of target cells infected with lymphocytic choriomeningitis virus: Different cytotoxic T cell specificities are associated with structures from H-2K or H-2D. J Exp Med 1975;141:1427–1436.
4 Bevan M: The major histocompatibility complex determines susceptibility to cytotoxic T cells directed against minor histocompatibility antigens. J Exp Med 1975;142:1349–1364.
5 Shearer G, Rehn T, Garbarino C: Cell-mediated lympholysis of trinitrophenylated-modified autologous lymphocytes. Effector cell specificity to modified cell surface components controlled by the H-2K and H-2D serological regions of the major histocompatibility complex. J Exp Med 1975;141:1348–1364.
6 Gordon R, Simpson E, Samelson L: In vitro cell-mediated immune response to male specific (H-Y) antigen in mice. J Exp Med 1975;142:1108–1120.
7 Kappler J, Marrack P: Helper T cells recognize antigen and macrophage surface components simultaneously. Nature 1976;262:797–799.
8 Waldmann H, Pope H, Brent L, et al: Influence of the major histocompatibility complex on lymphocyte interactions in antibody formation. Nature 1978;74: 166–168.
9 Ziegler K, Unanue E: *Listeria monocytogenes* immune T cells to macrophages. I. Quantitation and role of H-2 gene products. J Exp Med 1979;150:1143–1160.

10 Ziegler K, Unanue E: Decrease in macrophage antigen catabolism caused by ammonia and chloroquine is associated with inhibition of antigen presentation. Proc Natl Acad Sci USA 1982;79:175–179.
11 Grey H, Colon S, Chestnut R: Requirements for the processing of antigen by antigen-presenting B cells. II. Biochemical comparison of the fate of antigen in B cell tumor and macrophages. J Immunol 1982;129:2389–2395.
12 Shimonkevitz R, Kappler J, Marrack P, et al: Antigen recognition by H-2 restricted T cells. I. Cell-free antigen processing. J Exp Med 1983;158:303–316.
13 Unanue E: Antigen-presenting function of the macrophage. Ann Rev Immunol 1984; 2:395–428.
14 Chestnut R, Endres R, Grey H: Antigen recognition by T cells and B cells: recognition of cross-reactivity between native and denatured forms of globular antigens. Clin Immunol Immunopathol 1980;15:397–408.
15 Babbitt B, Allen P, Matsueda G, et al: Binding of immunogenic peptides to Ia histocompatibility molecules. Nature 1985;317:359–361.
16 Buus S, Sette A, Colon S, et al: The relation between major histocompatibility complex (MHC) restriction and the capacity of Ia to bind immunogenic peptides. Science 1987;235:1353–1358.
17 Townsend A, Rothbard J, Gotch F, et al: The epitopes of influenza nucleoprotein recognized by cytotoxic T lymphocytes can be defined with short synthetic peptides. Cell 1986;44:959–968.
18 Bjorkman PJ, Saper MA, Samraoui B, et al: Structure of the human class I histocompatibility antigen, HLA-A2. Nature 1987;329:506–512.
19 Morrison LA, Zukacher A, Braciale VL, et al: Differences in antigen presentation to MHC class I and class II-restricted influenza virus specific cytolytic T cell clones. J Exp Med 1976;163:903–920.
20 Kappler JW, Hunter PC, Jacobs D, et al: Functional heterogeneity among the T-derived lymphocytes of the mouse. I. Analysis by adult thymectomy. J Immunol 1974;113:27.
21 Heber-Katz E, Hansburg D, Schwartz R: The Ia molecule of the antigen presenting cell plays a critical role in the immune response gene regulation of T cell activation. J Mol Cell Immunol 1983;1:3–14.
22 Hunig T, Bevan M: Antigen recognition of cloned cytotoxic T lymphocytes follows rules predicted by the altered self hypothesis. J Exp Med 1982;155:111–125.
23 Allison J, McIntyre B, Bloch D: Tumor-specific antigen of murine T-lymphoma defined with monoclonal antibody. J Immunol 1982;129:2293–2300.
24 Meuer S, Fitzgerald K, Hussey R, et al: Clonotypic structures involved in antigen-specific human T cell function. Relationship to the T3 molecular complex. J Exp Med 1983;157:705–719.
25 Haskins K, Kubo R, White J, et al: The major histocompatibility complex-restricted antigen receptor in T cells. I. Isolation with a monoclonal antibody. J Exp Med 1983;157:1149–1169.
26 Samelson L, Schwartz R: The use of antisera and monoclonal antibodies to identify the antigen-specific T cell receptor from pigeon cytochrome c-specific T cell hybrids. Immunol Rev 1983;76:59–78.
27 Staerz U, Pasternack M, Klein J, et al: Monoclonal antibodies specific for a murine cytotoxic T lymphocyte clone. Proc Natl Acad Sci USA 1984;81:1799.
28 Kaye J, Porcelli S, Tite J, et al: Both a monoclonal antibody and antisera specific for determinants unique to individual cloned helper T cell lines can substitute for

antigen and antigen-presenting cells in the activation of T cells. J Exp Med 1983; 158:836–856.
29 Hedrick S, Cohen D, Nielsen E, et al: Isolation of cDNA clones encoding T cell-specific membrane-associated proteins. Nature 1984;308:149–153.
30 Yanagi Y, Yoshikai Y, Leggett K, et al: A human T cell-specific cDNA clone encodes a protein having extensive homology to immunoglobulin chains. Nature 1984;308:145–149.
31 Saito H, Kranz D, Takagaki Y, et al: A third rearranged and expressed gene in a clone of cytotoxic T lymphocytes. Nature 1984;312:36–40.
32 Chien Y, Becker D, Lindsten T, et al: A third type of murine T-cell receptor gene. Nature 1984;312:31–35.
33 Sim G, Yagüe J, Nelson J, et al: Primary structure of human T-cell receptor α-chain. Nature 1984;312:771–775.
34 Tonegawa S: Somatic generation of antibody diversity. Nature 1983;303:575–581.
35 Hayday A, Diamond D, Tanigawa G, et al: Unusual organization and diversity of T cell receptor α chain genes. Nature 1985;316:828–832.
36 Chien Y, Gascoigne N, Kavaler J, et al: Somatic recombination in a murine T cell receptor gene. Nature 1984;309:322–326.
37 Gascoigne N, Chien Y, Becker D, et al: Genomic organization and sequence of T cell receptor β-chain constant and joining region genes. Nature 1984;310:387–391.
38 Kavaler J, Davis M, Chien Y: Localization of a T cell receptor diversity-region element. Nature 1984;310:421–423.
39 Arden B, Klotz J, Siu G, et al: Diversity and structure of genes of the *a* family of mouse T cell antigen receptor. Nature 1985;316:783–787.
40 Clark S, Yoshikai Y, Taylor S, et al: Identification of a diversity segment of human T-cell receptor β-chain, and comparison with the analogous murine element. Nature 1984;311:387–389.
41 Malissen M, Minard K, Mjolsness S, et al: Mouse T cell antigen receptor. Structure and organization of constant and joining gene segments encoding the β polypeptide. Cell 1984;37:1101–1110.
42 Siu G, Clark S, Yoshikai Y, et al: The human T cell antigen receptor is encoded by variable, diversity and joining gene segments that rearrange to generate a complete V gene. Cell 1984;37:393–401.
43 Clark S, Yoshikai Y, Taylor S, et al: Identification of a diversity segment of human T-cell receptor β-chain, and comparison with the analogous murine element. Nature 1984;311:387–389.
44 Toyonaga B, Yoshikai Y, Vadasz V, et al: Organization and sequences of the diversity, joining, and constant region genes of the human T cell receptor β chain. Proc Natl Acad Sci USA 1985;82:8624–8628.
45 Born W, Yagüe J, Palmer E, et al: Rearrangement of T cell receptor β chain genes during T cell development. Proc Natl Acad Sci USA 1985;82:2925.
46 Fink P, Matis L, McElligott D, et al: Correlations between T cell specificity and structure of the antigen receptor. Nature 1986;321:219–226.
47 Kappler JW, Staerz U, White J, et al: Self-tolerance eliminates T cells specific for Mls-modified products of the major histocompatibility complex. Nature 1988;332: 35–40.
48 MacDonald HR, Schneider R, Lees RK, et al: T-cell receptor Vβ use predicts reactivity and tolerance to Mls^a-encoded antigens. Nature 1988;332:40–45.

49 White J, Herman A, Pullen AM, et al: The Vβ-specific superantigen Staphylococcal enterotoxin B. Stimulation of mature T cells and clonal deletion in neonatal mice. Cell 1989;56:27–35.

50 Kappler J, Roehm N, Marrack P: T cell tolerance by clonal elimination in the thymus. Cell 1987;49:273–280.

51 Hung ST, Kisielow P, Scott B, et al: Thymic major histocompatibility-complex antigens and the ab T cell receptor determine the CD4/CD8 phenotype of T cells. Nature 1988;335:229–233.

52 Kisielow P, Bluthmann H, Staerz UD, et al: Tolerance in T-cell-receptor transgenic mice involves deletion of nonmature $CD4^+8^+$ thymocytes. Nature 1988;333:742–746.

53 Fowlkes B, Schwartz P, Pardoll D: Deletion of self-reactive thymocytes occurs at $CD4^+CD8^+$ precursor stage. Nature 1988;334:620–623.

54 Jenkins MK, Schwartz RH: Antigen presentation by chemically modified splenocytes induces antigen-specific T cell unresponsiveness in vitro and in vivo. J Exp Med 1987;165:302–319.

55 Burkley L, Lo D: Personal commun.

56 Bevan M: In a radiation chimera host H-2 antigens determine the immune responsiveness of donor cytotoxic cells. Nature 1977;269:417–419.

57 Zinkernagel R, Callahan G, Althage A, et al: On the thymus in the differentiation of H-2 self-recognition by T cells. Evidence for dual recognition? J Exp Med 1978;147: 882–896.

58 Lo D, Ron Y, Sprent J: Induction of MHC-restricted specificity and tolerance in the thymus. Immunol Res 1986;5:221–232.

59 Marrack P, McCormack J, Kappler J: Presentation of antigen, foreign major histocompatibility complex proteins and self by thymus cortical epithelium. Nature 1989;338:503–505.

60 Murphy DB, Lo D, Roth S, et al: A novel MHC class II epitope expressed in thymic medulla but not cortex. Nature 1989;338:765–768.

61 Saito H, Kranz D, Takagaki Y, et al: Complete primary structure of a heterodimeric T cell receptor deduced from cDNA sequences. Nature 1984;309:757–762.

62 Hayday A, Saito H, Gilles S, et al: Structure, organization, and somatic rearrangement of T cell γ genes. Cell 1985;40:259–268.

63 Raulet D, Garman R, Saito H, et al: Developmental regulation of T cell receptor gene expression. Nature 1985;314:103–107.

64 Chien Y, Iwashima M, Kaplan KB, et al: A new T-cell receptor gene located within the alpha locus and expressed early in T-cell differentiation. Nature 1987;327:677–682.

65 Havran WL, Grell S, Duwe G, et al: Limited diversity of TcR γ chain expression of murine thy-1^+ dendritic epidermal cells revealed by Vγ3-specific monoclonal antibody. Proc Natl Acad Sci USA, in press.

66 Asarnow DM, Kuziel WA, Bonyhadi M, et al: Limited diversity of γδ antigen receptor genes of thy-1^+ dendritic epidermal cells. Cell 1988;55:837–847.

67 McConnell TJ, Yokoyama WM, Kikuchi GE, et al: σ-Chains of dendritic epidermal T cell receptors are diverse but pair with γ-chains in a restricted manner. J Immunol 1989;142:2924–2931.

68 Bonneville M, Janeway CA Jr, Ito K, et al: Interstitial intraepithelial lymphocytes are a distinct set of γδ T cells. Nature 1988;336:479–481.

69 Korman AJ, Marusic-Galesic S, Spencer D, et al: Predominant variable region gene usage by γ/δ T-cell receptor-bearing cells in the adult thymus. J Exp Med 1988;168:1021–1040.

70 Takagaki Y, Nakanishi N, Ishida I, et al: T cell receptor-γ and -δ genes preferentially utilized by adult thymocytes for the surface expression. J Immunol 1989;142:2112–2121.

71 Havran WL, Allison JP: Developmentally ordered appearance of thymocytes expressing different T-cell antigen receptors. Nature 1988;355:443–445.

72 Matis LA, Cron R, Bluestone JA: Major histocompatibility-linked specificity of γδ receptor-bearing T lymphocytes. Nature 1987;330:262–264.

73 Bluestone JA, Cron RQ, Cotterman M, et al: Structure and specificity of T cell receptor γ/δ on major histocompatibility complex antigen-specific $CD3^+$, $CD4^-$, $CD8^-$ T lymphocytes. J Exp Med 1988;168:1899–1916.

74 Matis LA, Fry AM, Cron RQ, et al: Structure and specificity of a class II MHC alloreactive γδ T cell receptor heterodimer. Science, submitted.

75 Rivas A, Koide J, Cleary ML, et al: Evidence for involvement of the γ, δ T cell antigen receptor in cytotoxicity mediated by human allo-antigen-specific T cell clones. J Immunol 1989;142:1840–1846.

76 O'Brien RL, Happ MP, Dallas A, et al: Stimulation of a major subset of lymphocytes expressing T cell receptor τσ by an antigen derived from *Mycobacterium tuberculosis*. Cell, in press.

77 Janis EM, Kaufmann SHE, Schwartz RH, et al: Activation of γδ T cells in the primary immune response to *Mycobacterium tuberculosis*. Science, in press.

78 Holoshitz J, Koning F, Coligan JE, et al: Isolation of $CD4^-$ $CD8^-$ mycobacteria-reactive T lymphocyte clones from rheumatoid arthritic synovial fluid. Nature, in press.

79 Modlin RL, Pirmez C, Hoffman FM, et al: Antigen-specific T cell receptor γδ bearing lymphocytes accumulate in human infections disease lesions. Nature, submitted.

Philippa Marrack, MD, Department of Medicine, Howard Hughes Medical Institute
at National Jewish Center for Immunology and Respiratory Medicine,
Denver, CO 80206 (USA)

Waksman BH (ed): 1939–1989: Fifty Years Progress in Allergy.
Chem Immunol. Basel, Karger, 1990, vol 49, pp 82–89

Lymphokines, Monokines, Cytokines

Clemens Sorg

Institute of Experimental Dermatology, University of Münster, FRG

Historical Note

At the beginning of this century one of the most puzzling questions for immunologists was how does a delayed-type hypersensitivity or cell-mediated immune response work. The importance of this reaction in health and disease had long been appreciated before Landsteiner and Chase [1] could demonstrate in 1942 that delayed-type hypersensitivity could be transferred to an unsensitized animal by lymphocytes from a sensitized animal.

Rich and Lewis [2] attempted to understand the mechanisms of cell-mediated immunity by studying tissue explants in vitro. In these studies the migration of mononuclear cells out of spleen fragments was inhibited in the presence of the sensitizing antigens. This was taken as a parameter of the delayed-type hypersensitivity state of the animal. Even though great efforts were put into standardizing this technique, only the simplification, using the migration of peritoneal exudate cells out of a capillary tube, the so-called macrophage migration assay introduced by George and Vaughan [3], provided a more quantitative and reproducible test system and allowed the dissection of the phenomenon itself.

The Discovery of the Macrophage Migration Inhibitory Factor

This was the stage setting in 1966 when Bloom and Bennett [4] and David [5] found that sensitized lymphocytes, when stimulated by antigen released a soluble material which inhibited the migration of peritoneal exudate macrophages. This material was called the 'macrophage migration inhibitory factor' (MIF). The MIF soon became very popular as it was the first known molecular equivalent of cell-mediated immunity and was widely applied in experimental and clinical immunological studies.

The Boom Times of Lymphokine Research

The publication of these observations triggered a whole avalanche of similar discoveries and soon an ever-growing list of factors became known, designated either as mediators of cell-mediated immunity, or lymphocyte activation products (LAP), or products of activated lymphocytes (PAL). In 1969 Dumonde et al. [6] suggested the term 'lymphokines' for all these factors, which is the most commonly used trade name for all nonantibody mediators released by lymphoid cells. There was always a discussion on terminology and nomenclature of factors, particularly on how far the term lymphokine should be stretched, whether it should also include low molecular weight, nonprotein cellular products.

The list of lymphokines grew rapidly and in a survey by Waksman and Namba [7] in 1976 the number was close to 100. Of course, since a characterization of the various factors was not feasible and, therefore, no comparative chemical data were available, many acronyms probably designated the same product. The desire to bring some order to the chaos first tried to work through nomenclature. It was first suggested to separate the factors into lymphokines and monokines, the latter being produced by monocytes and macrophages. As it became more and more apparent that lymphokine-like factors were not only produced by lymphoid cells but also under certain circumstances by a variety of nonlymphoid cells, Cohen et al. [8]. proposed to use the term 'cytokines' for the entire class of mediators. It was also discovered that cytokines not only participate in regulating immunological and inflammatory processes, but can also contribute to repair processes and the regulation of normal cell growth and differentiation. In fact, the development of the field until today makes this more comprehensive term to be the most appropriate and should be used generally.

In 1979 at the 2nd International Lymphokine Workshop at Ermatingen, Switzerland, it was discussed that many biological activities produced by lymphocytes, macrophages or other cells could be attributed to the same molecules. These molecules were renamed interleukin 1 and interleukin 2. Up to now several lymphokines have been renamed and numbered from interleukin 1 to 7 [9, 10].

The Depression Years

After the discovery of a novel biological activity, usually defined by an in vitro assay it was an obvious task to determine its molecular characteristics. If we look back at the year 1966 and the techniques available for protein purification the problems encompassed were insurmountable. It very soon became clear that lymphokines act in minute amounts, that is, as we know today, in the nanogram or femtomol range. This posed another enormous logistic problem. Where to obtain enough material for the purification procedures with macrotechniques? At that time cell culture techniques just started to become introduced in immunologist's laboratories. It was considered an art, which only a few mastered. Producing large quantities of lymphokine-rich supernatant for preparative purification was wishful thinking. All who where engaged in lymphokine research tried purification of a factor with ridiculously small samples of up to 100 ml. Retrospectively, it was a rather hopeless undertaking. Nevertheless, some raw data emerged and at least occasionally could even convince the skeptic that lymphokines were real [11]. The time and its technical possibilities were not ready for significant progress in the lymphokine field. A lot could be learned from the frustrating experiences with interferon which had been discovered in 1957 [12], almost 10 years before MIF was discovered, and so far had resisted all purification attempts. As is known, the structure of interferon-β was first described much later by a new breed of scientists, actually outsiders of the field, using the revolutionary gene technology [13].

At the time, 20 years ago, technology was not ready for the lymphokine field. Key amongst the problems that beset the infant speciality was the unreliability and inadequacy of source materials and primitive bioassays. After a sobering experience with the purification of a factor many quit the field which some out of frustration called 'lymphodreck'. And as much as it had been fashionable to work in lymphokines, it now became fashionable to turn its back on it.

The Comeback of Lymphokines

While the list of described lymphokine activities was steadily growing and despite some setbacks it became more and more obvious that this group of biological molecules was intertwined with virtually every aspect of mammalian cell biology. The need for communication and interaction of investi-

Table 1. International lymphokine workshops

Year	Location	Theme	Organizers
1976	Bethesda, Md.	Mechanisms of Action and Characterization of Lymphocyte Mediators	J. Oppenheim
1979	Ermatingen, Switzerland	Biochemical Characterization of Lymphokines	A.L. de Weck, F. Kristensen, M. Landy
1982	Haverford, Pa.	Interleukin, Lymphokines and Cytokines	J. Oppenheim S. Cohen
1984	Schloss Elmau, FRG	Cellular and Molecular Biology of Lymphokines	C. Sorg A. Schimpl
1986	Clearwater Beach, Fla.	Molecular Basis of Lymphokine Action	D.R. Webb, C.W. Pierce, S. Cohen
1988	Evian, France	Lymphokine Receptor Interaction	D. Fradelizzi J. Bertoglio

gators arose (table 1). In 1976 the first International Lymphokine Workshop was convened in Bethesda, Md. The second was held in Ermatingen, Switzerland, in 1979 where the number of participants had almost doubled. After the third workshop at Haverford College, Pennsylvania, in 1982, the conference series had firmly established itself. Interest in lymphokine research was again on the rise. What had happened? By 1980 a number of technical developments had reached the stage that made them applicable to lymphokine research. Not only did they prove decisive, they also transformed the lymphokine field and swiftly brought it into the modern era.

The problem of producing large quantities was solved by using established cell lines in mass culture. Presently, cell lines provide potent sources of all lymphokines and are eminently suitable for large-scale production. The monoclonal antibody technique which became available in 1976 had an enormous impact on lymphokine research. This technology was utilized with increasing sophistication to generate antibodies specific for a variety of lymphokines. Such antibodies have proved decisive in preparative isolation, for lymphokine identification, and, in recent years, in the molecular cloning of cDNA. Separation technology has also experienced tremendous progress.

High-performance liquid chromatography provided a major step forward in speed and reproducibility of protein purification. Another quantum leap was performed in amino acid sequencing. Ten to fifteen years ago even a partial amino acid sequence required milligram quantities of precious material. Utilization of gas phase protein sequencing enables that task to be completed with a few micrograms. The lymphokine field, however, was most dramatically changed by the application of recombinant DNA technology. Not only did structural analysis by DNA sequencing and translation into amino acid sequence become a relatively simple technique, but also production of recombinant proteins solved one of the most difficult problems of lymphokine research and that was the lack of sufficient material for biological experiments.

The speed of progress and announcement of breakthroughs in the molecular cloning of several lymphokines was characteristic for the year 1984 and generated a unique atmosphere at the 4th International Lymphokine Workshop at Schloss Elmau in West Germany, a peculiar mixture of eureka fever and gold rush.

The Years After

Already before 1984 the lymphokine field had got 'company'. Newly founded biotechnology and gene companies followed the lymphokine field like a shadow and there were good reasons for that. It was the lymphokine field which provided projects and promising profits. They felt particularly encouraged by one of the first spectacular successes, that is cloning of interferon, which up to then had resisted all attempts of structural analysis.

As outlined above, the problems of lymphokine research were tailor-made for this new technology to solve. All of a sudden recombinant lymphokines were available from gene companies – mostly provided free of charge – for the search of biological effects or, even better, of clinical indications. The hopes run high for the latter. In an article in 'Business Week' from October 8, 1984, it was announced that the next revolution in medicine is almost there. Today, 5 years later, where are the children of this revolution?

Currently, the list of interleukins has not even doubled. The definition of interleukin 8 is still discussed. Instead, major efforts have been put in the elucidation of biological functions of the few known lymphokines, mostly interleukin 1 and interleukin 2. It is tempting to speculate why so many human and material resources are spent on so few factors. Very likely, one reason is that availability of materials and the marketing efforts of interested companies have focused interest. Even though no final clinical indication for therapeutic or diagnostic use has been established, we have learned a great deal about their molecular and cellular biology. We have learned that all known lymphokines have pleiotropic effects, that is all lymphokines act on different types of cells inducing different, even opposite effects. The corresponding lymphokine receptors have been identified and characterized in most instances and many researchers are engaged in investigating the subcellular mechanisms of signal transduction which integrates cellular and molecular biology in the lymphokine field.

If we look, however, at the applications of lymphokines in clinical diagnosis and therapy, the promises have not yet been fulfilled. If we consider the tremendous efforts spent on, for example, interferons to develop a clinical indication, one has to admit that the results are rather meager. Treatment of hairy cell leukemia by interferon-α is now established [14]. On the other hand, interferon-β and interferon-γ could not yet be assigned to a special treatment or diagnosis. Factors like GM-CSF or G-CSF or erythropoietin are sure to have a place in the list of therapeutics of radiation damage [15]. Interleukin 2 has been expected to find a broad application in tumor therapy, but apparently its efficacy has no relation to its side effects and cost of treatment [15]. Another factor is interleukin 1, which is pleiotropic, that is an absolutely bewildering array of biological effects has been described mostly on cultured cells which make it impossible to see a common denominator for a biological concept. The notion that interleukin 1 is the principal mediator in chronic inflammatory lesions is difficult to conceive, as no convincing data demonstrate interleukin 1 production by tissue macrophages in situ [17].

This brings up another issue. Immunology, in particular the lymphokine field, has been shaped by reductionism, that is the dissection of a complex biological system into its smallest component. Unfortunately, this approach has its severe drawbacks which arc manifest in the application of lymphokines in diagnosis and therapy.

It is a long known fact that cells which are thought to function on a special milieu in the tissue will express a different program in order to survive in a strange environment. This is particularly so with macrophages, but all other cells behave similarly or they die. In the new environment surviving cells may express gene products which are suppressed in the normal phenotype. Therefore, it is difficult if not impossible to make a conclusion, from the behavior in vitro to the one in vivo. Furthermore, in vitro bioassays barely reflect the complex in vivo situation and some research areas in the cytokine field just experience how misleading biossays in vitro can be.

If one reads the ever increasing literature on lymphokines and evaluates 'sine ira et studio' results of clinical trials one cannot help to conclude that the lymphokine field is entering another phase of depression. Why? Certainly not because of technical shortcomings. This time it is rather ideological. The influence of business is dominating and the longing for a quick profit is too strong. Shortcuts, therefore, are the preferred choices. As soon as a recombinant lymphokine is available, it is sent to clinicians, who screen it on all sorts of diseases, no matter how different the pathomechanisms might be. The success rate, therefore, is alarmingly low. It is difficult to conceive whether companies can continue for long to 'shepherd' lymphokines to market without clear-cut indications for diagnosis and therapy. What is widely missing in the lymphokine field is the solid assessment of the role of a particular lymphokine in a particular disease. This means that with current technology (antibodies, nucleotide probes) it should be possible to detect a lymphokine in a lesion, identify the producing cell, and explore the regulating mechanisms.

Fortunately, there are clear-cut indications that we are learning again to appreciate the power and unequivocal answers of histological and pathological analyses. If the lymphokine field will also integrate molecular pathology, and continue what has once started by Virchow as cellular pathology, it will continue to be one of the most exciting research ventures of our time.

References

1 Landsteiner K, Chase MW: Experiments on transfer of cutaneous sensitivity to simple compounds. Proc Soc Exp Biol Med 1942;49:688.

2 Rich AR, Lewis MR: The nature of allergy in tuberculosis as revealed by tissue culture studies. Bull Johns Hopkins Hosp 1932;50:115.

3 George M, Vaughan JH: In vitro cell migration as a model for delayed hypersensitivity. Proc Soc Exp Biol Med 1962;111:514.

4 Bloom BR, Bennett B: Mechanism of a reaction in vitro associated with delayed hypersensitivity. Science 1966;153:80.

5 David JR: Delayed hypersensitivity in vitro: its mediation by cell-free substances formed by lymphoid cell interaction. Proc Natl Acad Sci USA 1966;56:72.

6 Dumonde DC, Wolstencroft RA, Panayi GS, et al: 'Lymphokines': non-antibody mediators of cellular immunity generated by lymphocyte activation. Nature 1969; 224:38.

7 Waksman BH, Namba Y: On soluble mediators of immunologic regulation. Cell Immunol 1976;21:161.

8 Cohen S, Ward PA, Bigazzi PE: Cell cooperation in cell-mediated immunity; in McCluskey, Cohen, Mechanisms of Cell-Mediated Immunity, New York, Wiley, 1974, p 331.
9 O'Garra A, Umland S, De France T, et al: B-cell factors' are pleiotropic. Immunol Today 1988;9:33.
10 Henney CS: Interleukin 7: effects on early events in lymphopoiesis. Immunol Today 1989;10:170.
11 Sorg C, Bloom BR: Products of activated lymphocytes. I. The use of radiolabelling techniques in the characterization and partial purification of the migration inhibitory factor of the guinea pig. J exp Med 1973;137:148.
12 Isaacs A, Lindemann J: Virus interference. I. The interferon. Proc R Soc Lond 1957; 147:258.
13 Taniguchi T, Mantei N, Schwarzstein M, et al: Human leukocyte and fibroblast interferons are structurally related. Nature 1980;285:547.
14 Quesada JR, Reuben J, Manning JT, et al: Alpha interferon for induction of remission in hairy-cell leukemia. N Engl J Med 1984;310:15.
15 Cosman D: Colony-stimulating factors in vivo and in vitro. Immunol Today 1988; 9:97.
16 O'Garra A: Interleukins and the immune system 1. Lancet 1989; April 29:943.
17 Dinarello CA: Interleukin-1 and its related cytokines. Cytokines 1989;1:105.

Prof. Dr. C. Sorg, Institute of Experimental Dermatology,
University of Münster, D-4400 Münster (FRG)

Waksman BH (ed): 1939–1989: Fifty Years Progress in Allergy.
Chem Immunol. Basel, Karger, 1990, vol 49, pp 90–120

From Prausnitz-Küstner to Passive Cutaneous Anaphylaxis and Beyond[1]

Zoltan Ovary

Department of Pathology, New York University Medical School and Kaplan Cancer Center, New York, N.Y., USA

In April 1988 I received a letter from Paul Kallós dated March 27, 1988, inviting me to write a review for the 50th anniversary volume of International Archives of Allergy and Applied Immunology. He proposed to me the following title: 'From P-K to PCA'. I accepted with pleasure and wrote back to him, that I would like to modify somewhat the title to be: 'From P-K to PCA and Beyond', if he does not think that it is too presumptuous. I am including here a photocopy of his answer.

June 26,1988

Professor Zoltan Ovary
New York.

Dear Zoltan:

Thank you for your letter of June 15. It is a historic fact that PCA is a development of the PK-reaction of great importance. Prausnitz himself admired your work. I cannot think that it would be pretentious or inappropriate to use the title 'From Prausnitz-Küstner to PCA and beyond' !!I am very glad that you will contribute to the Anniversary Volume.

Liselotte joins me in sending you fondest regards and best wishes.

Yours as ever,

Paul

Paul Kallős

[1] Supported by NIH Grant AI-03075.

Passive Cutaneous Anaphylaxis: A Tool for Quantification and Determination of Biological Activities of Antibodies

Investigations of human allergic reactions were made possible when, in 1922, Prausnitz transmitted passive hypersensitivity to himself by injecting the serum of his colleague Kustner, who was highly sensitive to fish [1]. This occurred about 20 years after the publication by Portier and Richet [2] in which these authors described *anaphylaxis*.

Anaphylactic reactions had by then been actively studied in animals. In 1907 Maurice Nicolle [3] showed that anaphylaxis can be transmitted passively in animals, and that the agent responsible is an antibody . In 1910, Sir Henry Dale showed that one of the main substances liberated during anaphylactic reactions is histamine [4, 5]. Histamine has many different actions in vivo, the two most important being the contraction of smooth muscles (utilized by Dale) and the increase in permeability of the small venules.

Studies in animals showed that when serum from immunized animals is mixed with antigen in vitro, then this mixture is toxic and provokes symptoms of anaphylaxis when injected into normal animals [6]. Underlined here is a very subtle difference: 'anaphylaxis' was not written but 'symptoms of anaphylaxis', and it will be seen why this subtle difference is important. It had been shown by Bordet [7] in 1913 that when agar is added to normal fresh serum, this mixture becomes toxic and could kill the animal when injected (even if the agar was removed from the mixture) and the symptoms were those of anaphylaxis. The substance(s) generated were named 'anaphylatoxins'. These facts were confusing and different possibilities were proposed to explain them. One explanation was that the colloidal stability of the serum was altered during anaphylaxis. This perturbation was named 'colloidoclasie' and the famous clinician Fernand Widal [8] in 1921 even published an article on *Colloidoclasie*. Today we know that this explanation is not much more than a dream-desire. It is not necessary to analyze here the publications concerning the problems generated by the 'anaphylatoxins'. Suffice it to say that it was only in 1959 that, with Osler, we showed that the 'anaphylatoxins' are fragments of complement [9]. In 1959, only 4 components of complement were known, and it was not even envisaged that others might exist. We demonstrated that 'anaphylatoxins' are liberated only when the last component of the cascade (the preparations containing C3) were added. What were called anaphylatoxins are in reality the small fragments detached from the alpha-chains of C3 or C5 (C3a or C5a). C3a or C5a are potent histamine liberators from mast cells. The pathways of histamine liberation by the anaphylatoxins and real anaphylactic reactions are different, but the end result (histamine liberation) is the same. Unfortunately, the misnomer 'anaphylatoxin' has been used in the literature for so long that it cannot now be changed.

The Prausnitz-Kustner or PK reaction was widely used to investigate human allergies and we had learned from these studies that very small amounts of antibodies are needed. The PK reaction was the only way to standardize the allergens and thus, in addition to its utility in diagnosis, it also had a tremendous practical importance.

Antibodies were supposed to be a sort of propriety of immune sera, but the nature of antibodies was not known. Antibodies were described according to their functions, such as: precipitins, lysines, antitoxins, and agglutinins. It was thought that each function was a different propriety of the immune serum or perhaps a different substance. Heidelberger and Kendall [10], in the mid-1930s developed the quantitative microprecipitin reaction and showed that antibodies are proteins and can be weighed. However, property cannot in fact be weighed. The quantitative micropreciptin reaction became one of the most important tools of immunology at that time as great emphasis was placed on quantitation. When ultracentrifugation became available Heidelberger sent his student Elvin Kabat to Tiselius and they showed that antibodies are gamma globulins [11].

In the late 1930s Bovet and Staub [12] synthesized the first antihistamines and with these products a new therapy became available for treatment of allergic reactions. We know that these antihistamines act by competitive inhibition on the H_1 receptors [13], but they do not have any impact on the fixation of histamine to H_2 receptors and thus do not inhibit secretion of the gastric glands.

A turning point came in the early 1940s when Chase [14] showed that anaphylactic-type hypersensitivities could be transmitted with antibodies (immune serum) but the tuberculin-type reactions required living cells. This was then the basis of the classification of hypersensitivity reactions into four types by Gell and Coombs [15].

Mentioned above was that Sir Henry Dale used the action of histamine on smooth muscles for his investigations. Nicolas Jancso, the Hungarian pharmacologist, used the action of histamine on the permeability of the small venules [16]. He introduced a new technique: the intravenous injection of indian ink to visualize the permeability of the small venules and the phagocytosis of the carbon particles by the endothelial cells of these venules.

With this method he could easily visualize the increased permeabil. produced by injected or liberated histamine in vivo. The technique of Jancsc was used by the writer with Biozzi at the Clinica Medica of the University of Rome [17]. Biozzi and Benacerraf used it in Paris for the important work they published on the reticulo-endothelial system [18]. Later, in our work Evans blue dye was substituted for india ink, as it was easier to manipulate. With this technique, it was discovered that very minute amounts of intradermally injected rabbit antibody into guinea pigs could be detected and this technique was named passive cutaneous anaphylaxis or PCA [19]. With PCA nanogram amounts of rabbit antibody were detected in guinea pigs, which was remarkable since, with other methods, the minimal detectable amount was about 50 µg. However, antibodies from chicken or horse could not sensitize the guinea pig for PCA reactions.

The reaction where the antigen was injected before the antibody and the antibody was used for the challenge was called *reverse PCA*. Reverse PCA reactions could be elicited only with gamma globulins and only gamma globulins from those species which could sensitize the animals for PCA reactions. For example, in guinea pigs, good reverse PCA reactions could be obtained with rabbit IgG molecules and antibodies against rabbit gamma globulins. Every antirabbit gamma globulin was effective even if produced in chickens or horse and, as had been seen, immunoglobulins from these species do not sensitize the guinea pig for PCA reactions [20].

Fred Karush and the writer showed that monovalent haptens are inefficient as challenging antigens for the PCA reaction [21]. It was possible to show with anti-dinitrophenyl antibodies, that monovalent haptens are ineffective, but bivalent haptens are effective as challenging antigens for PCA reactions [22].

An important breakthrough came at this time (1959): Porter [23] published his studies on papain digestion of rabbit gamma globulin. Rabbit gamma globulin is a 150,000 kdalton protein. Porter reasoned that it would be easier to analyze this molecule if smaller fragments could be obtained for analysis. He showed that when rabbit gamma globulin is digested with papain two types of fragment are obtained: one which can still combine with the antigen (he named it Fab) and the other which cannot combine with the antigen he named Fc as it was crystallizable. Nobody guessed then that this was because the Fc fragment was the same in all rabbit gamma globulin molecules, but the Fab fragment varied from one molecule to another. Yet the fact that a fragment was crystallizable already established that in all rabbit IgG molecules there is a portion which is constant.

Very soon after the publication of Porter we showed, with Fred Karush, that the Fab fragment from an antilactoside antibody could not sensitize guinea pigs for PCA, though the intact antibody was very effective. However, the Fc fragment, even from a 'nonspecific' (normal) gamma globulin could sensitize for reverse PCA reactions [24]. We had also shown that without the Fc fragment, complement fixation did not occur [25]. Of course, complement fixation by the classical pathway is referred to as the alternative pathway had not yet been discovered.

Later, Nisonoff [26] digested the rabbit gamma globulin with pepsin and obtained a bivalent fragment $F(ab')_2$ which could precipitate the polyvalent antigen. We showed that this fragment, although it combines with the antigen, is still unable to sensitize the guinea pig for PCA reaction and cannot fix complement by the classical pathway. Our conclusion was that the Fc fragment carries the biological activities such as the structures necessary for complement fixation by the classical pathway, and the structure necessary for fixation to cellular receptors on mast cells. Figure 1 shows a typical PCA reaction in a guinea pig by the native rabbit antihuman gamma globulin antibody and its inhibition by an excess of the pepsin digested $F(ab')_2$ fragment of the same antibody. In retrospect it must be underlined how fortunate we were that the rabbit IgG can sensitize guinea pigs – that it can fix to receptors on guinea pig mast cells.

The fact that injection of antigen immediately after the injection of the antibody is ineffective for PCA reactions was already showing that the antibody must become fixed to some receptors (receptors on mast cells). The time necessary between the injection of the antibody and the challenging injection of the antigen was called the sensitization period. In figure 1 it is marked as LP, latent period, but this term was abandoned as sensitization period better describes the phenomenon. With the experiments quoted above, we showed that this receptor fixes to a structure on the Fc fragment. We already knew that for PCA reactions we need at least a bivalent antigen.

All these facts led to the proposal of the cross-linking or bridging hypothesis in 1961 [22], which states that 'a bridging of two different combining sites of two different antibody molecules by two antigenic determinants on the same antigenic molecule is necessary and sufficient to obtain a PCA reaction [27]. This 'cross-linking' or 'bridging' was confirmed by many investigators and was later also proposed for many other systems. The most elegant demonstration of the bridging hypothesis was done by Ishizaka and collaborators who showed that cross-linking of the high-affinity mast cell IgE

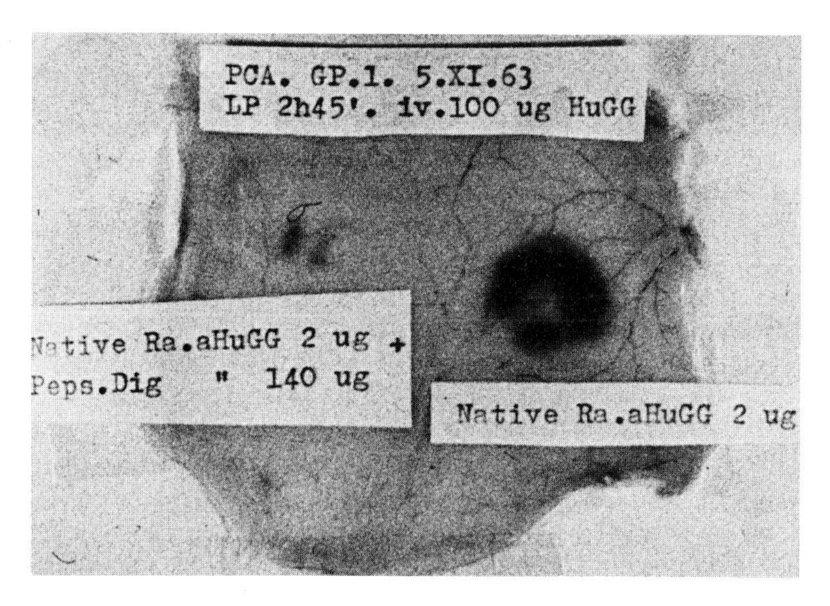

Fig. 1. On the right side 2 μg antibody from a native anti-human gamma globulin was injected, on the left side the same dilution of the same antibody was injected, mixed with 140 μg pepsin digested sample from the same antibody. Two hours and 45 min later the guinea pig was injected i.v. with 100 μg human gamma globulin in 0.5% Evans blue dye. The reactions were read on the reflected skin 30 min later.

receptor (FcεRI) by an antireceptor antibody is sufficient to degranulate the mast cells and liberate the important active substances which they contain [28]. The mast cell is the important cell in anaphylactic reactions. It contains very active substances either preformed or in a precursor state. The anaphylactic reaction is a typical autopharmacological reaction. The mast cell can be considered a chest of poison, a sort of Pandora's box normally well closed, that opens only as a result of the bridging of its FcεRI or after the action of C3a or C5a. The most important preformed substances are histamine (serotonin in rodents), heparin, two tetrapeptides called the eosinophilic chemotactic factors of anaphylaxis (EcF-A), heparin and some enzymes. The most important generated substances are the arachidonic acid metabolites (by the lipoxygenase pathway) formerly called the slow-reacting substance of anaphylaxis or SRS-A, now identified by Piper et al. [29] and Samuelson [30] as the leukotrienes LTC_4, LTD_4 and LTB_4. LTC_4 and especially LTD_4 are a thousand times more potent than histamine as vasoconstrictors. LTB_4 is a potent chemotactic for neutrophils and eosinophils. The action of the platelet-activating factor, PAF, in the acute phase of the anaphylactic reaction is insignificant compared to that of histamine or the leukotrienes.

The degranulation of mast cells following the signal from the FcεRI was intensively investigated by Austen's group [31], Lichtenstein, Ishizaka and co-workers [32, reviewed in 33] to mention some of the pioneers in this field. The cascade of events which follows the bridging of the FcεRI starts with the activation of a serine esterase [31] and is rapidly followed by activation of adenylate cyclase, Ca^{++} influx, activation of methyltransferases and many other events, the last of which is histamine release (or serotonin in rodents). The generation of leukotrienes follows these first events in a few minutes.

Porter [34] continued his investigation of the rabbit IgG molecule with other methods and showed that each molecule is made up of two identical A and two identical B chains (the classic 4 chain model of Porter). Later, the chains were named H (heavy) and L (light). This wonderful work was rewarded with the Nobel prize. Shortly after Porter's work Hilschmann and Craig [35] showed that the L chain is divided into two portions of about the same size: the N-terminal half is variable from one L chain to the other (V), but the C terminal half is constant (C) and is identical from one L chain to the other of the same group. It cannot be emphasized enough what a tremendous impact this discovery had on future investigations, not only on immunoglobulins, but also on other molecules, such as class I and class II molecules coded for by the genes of the major histocompatibility complex and even on the T cell receptors and others. Shortly after the discovery by Hilschmann, it was also shown that the H chain had an N-terminal variable and a C-terminal constant portion, but that the constant portion of the H chain is three times longer than the V portion. In retrospect, as noted above, the fact that the rabbit Fc fragment was crystallizable should have led us to an awareness of its constant structure.

Intrachain disulfide bonds delimit loops in each chain, called domains. It is interesting to note that many proteins involved in immunological interactions have domain structures: immunoglobulins, T cell receptors, MHC class I and II molecules, to mention only some of them. These molecules were later grouped under the name of immunoglobulin superfamily [36] and it is thought that they all derive from a primitive domain molecule. It was then investigated whether some functions could be assigned to separate domains. The combining site (later also called paratope) which reacts with the antigen is in the first domain. As expected, in view of the enormous variability of the antigenic determinants (later also called epitopes), the first domain is made up by the variable segment of each chain.

At that time the writer was concerned especially with the biological activities of the antibody molecules. We and others showed that the first domain of the Fc fragment carries the structures reacting with the first member of the first component of complement, namely with C1Q [25]. However, the domain concept does not hold for some of the other reactions. For example, the fixation to macrophages and to the FcεRI on mast cells necessitates the integrity of the Fc fragment. We had shown with Lamm that when the last domain of the rabbit gamma globulin is separated, neither the remainder nor the separated CH3 domain has any activity. Moreover, the mixture of the two separated fragments is unable to inhibit the fixation of the intact molecule to the receptor on macrophages or to the high-affinity FcεRI on mast cells [37]. Quite recently, Geha and colleagues, in wonderful experiments, showed that the same is true for the human IgE antibody fixation to the FcεRI on human mast cells. Geha showed that only those peptides in which the junction of the 2nd and 3rd domains were intact could inhibit the fixation of the intact IgE molecule [38]. Glovsky obtained the same result [39].

From the studies of the immunoglobulin chains came better knowledge concerning allotypes. From the study of the C portions of the H chains came the division of immunoglobulin molecules in classes and subclasses (isotypes).

Antibody Isotypes, the Prausnitz-Küstner Reaction for IgE Studies. Monoclonal Antibodies, in vitro Methods for IgE Determinations

In 1962 at NYU in collaboration with Benacerraf and Bloch, we made an observation studying guinea pig antibodies: the guinea pig IgG could be subdvided into two classes. One, the electrophoretically faster moving IgG1 can sensitize the guinea pig for anaphylactic reactions, but cannot fix complement by the classical pathway. The other, the electrophoretically slower-moving IgG2 does not sensitize the guinea pig for PCA reactions, but fixes complement by the classical pathway [40–43]. Figure 2 shows the separation of guinea pig IgG1 and IgG2 by starch block electrophoresis [43].

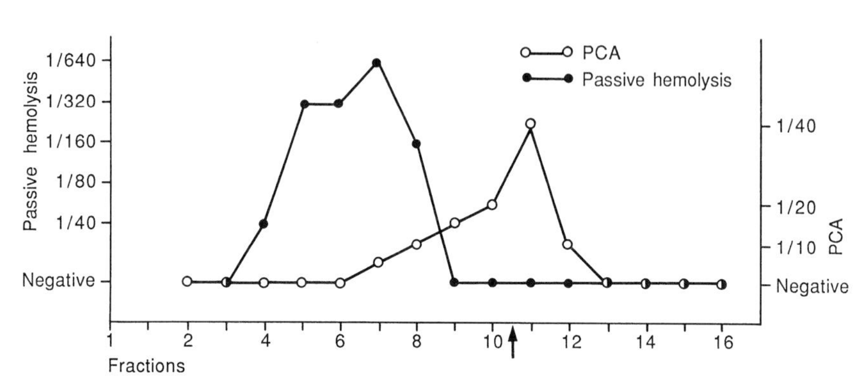

Fig. 2. Titration of passive hemolytic and PCA activities of eluates from starch block electrophoresis of purified guinea pig antipicryl-guinea pig albumin antibodies. Arrow indicates point of application of antibodies on strach block.

These studies were confirmed by many investigators and were an impetus to study the heterogeneity of immunoglobulins in other species too. Different immunoglobulin isotypes were found in many species and it was confirmed that different isotypes may carry different biological activities. We know that the differences between the isotypes in a species are due to difference in the amino acid sequences of the constant regions of the H chains. As the Fc fragment is made of the last domains of the two heavy chain constant regions it is expected that different isotypes might carry different biological activities. These first observations made with guinea pig IgG1 and IgG2 were thus important in view of the fact that biological functions of antibodies could be better understood, and called attention to the importance of the different immunoglobulin classes.

We had just described two important problems: fixation of antibodies to high-affinity mast cell receptors (this process was called the sensitization of mast cells by the antibody) and the different biological properties of different antibody isotypes. Both problems were solved with the help of PCA, or in the case of human antibodies, by the P-K reaction.

We had shown that rabbit IgG can sensitize the guinea pig for PCA reactions, but not the rabbit for anaphylactic reactions and guinea pig IgG1 can sensitize the guinea pig, but not the rabbit or other species. At that time it was still not clear which isotype was involved in human allergic reactions. It was first thought that it was the IgA isotype, but then Mary Loveless presented data on patients who were agammaglobulinemic for IgA, but still had tremendous allergic reactions [44]. The Ishizakas, with great effort and insight and marvellous work, showed that the sensitizing antibody is not of the IgG isotype and called it IgE [45, 46]. Shortly after the studies of the Ishizakas, Johannsson and Bennich [47] in Sweden found an unusual human myeloma protein. The Ishizakas then found that it was of the same isotype as the human antibodies capable of transmitting the P-K reaction. The existence of a human IgE myeloma was important, not only for the confirmation of the existence of the human IgE immunoglobulin class, but also made it possible to develop in vitro methods of human IgE determinations (such as the commerically available Pharmacia Phadebas Rast and Phadebas IgE tests). The importance of these in vitro tests cannot be overestimated, as today it would be impossible to use the P-K test in view of the possibility of transmission of the hepatitis and AIDS viruses. IgE is present in very small amounts in the serum (in nanogram amounts). This is the principal reason why it was not discovered earlier. Figure 3 shows a P-K reaction. Prof. Tomio Tada, the passive participant in these studies, is thanked for this photograph. It is an historical picture as it shows the original observation of the Ishizakas demonstrating that an antibody against IgE can neutralize human reaginic antibody.

IgE is also present in the sera of animals. Here too it is present in nanogram amounts and here too it sensitizes for anaphylactic reactions. However, we had seen that in addition to IgE in experimental animals there is often a class (or subclass) of IgG (IgG1 in the guinea pig) which is present in milligram amounts and can also sensitize for anaphylactic reactions. From IgE, generally, a few nanograms can give threshold PCA reactions, from the IgG classes usually, 50 or 100 times more is needed to obtain similar threshold PCA reactions.

IgE is heat labile in every species, which easily permits its inactivation in vitro (56 °C for 1 h). For some time this heat lability was a very practical tool to demonstrate its presence. IgE is heavier than IgG (the MW is about 180,000 daltons) as it has 4 constant domains rather than 3, like IgG.

In mice where no IgE myeloma has been found for 20 years the PCA reaction was the only method by which IgE could be detected and quantitated. In 1975 Kohler and Milstein [48] constructed the first monoclonal antibody-producing hybridoma. This work had a tremendous effect on the development of immunology and also on other branches of biological research. It was now possible to use this technique for precise analysis of many different immunological reactions and reactants. In 1980 we constructed the first hybridoma-producing monoclonal anti-dinitrophenyl antibody (DNP) derived from cells of BALB/c mice [49]. At nearly the same time

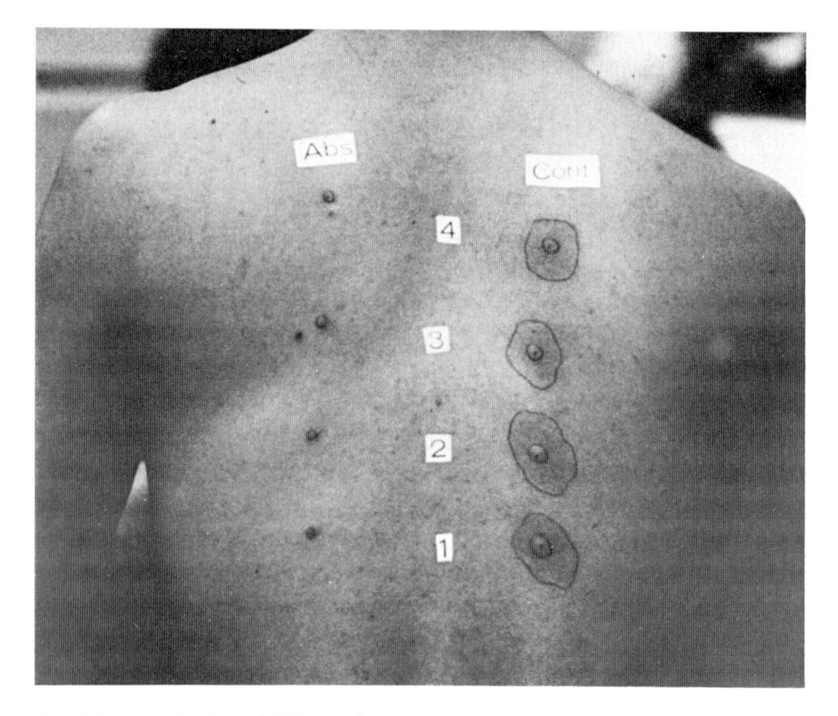

Fig. 3. Neutralization of PK reactions with rabbit anti-IgE antibody. On the right side different dilutions of human sera in sterile saline, and on the left side the same dilutions mixed with rabbit anti-IgE antibody. All injections were done 48 h before challenge. Challenge = The respective antigen dilutions injected into the sites where the sera were previously injected. Photographs taken 20 min later.

Eshhar et al. [50] and Katz and his group [51] also constructed anti-DNP IgE-producing hybridomas from cells derived from different strains of mice.

It was now possible to use radioimmunoassays and ELISA [52] instead of PCA for many investigations. The construction of monoclonal antimurine IgE antibodies [53, 54] made ELISA a very sensitive, reliable method for IgE antibody determinations. Quite recently we showed that, to determine the amount of antibody specific for a given epitope, the order of addition of the reactants is critical. Generally, the wells are first coated with the antigen. If sera from immunized mice are titrated for IgE antibody specific for a given epitope, the first reactant to be added must be the monoclonal anti-IgE antibody. The reason is that in sera of immunized mice, the amount of specific IgE antibody is much less than antibody with the same specificity from other isotypes and these other isotypes may compete very effectively with the antibody of the isotype IgE. However, if the anti-IgE antibody captures the IgE molecules first, then, when the specific antigen is added there is no more competition as the other isotypes were washed away [55]. Figure 4 illustrates the importance of the order of addition of the reactants in ELISA.

It must be underlined, however, that great progress was accomplished using the PCA reaction and, in fact, progress was made because of the use of PCA. For certain investigations even today PCA is very useful. It was with PCA reactions and monoclonal IgE antibodies that, in collaboration with Tada's group in Tokyo, we could determine the allotype of IgE (allotype 7 [56]).

Monoclonal antibodies made it possible to determine the half-life of the IgE antibodies. The half-life of circulating murine IgE is very short: 6–8 h for circulating IgE in comparison with circulating IgGl with a half-life of 12 days. However, when IgE is fixed to the mast cell FcεRI the half-life is 25 times longer [57, 58].

The bridging or cross-linking hypothesis [22, 27] was not the only consequence of the work done with PCA as a tool. Another very important problem was the anchorage of antibody molecules to cellular receptors. The receptors on cells which bind to the Fc fragment have been studied by many investigators. Metzger et al. [59] showed that the mast cell IgE receptor is a multichain molecule and is made up of one α-, one β- and two γ-chains. It is the first domain of the α-chain which reacts with the structure between the CH2 and CH3 domains of the Fc fragment mentioned above [37–39]. The chains have been sequenced and though we do not yet know the crystallographic structure of this high-affinity FcεRI molecule (affinity of 5×10^8 to 1×10^{10}), a model has been presented by Metzger et al. [60].

A second type of receptor for the IgE molecules has been described. This is the low-affinity IgE receptor also called FcεRII. Its affinity is of several orders lower than that of the FcεRI. The gene encoding the FcεRII was cloned simultaneously by Kishimoto's [61] and Yodoi's [62] groups. It is composed of a single chain. It turned out that this molecule had already been identified on the surface of B cells and was called CD23. Interestingly, there are two different forms of human FcεRII. The C-terminal portion is the same in both molecules, but the N-terminal, intracellular portion is different. The FcεRIIa is expressed constitutively only on B cell lines, FcεRIIb is detectable also on other cell types such as monocytes and eosinophils. Normally, FcεRIIb is undetectable on B cells or monocytes; however, it can be induced in both cells by interleukin 4 [63].

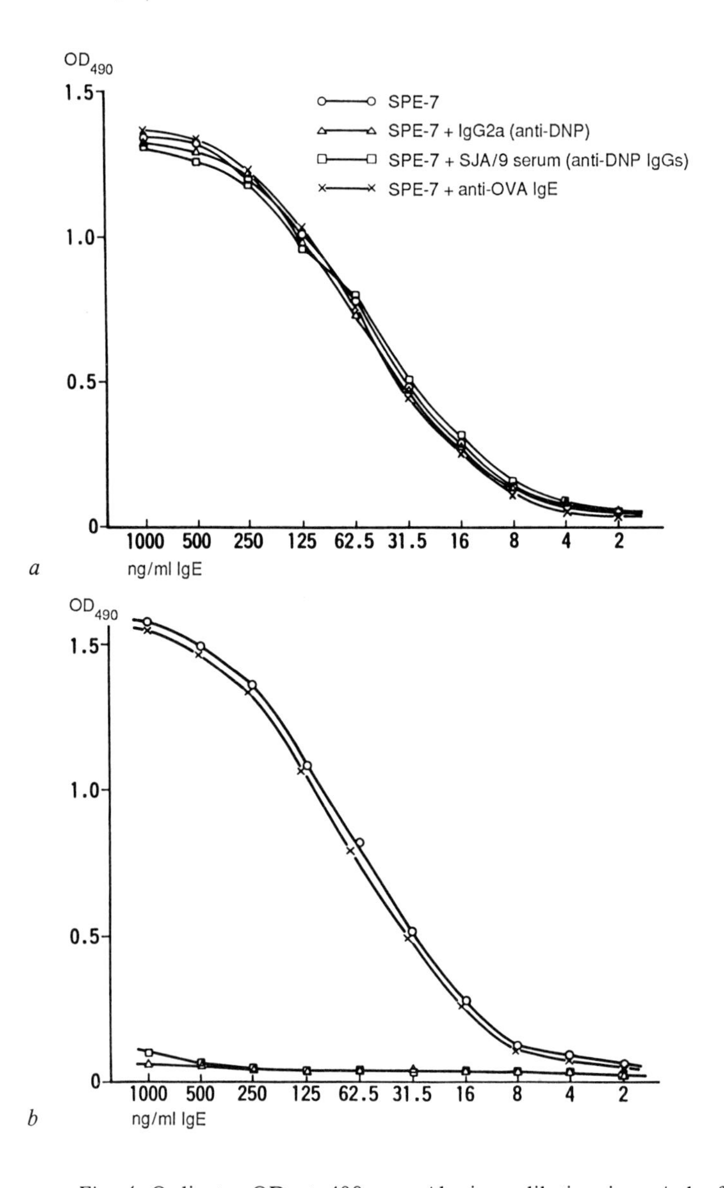

Fig. 4. Ordinate: OD at 490 nm. Abscissa: dilution in ng/ml of SPE-7. SPE-7=Monoclonal anti-dinitrophenyl IgE antibody diluted according to figures on the abscissa IgG2a = constant dilution (1μg/ml) of anti-dinitrophenyl IgG2a monoclonal antibody; SJA/9 = constant dilution (1/100) of serum pool from SJA/9 mice containing anti-dinitrophenyl antibodies of other isotypes than IgE; anti-OVA IgE=constant dilution of monoclonal anti-OVA IgE (1 μg/ml). *a* Wells coated first with monoclonal anti-IgE antibody. *b* Wells coated first with dinitrophenylated bovine serum albumin.

Finally, with Lynch and Mathur we described a third type of IgE binding receptor on Lyt-2 cells [64]. The phenotype of this T cell is that of the so-called cytotoxic/suppressor phenotype, i.e. Lyt-2^+3^+, also called CD8 T cell. Mathur showed that these T cells suppress the synthesis of the messenger RNA for the ε-chain, even in the IgE-producing hybridoma cells [65].

Receptors for IgE (or shed receptors or similar products secreted by cells with these receptors), excluding the high-affinity receptor on mast cells (FcεRI), might have a regulatory role in IgE production. We have seen an example in the Lyt-2^+ T cells, which could inhibit the synthesis of the messenger RNA for the ε-chain. The regulation of IgE production by IgE receptors and IgE-binding factors have been extensively studied in the laboratories of Ishizaka and Katz [reviewed in 66, 67]. A review concerning the structures and functions of Fc receptors for IgE on lymphocytes, monocytes, and macrophages was published by Spiegelberg [68] in 1984. Of course this review preceded the recent discoveries mentioned above [60–65].

The necessity of receptors for IgE fixation for elicitation of anaphylactic reactions has been demonstrated by the observation that injection of antigen immediately after the injection of antibody is ineffective for challenging the PCA reaction. Boyden and Sorkin showed that antibody molecules might fix to lymphoid or other cells [69, 70]. These authors coined the term 'cytophilic' for those antibodies which were capable of 'passively sensitizing' cells. Berken and Benacerraf [71] showed that only guinea pig IgG2 and not IgG1 is cytophilic for guinea pig macrophages. With Tigelaar and Vaz, we showed that mouse mast cells possess receptors for IgG molecules [72]. The great interest concerning the Fc receptors is shown by the fact that Unkeless et al. [73] wrote a chapter in volume 6 of *Annual Review of Immunology* which reviews the Fc receptor.

The Carrier Effect, B and T Cells and Molecules Coded for the I Region of the Murine Major Histocompatibility Complex, Antigen Processing: The T Cell Receptor

In 1962 we observed that when haptenated proteins are used as antigens for secondary antihapten antibody production, it is necessary that the hapten should be coupled to the same carrier that was used to induce the primary response. This was called the 'carrier effect' [74]. About 7 years later Claman et al. [75], Miller et al. [76] and Szenberg and Warner [77] showed that the lymphocytes involved in the antibody production are divided into two distinct families: the B cell which actually produces the antibody (in our case against the hapten as epitope) and the T cells which do not produce antibodies. Mitchinson [78] and Rajewsky [79] then showed that a subpopulation of T cells (generally L3T4$^+$ in the mouse, and T4$^+$ in the human) later called T helper cells (Th) recognize the carrier protein and then help the B cells to produce antibodies. The carrier effect is thus due to the recognition by Th cells of the carrier protein.

Many scientists worked to discover how B cells recognize the epitope (or the hapten) and the T cells, the carrier protein. It was readily understood that the immunoglobulin, anchored to the surface of the B cell, recognizes the antigen by its specific epitope (in our case the hapten). This involves the recognition of the surface configuration in natural proteins, therefore it can be recognition of tertiary or quaternary structures which may not exist on the unfolded protein. However, the recognition mechanism used by T cells was for a long time elusive. Finally the group of Davis and Hedrick [80] were able to identify a gene encoding one of the chains of the T cell receptor by subtractive hybridization. It turned out that the T cell receptor for antigen is also a two-chain molecule and has domains similar to those of the immunoglobulin molecules.

To understand the recognition of the antigen by T cells, it is necessary to go back in time and describe very briefly what we learned from the study of different mouse strains and their response to antigens.

It was known for a very long time that some substances are not antigenic and some are poor antigens. It was also known that antibody production could be enhanced by adjuvants. The first clearly formulated question was: Why is it that some substances are not antigenic in a given strain or species? This question was already asked before the beginning of this century, when immunization and vaccination were experimentally investigated. Autologous proteins are generally not antigenic, as Ehrlich [81] claimed at the end of the last century by the principle of 'horror autotoxicus'. We should add that today this dogma has turned out to be not so absolute as Ehrlich had thought. Indeed, it is well documented that auto-antibodies can be produced and below we discuss this problem briefly, when we present some work done in our laboratory.

When inbred strains of mice became available [82], it was observed that well-defined synthetic antigens were antigenic in some but not in other strains. When the major histocompatibility complex (MHC) of mice (*H-2*) on chromosome 17 was investigated, it was noted that there was a good correlation between antigenicity and *H-2* [83]. The MHC is a remarkable region of this chromosome as it codes for several very important proteins including one chain of the class I and both chains of the class II proteins. The best-defined class I antigens called K and D are those which are present in all (or nearly all cells). These proteins are those which are responsible for graft rejection and which gave the MHC its name. The class II antigens are coded for by the I region located between the K and D regions. The designation I was chosen because the immune response for different compunds and the MHC resctriction of this region was firmly established and the genes coded for by the I regions were called the immune response genes or Ir genes [84]. The proteins coded fo by the Ir genes are the class II MHC proteins, also called Ia molecules. These molecules are present on the surface of B cells and macrophages [85, reviewed in 86]. The Ia molecules are made up of two chains, each chain has two extracellular domains (determined by two intra-chain cysteins), a short transmembrane and a short intracellular portion [86].

The studies of Marrack and co-workers [87], Rock et al. [88], Lanzavecchia [89], Benjamin et al. [90], Berkower et al. [91], Sette et al. [92] and Allen et al. [93], to mention some of the studies on antigen processing [94], have shown that T cell recognition is a complex phenomenon. This discussion is limited to the antigen processing for T helper cell recognition. For effective antibody production for all the isotypes except IgM, T cell help is needed. However, the T cell does not recognize the antigen if it is not processed and presented in the context of class II MHC molecule: the Ia molecule on the surface of the antigen-presenting cell. The first step in this process is the internalization of the entire antigenic molecule, the second step is the unfolding and partial breakdown of the polypeptide chain of the antigen (now inside the antigen presenting cell), the third is the exposition of the processed molecule on the membrane of the antigen-presenting cell in context with an Ia molecule.

The macrophage-type antigen-presenting cell can internalize any antigenic molecule; when the antigen is presented to a B cell, then only those antigenic molecules which have epitopes complementary to the surface-anchored antibody molecule of the B cell can be recognized. Thus, the B cell has specificity in the recognition process, the macrophage-type antigen-presenting cell has no specificity. The hapten fixed to the carrier molecule is the epitope which the B cell recognizes. In native proteins the epitope is a surface structure. As proteins are folded polypeptide chains, it is possible that the epitope is formed by juxtaposition of amino acids, which are far away from each other in the polypeptide chain.

The exact mechanism of the 'antigen processing' is not yet known. It is thought that the processing is necessary because it unfolds the protein (and perhaps breaks into smaller entities) and T cells unlike B cells recognize (sequential) contiguous amino acid structures. The unfolding of the protein (the carrier moiety of the antigen) exposes amino acids which are recognized by the T cell receptors and which might also interact with the Ia molecule; it is also possible that some of the amino acids interact only with amino acids of the Ia molecule or only with amino acids of the variable portions of the T cell receptor [90–94]. B cells might recognize tertiary or quaternary structures, T cells recognize portions of the unfolded polypeptide chain of the carrier protein. Therefore, the portion of the antigen (carrier) recognized by the T cell receptors should not be called epitope, as it is not exposed on the surface of the antigenic molecule.

A better designation would be *aphantotope*. The Greek word 'aphantos' means 'hidden', and the structure recognized by the T cells is not readily available for recognition, as it is not on the surface of the molecule.

It is now much easier to understand the carrier effect and the Ir genes. The Ir genes are simply Ia molecules coded for by the I regions of the MHC complex. They regulate T cell recognition and thus indirectly influence antibody production by the B cells. It has been established that some amino acids of antigenic proteins may react with specific amino acids of the Ia molecule and some others with the T cell receptor variable regions or more precisely with the variable portions of its α- and/or β-chains [90–94]. It is postulated that the variable portion of the α- and/or β-chain of the T cell receptor directly recognizes some portion of the Ia molecule, but we do not yet have any direct evidence to know which are these amino acids.

When L3T4(CD4) T cells recognize the processed antigen fragment + the MHC class II (Ia) molecule on the surface of the antigen-presenting cell, the T cell is 'activated' and one of the important consequences of this activation is the production and secretion of interleukins. We still do not know how these processes come about.

These developments have of course been progressive, but have occurred in a relatively short time. The antigen processing is such an important problem of actuality that an entire issue of *Immunological Review* was dedicated to 'Antigen Processing' [94]. This example shows how we progressed from the simple observation of the 'carrier effect' to a much better understanding of antibody production.

We had seen that T helper cells recognize sequential amino acid configurations of protein antigens and we shall return below to other important functions of T helper cells. These T helper cells have a marker: the L3T4 molecule (or CD4 molecule). There are other T cells which have a different marker, namely Lyt-2/3 (or CD8 molecule). These are the so-called cytotoxic or Tc cells, also called cytotoxic/suppressor type T cells. These CD8 T cells recognize processed antigen on class I molecules.

The class I molecules have recently been crystallized [95, 96] and the form suggests that a peptide might fit into a groove formed by the first two domaines of the Class I chain. By analogy, it is proposed that the class II molecule would similarly accommodate a peptide derived from the processed antigen [97].

Two great breakthroughs came in the mid-1970s which contributed to shaping our thinking and helped to perform much successful research in the field of immunology. One was the construction of monoclonal antibodies by Kohler and Milstein [48] in 1975, the second was the discovery of the immunoglobulin genes and their organization in the germline by Tonegawa [98] in 1976. That the molecule of immunoglobulin gene is made up of DNA segments (exons) separated by other DNA elements (introns) and that the rearrangement of the 'germline' DNA is necessary for the actual protein producing template or 'rearranged' DNA was a tremendous novelty. These facts opened new horizons for the explanation of antibody diversity and also for the study of the stages of the formation of the immunoglobulins.

It turned out that those DNA elements which do not code for the actual amino acid chains of the antibody are nonetheless very important elements, as these introns contain the promoters, enhancers and possibly other DNA elements necessary for the production of the antibody molecule.

Adjuvants and IgE Regulation

Some problems related to antigenicity were discussed above, and we mentioned that antibody production can be enhanced by adjuvants. The current status of immunological adjuvants was recently described by Warren et al. [99].

The use of adjuvants goes back to the work of Ramon, when he was director of the Garches Annex of the Pasteur Institute of Paris and was actively engaged in antisera production. He observed that those horses which had a slight inflammation at the site of the antigen injection produced more antibody than others which had no inflammation. He then investigated various substances (tapioca, agar, etc.) to boost the antibody production [100].

One of the most used adjuvants is Freund's complete adjuvant (FCA), where the antigen is in a water in oil emulsion and contains dead *Mycobacterium tuberculosis* or acid-fast bacilli [101]. Freund was aware of the work of Coulaud and Saenz, both collaborators of Monsieur Calmette at the Pasteur Institute in Paris on the adjuvant effect of acid-fast bacteria, and discussed their findings when he developed the FCA [101]. Other adjuvants, such as alum and *Bordetella pertussis* vaccine are also frequently used [99] especially if IgE antibody is to be produced [102].

IgE antibody production is potentiated by parasitic infections, especially by helminths. This was exploited by Ogilvie and Jones [103], Orr and Blair [104],and Jarret [105] in England; and Bloch and Wilson [106], Kojima and Ovary [107], Ishizaka et al. [108], and Katona et al. [109] in this country.

How do all these adjuvants work? It is believed that they may attract cells involved in antibody formation and in immunological processes such as cytolysis, graft rejection, delayed-type hypersensitivity and others.

When we investigated the allotype 7 (that of IgE) we had to immunize mice of many different strains [56]. It was surprising that one of the strains, the SJA/9 strain, MHC congenic to the SJL strain, did not produce any IgE antibody. Using adoptive transfer and examining the B or T cells from this strain, we found that both cells can collaborate with B or T cells of the congenic SJL mice. It was puzzling, as nothing seemed to be wrong with either the B or the T cells of the SJA/9 strain and yet in the intact SJA/9 animal they could not collaborate for IgE production. However, when the B cell donors were injected with unprimed T cells from the SJL strain or infested 14 days after priming with *Nippostrongylus brasiliensis*, good IgE antibody production was obtained against the immunizing antigen. From these experiments it was obvious that in ordinary conditions the SJA/9 mice lacked some sort of T cell activity.

We proposed a hypothesis based on the above experiments which postulated that for IgE production, intervention of several sequential T cell actions are necessary and IgE antibody is secreted only after the last intervention of T helper cells and antigen [110].'Or the product of some T helper cells' could now be added.

Interleukines: Their Roles in B and T Cell Activities, Positive and Negative T Cell Selection

Other, very interesting discoveries, first thought to be involved mainly with B cell functions, might shed some light on the role of T cells in antibody production and immunity in general. These discoveries concerned products of lymphocytes and of mononuclear cells. Several of these 'factors' were described by different authors. It was not easy to keep up with the different factors described by the different authors. Sometimes the same products were described under different names. These products, synthesized by lymphocytes, were called lymphokines and later, when they were cloned, the word 'interleukin' was adopted for many of them with a numerical suffix according to the order of definitive identification. However, it must be noted that interleukin 1 (IL-1) is produced by macrophages. From several works, only a few are quoted, such as those written by Howard and Paul [111] on B cell growth and differentiation factors, by Smith [112] on interleukin 2 (IL-2), and its receptor by Greene and Leonard [113], on IL-1 by Durum et al. [114], on interleukin 3 by Schrader [115], on interleukin-4 by Paul and Ohara [116], and on interleukin-5 and interleukin-6 by Kishimoto and Hirano [117].

In the mid-1980s it became clear that IL-4 is essential for the production of IgE and IgG1 [118]. It was therefore of great interest to study the IL-4 producing T cells. From the DNAX laboratory Mosmann, Coffman and collaborators [119–121] cloned murine L3T4 (T helper) cell lines and showed that there were two different types of cloned L3T4 cells: one which produced, among other lymphokines, IL-2 and interferon-γ, but not IL-4 and IL-5. This was called the T helper 1, and the other, which does not produce IL-2 and interferon-γ, but produces IL-4 and IL-5 was called T helper 2. The T helper 2 and at least one of its products, IL-4, is essential for the production of IgE. We have shown that lipopolysaccharide-stimulated normal spleen cells from SJA/9 and even from nude mice can produce IgE, when IL-4 is added to the cultures [122]. It was shown long ago that IgE production is absolutely T cell dependent [123]. Our earlier hypothesis was that sequential T cell interventions are necessary for IgE production [110] and these results were in line with this hypothesis. Now we could formulate this hypothesis in a somewhat modified form: one of the T helper cells (and/or its product) sets the B cell apparatus for antibody production in 'motion'. This first T cell intervention might even be produced by a T helper 1 cell producing IL-2 or it might already be a T helper 2 cell producing IL-4. A second interaction, this time certainly by a T helper 2 cell and/or IL-4 will lead to IgE secretion. This is of course only a very simplified model and it is quite probable that in reality the events are more complicated.

It is tempting to propose that one of the actions of the different adjuvants is some sort of chemotaxis and/or activation of the Thl and/or Th2 cells. Freund's complete adjuvant seems to produce preferentially delayed-type hypersensitivity and hinders the production of the IgE isotype [102]. It might be possible that the attraction and/or activation of the Thl is favored by this adjuvant. This fits the fact that Thl is the cell involved in the delayed-type hypersensitivity reaction [124] and also that for IgE production Th2 seems to be important. It is possible that parasitic infections produce a substance which attracts or favors the functions of Th2 cells producing IL-4 and this also fits with the observation that parasitic infections favor IgE production.

Discussed above were antigen processing and the necessity for the processed antigen to be exposed on an Ia molecule to be recognized by the T helper cell. We also mentioned that the other family of T cells carries not the CD4 (L3T4) molecule but the CD8 (Lyt-2^+3^+) molecule and are also called cytotoxic/suppressor T cells. These CD8 cells recognize the antigen complexed to a self class I antigen (generally a K or D molecule in the murine system) and these T cells are those which are responsible for graft rejections [125]. It was shown that a virus-infected cell is killed by immune cytotoxic T cells only if both the target (infected) cell and the killer (cytotoxic) cell have the same MHC [125]. Quite recently it was shown that, also in this case, the antigen must be processed [126]. Practically nothing is known of how the antigen is processed and complexed to the class I molecule. However, the fact that cytotoxic T cells recognize antigen only in context with self class I molecule has been verified by innumerable scientists. The crystallographic studies of Bjorkman et al. [95, 96] mentioned above, showed the groove on a human class I molecule where the processed antigen could fit.

A positive and a negative selection were proposed to explain the fact that the cytotoxic cell must have the same MHC as the target, but they do not react (do not become activated) against self MHC. According to this proposal, in the positive selection, the progenitors of mature T cells are instructed (educated) in the thymus (1) to recognize 'self' MHC molecules; and (2) to recognize antigenic molecules only on a self MHC molecule. In the negative selection, three possibilities might be effective: (1) inactivation of self-reactive lymphocytes (clonal anergy); (2) suppression (clonal suppression), or (3) elimination (clonal deletion [127]). However, we had to wait until Kappler et al. [128] showed that T cell tolerance is obtained by clonal elimination. Clonal deletion was also confirmed by other experiments [129, 130].

It is thought that there are plenty of B cells to synthetize a great number of antibodies with different specificity, but the T helper cells to give the necessary help are missing. An example is provided by the auto anti-IgE antibody production by A/J mice, if these mice were immunized with A/J-derived monoclonal IgE antibodies coupled with keyhole limpet hemocyanin [131]. The T helper cells specific for the coupled protein, in this case the hemocyanin, gave the necessary help and large amounts of anti-IgE antibodies could be produced. Obviously, the A/J mice lack the necessary T helper cells recognizing syngenic proteins.

Quite recently, positive selection was also demonstrated [132, 133]. Moreover, the selection is operative on very young thymocytes which have both the CD4 and CD8 molecules [134]. Some of these experiments were performed with the use of transgenic mice, a technique used increasingly to investigate immunological problems [135].

Auto-Antibodies

We mentioned above that auto-antibodies are now well documented. One of the most important works on auto-antibodies was that of Jerne. When Jerne developed his network theory, based on the production of auto-antibodies, he called the anti-idiotypic antibodies Ab_2 and divided these into three categories: $Ab_{2\alpha}$, which recognizes idiotopes remote from the combining site of the immunizing idiotypic antibody (Ab_1), $Ab_{2\gamma}$ which recognizes idiotopes near the binding sites of Ab_1 and its binding to Ab_1 can be inhibited by the antigen, and $Ab_{2\beta}$ which mimics the original antigen, and called this $Ab_{2\beta}$, the *internal image* [136]. Furusawa et al. [137] have shown that the $Ab_{2\beta}$ in reality is better described, not as the internal image of the antigen, but as the *mirror image of the paratope* because when the paratope cross-reacts with a similar but not identical epitope, then the $Ab_{2\beta}$ has the same cross-reacting specificity. For example, if Ab_1 was produced against the dinitrophenyl (DNP) epitope, but cross-reacted with the trinitrophenyl (TNP) and other similar epitopes, then $Ab_{2\beta}$ reacted not only with the antibody produced against the epitope of the immunizing antigen, but also with antibodies produced by immunization with crossreacting epitopes present on the immunizing antigen, in exactly the same way as did Ab_1.

Investigating these anti-idiotypic antibodies we developed a new method to test the heterocliticity of antibodies using the % inhibition by haptens

[138]. This new method is faster, simpler and does not need radiolabeled reagents like the method previously employed [139].

Suppression

Antibody production is sometimes rapidly terminated. It was in the 1970s that Gershon and Kondo [140] called attention to the suppression of antibody production and hypothesized that not only helper T cells, but also T suppressor cells exist. Usually, the phenotype of these cells is different from the phenotype of the T helper cells. In the human system the cytotoxic/suppressor T cell (Tc or Ts) is characterized by the presence of the $CD8^+$ antigen; in the mouse by the $Lyt\text{-}2^+3^+$ molecule.

Suppression is a most complex phenomenon. In the early 1970s it was thought that the T cell responsible for antibody suppression was $Lyt\text{-}2^+3^+$ in the murine system or $CD8^+$ in the human system. Today, many investigators propose that a complex interaction between different T cells or their products is necessary to bring about suppression.

Tada et al. [141] were the first to show that in the rat a radiosensitive cell can suppress IgE production. Our experiments were the first to show that in the SJL strain of mice, a $Lyt\text{-}1^+$ cell must interact with other T cells to produce suppression of IgE production [142]. At that time the marker L3T4 had not yet been described. The great complication, however, is that the T cell receptor of the *suppressor* cell has not yet been identified. It was not easy to identify the T cell receptor of the helper or cytotoxic T cells, but the progress in that field was phenomenally fast. Antibody suppression was described more than 15 years ago and we still do not know the structure of the T cell receptor on suppressor cells.

This presents different possibilities. The T suppressor cell might use a different recognition system. An alternative possibility will also be discussed below. In any case, the phenomenon of antibody suppression is a real one and this is not to be questioned. By what mechanism it is obtained, and if there are cells whose only function throughout their lifetime is suppression, is open to debate.

We mentioned that in SJL mice we showed that $Lyt\text{-}1^+$ T cells were necessary to obtain suppression of antigen-specific IgE antibody production. At that time we did not know that T helper cells consist of at least two subpopulations mentioned above (Th1 and Th2). We now have other experiments on IgE suppression – this time suppression in vitro. When spleen cells from the SJL, BALB/c or even SJA/9 or BALB/c nude strains are cultured in vitro and first stimulated with lipopolysaccharide then recombinant IL-4 brings about the secretion of immunoglobulins of the IgG and IgE isotypes. This secretion is inhibited when, in addition to IL-4, IL-2 is also added from the start and during the entire period of culture of these cells [Hirano, Ueda and Ovary, in preparation]. It has already been shown that IgE production is inhibited by interferon-γ [120]. We could postulate, therefore, that suppression of IgE production could be obtained by product(s) of Th1 cells without any 'special suppressor cells'. For other isotypes, intervention of Th2 or perhaps both T helper cells or even other T cell types might be necessary. It is also possible that suppression of antibody production can be obtained by several different mechanisms and the different pathways proposed by different authors are not absolutely exclusive.

The modulation of IgE production is a very complex phenomenon. We had seen that in mice injected with IgE-producing hybridoma cells as long as the IgE secreted by the hybridoma is present in abundance in their sera, it is impossible to obtain IgE antibody production by the normal B cells of the host [143]. For example, mice injected with the anti-DNP IgE-producing hybridoma cells did not produce anti-ovalbumin antibody when injected with ovalbumin, as long as Lyt-2^+ lymphocytes bearing receptors for IgE were present in these mice. As soon as the hybridoma was eliminated the anti-ovalbumin IgE appeared in their sera. The synthesis of anti-ovalbumin IgE in these mice immunized with ovalbumin after elimination of the anti-DNP IgE-secreting hybridoma is very similar to what we described above concerning the presence of Lyt-2^+ cells with receptor for the Fc fragment of IgE in mice bearing IgE-producing hybridomas [64].

The first description of the PK reaction (1921) was separated by 28 years from the first description of the PCA reaction (1949). About the same time separated the first description of the PK from the PCA reactions (28 years) as the description of the PCA reaction and the present time (30 years). However, the progress during the second period was incomparably faster. The trend seems to follow this exponential progression. Much remains to be learned, and hopefully many problems will be solved in the future, even faster than in the past.

References

1 Prausnitz C, Küstner H: Studien über die Überempfindlichkeit. Zentralbl Bakt 1921;86:160–169.

2 Portier P, Richet C: De l'action anaphylactique de certains venins. C R Soc Biol 1902;54:170–172.
3 Nicolle M: Contributions a l'étude du phenomène d'Arthus. Ann Inst Pasteur, Paris 1907;21:128–137.
4 Dale HH, Laidlaw PP: The physiological action of Beta-iminazol-ε thylamine. J Physiol 1910–1911;41:318–344.
5 Dale HH: The anaphylactic reaction of plain muscle in the guinea pig. J Pharmacol 1913;4:167–223.
6 Friedberger E: Über die Beziehungen zwischen Überempfindlichkeit und Immunität. Berl Klin Wochenschr. 1910;47:1490–92.
7 Bordet J: Gelose et anaphylatoxine. C R Soc Biol 1913;74:877.
8 Widal F: Definition and classification of colloidoclasies. Prog Med 1921;36:48.
9 Osler AG, Randall HG, Hill BM, et al: Studies on the mechanism of hypersensitivity phenomena. III. The participation of complement in the formation of anaphylatoxin. J Exp Med 1959;110:311–339.
10 Heidelberger M, Kendall FE: A quantitative study of the precipitin reaction between type III pheumococcus polysaccharide and purified homologous antibody. J Exp Med 1929;50:809–823.
11 Tiselius A, Kabat EA: An electrophoretic study of immune sera and purified antibody preparations. J Exp Med 1939;69:119–131.
12 Bovet D, Staub AM: Action protectrice des ethers phenoliques au cours de l'intoxication histaminique. C R Soc Biol 1937;124:547–549.
13 Ash ASF, Schild HO: Receptors mediating some action of histamine. Br J Pharmacol 1966;27:427–439.
14 Chase MW: The cellular transfer of cutaneous hypersensitivity to tuberculin. Proc Soc Exper Biol Med 1945;49:134–135.
15 Gell PCH, Coombs RRA (eds): Clinical aspects of immunology. ed 2. Philadelphia, Davis, 1969.
16 Jancso N: Speicherung. Budapest, Akademiai Kiado, 1955.
17 Biozzi G, Mene G, Ovary Z: L'histamine et la granulopexie de l'endothelium vasculaire. La Rev Immunol 1948;12:320–334.
18 Halpern BN, Benacerraf B, Biozzi G, et al: Facteurs regissant la fonction phagocytaire du système reticuloendothelial. Rev Hematol 1954;9:621–642.
19 Ovary Z: Immediate reactions in the skin of experimental animals provoked by antigen-antibody interactions. Prog Allergy. Basel, Karger, 1958, vol 5, pp 460–508.
20 Ovary Z.: Reverse passive cutaneous anaphylaxis in the guinea pig with horse, sheep or hen antibodies. Immunology 1960;3:19–27.
21 Ovary Z, Karush F: Studies on the immunologic mechanism of anaphylaxis. I. Antibody-hapten interactions studied by passive cutaneous anaphylaxis in the guinea pig. J Immunology 1960;84:409–415.
22 Ovary Z: Activité des substance à faible poids moleculaire dans les reactions antigène-anticorps in vivo et in vitro. C R Acad Sci 1961;253:582–583.
23 Porter RR: The hydrolysis of rabbit gamma-globulin and antibodies with crystalline papain. Biochem J 1959;73:119–126.
24 Ovary Z, Karush F: Studies on the immunologic mechanism of anaphylaxis. II. Sensitizing and combining capacity in vivo of fractions separated from papain digests of antihapten antibody. J Immunol 1961;86:146–150.

25 Ovary Z, Taranta A: Passive cutaneous anaphylaxis with antibody fragments. Science 1963;140:193–195.
26 Nisonoff A, Wissler FC, Lipman LN, et al: Separation of univalent fragments from the bivalent rabbit antibody molecule by reduction of disulfide bond. Archs Biochem Biophys 1960;89:230–244.
27 Ovary Z: The mechanism of passive sensitization in PCA and RPCA in guinea pigs. Proc 3rd Int Pharmacol Meet, July 24–30, 1966. Immunopharmacology 1968; 11:15–22.
28 Ishizaka T, Hirata F, Sterk AR, et al: Bridging of IgE receptors activates phospholid methylation and adenylate cyclase in mast cell plasma membranes. Proc Natn Acad Sci USA 1981;78:6812–6816.
29 Piper PJ, Samhoun MN, Tippins JR, et al: SRS-A and SRS: Their structure, biosynthesis and actions. Int Archs Allergy Appl Immunol 1981;66(suppl 1):107–112.
30 Samuelsson B: Leukotrienes. Mediators of allergic reactions and inflammation. Int Archs Allergy Appl Immunol 1981;66(suppl 1):98–106.
31 Orange RP, Austen WG, Austen KF: Immunological release of histamine and slow reacting substance of anaphylaxis from human lung. J Exp Med 1961;34:136s–149s.
32 Ishizaka T, DeBertrand R, Tomioka H, et al: Identification of basophil granulocytes as a site of allergic histamine release. J Immunol 1972;108:1000–1008.
33 Ishizaka T: Biochemical analysis of mast-cell induced triggering by bridging IgE receptors; in Froese A, Paraskevas P (eds): Structure and Function of Fc Receptors, chap. 9. New York, Dekker, 1983, pp 157–176.
34 Porter RR: in Gellhorn A, Hirschberg E (eds): Symposium on Basic Problems in Neoplastic Disease. New York, Columbia University Press, 1962, pp 177–194.
35 Hilschmann N, Craig LC: Amino acid sequence studies with Bence-Jones proteins. Proc Natn Acad Sci USA 1965;53:1903–1909.
36 Hood L, Kronenberg M, Hunkapiller T: T cell antigen receptors and the immunoglobulin supergene family. Cell 1985;40:225–229.
37 Ovary Z, Saluk PH, Quijada L, et al: Biologic activities of rabbit immunoglobulin G in relation to domains of the Fc region. J Immunol 1976;116:1265–1271.
38 Helm B, Marsh P, Vercelli D, et al: The mast cell binding site on human immunoglobulin E. Nature 1988;331:180–183.
39 Kebo D, Glovsky MM, Helm B, et al: Recombinant IgE (301–376) inhibition of passive transfer on reagin and IgE to human basophils. Fed Proc 1988;5531.
40 Ovary Z, Benacerraf B: Separation of skin sensitizing from blocking 7S antibodies in guinea pig sera. Fed Proc 1962;21:2.
41 Benacerraf B, Ovary Z, Bloch KJ, et al: Properties of guinea pig 7S antibodies. I. Electrophoretic separation of two types of guinea pig 7S antibodies. J Exp Med 1963;117:937–949.
42 Ovary Z, Benacerraf B, Bloch KJ: Properties of guinea pig 7S antibodies. II. Identification of antibodies involved in passive cutaneous and systematic anaphylaxis. J Exp Med 1963;117:951–964.
43 Bloch KJ, Kourilsky FM, Ovary Z, et al: Properties of guinea pig 7S antibodies. III. Identification of antibodies involved in complement fixation and hemolysis. J Exp Med 1963;117:965–981.
44 Loveless MH: Reagin production in a healthy male who forms no detectable beta$_{2A}$ immunoglobulins. Fed Proc 1964;23:403.

45 Ishizaka K, Ishizaka T, Hornbrook M: Physicochemical properties of reaginic antibodies. V. Correlation of reaginic activity with gamma-E-globulin antibody. J Immunol 1966;92:840–853.
46 Ishizaka K, Ishizaka T: Identification of E-antibody as a carrier of reaginic activity. J Immunol 1967;99:1187–1196.
47 Johansson SGO, Bennich H: Immunological studies of an atypical (myeloma) immunoglobulin. Immunology 1967;12:381–394.
48 Kohler G, Milstein C: Continuous cultures of fused cells secreting antibody of predefined specificity. Nature 1975;256:495–497.
49 Bottcher I, Ulrich M, Hirayama N, et al: Production of monoclonal mouse IgE antibodies with DNP specificity by hybrid cell lines. Int Archs Allergy Appl Immun 1980;61:248–250.
50 Eshhar Z, Ofarim M, Waks T: Generation of hybridomas secreting murine reaginic antibodies of anti-DNP specificity. J Immunol 1980;124:775–780.
51 Liu F, Goh JW, Ferry EL, et al: Monoclonal dinitrophenyl-specific murine IgE antibody: preparation, isolation and characterization. J Immunol 1980;124:2728–2737.
52 Maekawa S, Ovary Z: Correlation of murine anti-dinitrophenyl antibody content as determined by ELISA, passive cutaneous anaphylaxis and passive hemolysis. J Immunol Methods 1984;71:229–239.
53 Baniyash M, Eshhar Z: Inhibition of IgE binding to mastcells and basophils by monoclonal antibodies to murine IgE. Eur J Immunol 1984;14:799–801.
54 Hirano T, Miyajima H, Kitagawa H, et al: Studies of murine IgE with monoclonal antibodies. I. Characterization of rat monoclonal anti-IgE antibodies and the use of these antibodies for determination of serum IgE levels and for anaphylactic reactions. Int Archs Allergy Appl Immunol 1988;85:47–54.
55 Hirano T, Yamakawa N, Myajima H, et al: An improved method for the detection of IgE antibody of defined specificity using rat monoclonal anti-IgE antibody. J Immunol Methods 1989;119:145–150.
56 Borges MS, Kumagai Y, Okumura K, et al: Allelic polymorphism of murine IgE controlled by the seventh immunoglobulin heavy chain allotype locus. Immunogenetics 1981;13:499–507.
57 Hirano T, Hom C, Ovary Z: Half-life of murine IgE antibodies in the mouse. Int Arch Allergy Appl Immunol 1983;71:182–184.
58 Haba S, Ovary Z, Nisonoff A: Clearance of IgE from serum of normal and hybridoma-bearing mice. J Immunol 1985;134:3291–3297.
59 Metzger H, Alcaraz G, Hohman R, et al: The receptor with high affinity for immunoglobulin E. Ann Rev Immunol 1986;4:419–470.
60 Blank U, Ra C, Miller L, et al: Complete structure and expression in transferred cells of high affinity IgE receptor. Nature 1989;337:187–189.
61 Kikutani H, Inui S, Sato R, et al: Molecular structure of human lymphocyte receptor of immunoglobulin E. Cell 1986;47:657–665.
62 Ikuta K, Takami M, Won Kim C, et al: Human lymphocyte Fc receptor for IgE: Sequence homology of its cloned cDNA with animal lectins. Proc Natn Acad Sci USA 1987;84:819–823.
63 Yokota A, Kikutani H, Tanaka T, et al: Two species of human Fc receptor II (FcεRII/CD23): Tissue-specific and Il-4-specific regulation of gene expression. Cell 1988;55:611–618.

64 Mathur A, Maekawa S, Ovary Z, et al: Increased T-ε cells in BALB/c mice with an IgE secreting hybridoma. Mol Immunol 1986;23:1193–1201.
65 Mathur A, Kamat DM, Van Ness BG, et al: Thymus-dependent in vivo suppression of IgE synthesis in a murine IgE-secreting hybridoma. J Immunol 1987;139:2865–2872.
66 Ishizaka K: Regulation of IgE synthesis. Ann Rev Immunol 1984;2:159–182.
67 Marcelletti JF, Katz DH: FcRε[+] lymphocytes and regulation of the IgE antibody system. I. A new class of molecules termed IgE-induced regulants (EIR), which modulate FcRε expression by lymphocytes. J Immunol 1984;133:2821–2828.
68 Spiegelberg HL: Structure and function of Fc receptors for IgE on lymphocytes, monocytes and macrophages. Adv Immunol 1984;35:6188.
69 Boyden SV, Sorkin E: The absorption of antibody and antigen by spleen cells in vitro. Some further experiments. Immunology 1961;4:244.
70 Boyden SV: Cytophilic antibody in guinea pigs with delayed type hypersensitivity. Immunology 1964;7:474–483.
71 Berken A, Benacerraf B: Properties of antibodies cytophilic for macrophages. J Exp Med 1966;123:119–144.
72 Tigelaar RE, Vaz NM, Ovary Z: Immunoglobulin receptors on mouse mast cells. J Immunol 1971;106:661–672.
73 Unkeless JC, Scigliano E, Freedman V: Structure and function of human and murine receptors for IgG. Ann Rev Immunol 1988;6:251–281.
74 Ovary Z, Benacerraf B: Immunological specificity of the secondary response with dinitrophenylated proteins. Proc Soc Exp Biol Med 1963;114:72–76.
75 Claman HN, Chaperon EA, Triplett RF: Immunocompetence of transferred thymus-marrow cell combinations. J Immunol 1966;97:828–832.
76 Miller JFAP, Mitchell GF, Weiss NS: Cellular basis of the immunological defect in thymectomized mice. Nature 1967;214:292–297.
77 Szenberg A, Warner NL: The role of antibody in delayed hypersensitivity. Brit Med Bull 1967;23:30–34.
78 Mitchinson NA: The carrier effect in the secondary response to hapten-protein conjugates. II. Cellular cooperation. Eur J Immunol 1971;1:18–27.
79 Rajewsky K: The carrier effect in the induction of antibodies. Proc R Soc Biol 1971;176:385–392.
80 Hedrick SM, Cohen DI, Neilsen EA, et al: Isolation of cDNA clones encoding T cell-specific membrane-associated proteins. Nature 1984;308:149–153.
81 Ehrlich P: On immunity with special reference to cell life (Croonian Lecture). Proc R Soc Lond 1900;66:424–448.
82 The Jackson Laboratory Staff: Biology of the Laboratory Mouse, ed 2, Green EL (ed.). London, MacGraw-Hill, 1966.
83 McDevitt HO, Chinitz A: Genetic control of the antibody response. Relationship between immune response and histocompatibility (H-2) type. Science 1969;163: 1207–1208.
84 Benacerraf B: The genetic control of specific immune responses. Harvey Lect 1973;67:109–141.
85 Klein J, Nagy ZA: MHC restriction and Ir genes. Adv Cancer Res 1982;37:233–317.
86 Klein J, Figueroa F, Nagy ZA: Genetics of the major histocompatibility complex: The final act. Ann Review Immunol 1983;1:119–142.
87 Shimonkevitz R, Kapler J, Marrack P, et al: Antigen recognition by H-2 restricted T cells. I. Cell-free antigen processing. J Exp Med 1983;158:303–316.

88 Rock KB, Benacerraf B, Abbas AK: Antigen presentation by hapten-specific B lymphocytes. I. Role of surface immunoglobulin receptors. J Exp Med 1984;160: 1102–1113.
89 Lanzavecchia A: Antigen-specific interaction between T and B cells. Nature 1985; 314:537–539.
90 Benjamin DC, Berzofsky JA, East IJ, et al: The antigenic structure of proteins. A reappraisal. Ann Rev Immunol 1984;2:67–101.
91 Berkower I,, Buckenmeyer GK, Berzofsky JA: Molecular mapping of a histocompatibility-restricted immunodominant T cell epitope with a synthetic and natural peptides. Implications for T cell antigenic structure. J Immunol 1986;136:2498–2503.
92 Sette A, Buus S, Colon S, et al: Structural characteristics of an antigen required for its interaction with Ia and recognition by T cells. Nature 1987;328:395–399.
93 Allen PM, Matsueda GM, Evans RJ, et al: Identification of the T-cell and Ia contact residues of a T-cell antigenic epitope. Nature 1987;327:713–715.
94 Antigen Processing: Immunological Review No 106, Möller G (ed). Copenhagen, Munksgaard.
95 Bjorkman PJ, Saper MA, Samraoui B, et al: Structure of the human class I histocompatibility antigen, HLA-A2. Nature 1987;329:506–512.
96 Bjorkman PJ, Saper MA, Samraoui B, et al: The foreign antigen binding site and T cell recognition regions of class I histocompatibility antigens. Nature 1987;329: 512–518.
97 Brown JH, Jardetzky T, Saper MA, et al: A hypothetical model of the foreign antigen binding site of class II histocompatibility molecules. Nature 1988;332:845–850.
98 Tonegawa S: Somatic generation of antibody diversity. Nature 1983;302:575–581.
99 Warren HS, Vogel FR, Chedid LA: Current status of immunological adjuvants. Ann Rev Immunol 1986;4:369–388.
100 Ramon G: Sur l'augmentation anormale de l'antitoxine chez les chevaux producteurs de sérum antidiphtérique. Bull Soc Centr Med Vet 1925;101:227.
101 Freund J, Casals J, Hismer EP: Sensitization and antibody formation after injection of tubercle bacilli and paraffin oil. Proc Soc Exp Biol Med 1937;37:509.
102 Revoltella R, Ovary Z: Reaginic antibody production in different mouse strains. Immunology 1969;17:45–54.
103 Ogilvie BM, Jones VE: Reaginic antibodies and helminth infection; in Movat HZ (ed): Cellular and Humoral Mechanism of Anaphylaxis and Allergy. Mechanism of Anaphylaxis and Allergy. Basel, Karger, 1970, p 13.
104 Orr TSC, Blair AMJ: Potentiated reagin response to egg-albumin and conalbumin in *N. brasiliensis* infected rat. Life Sciences 1969;8:1073–1077.
105 Jarrett EEE: Potentiation of reaginic antibody to ovalbumin in rat. Immunology 1972;22:1099–1101.
106 Bloch KJ, Wilson RJM: Homocytotropic antibody response in the rat infected with the nematode *Nippostrongylus brasiliensis*. J Immunol 1968;100:629–636.
107 Kojima S, Ovary Z: Effect of *Nippostrongylus brasiliensis* infection on anti-hapten IgE antibody response to a heterologous hapten-carrier conjugate. Cell Immunol 1975;17:383–391.
108 Ishizaka T, Urban JF, Ishizaka K: IgE formation in rat following infection with *Nippostrongylus brasiliensis*. Proliferation and differentiation of IgE bearing cells. Cell Immunol 1976;22:248–261.

109 Katona IM, Urban JF, Scher I, et al: Induction of an IgE response in mice by *Nippostrongylus brasiliensis*: characterization of lymphoid cells with intracellular or surface IgE. J Immunol 1983;130:350–356.
110 Hirano T, Kumagai Y, Okumura K, et al: Regulation of murine IgE production: Importance of a not-yet described T cell for regulation for IgE secretion demonstrated in SJA/9 mice. Proc Natl Acad Sci USA 1983;80:3435–3438.
111 Howard M, Paul WE: Regulation of B-cell growth and differentiation by soluble factors. Ann Rev Immunol 1983;1:307–333.
112 Smith KA: Interleukin 2. Ann Rev Immunol 1984;2:319–333.
113 Greene WC, Leonard WJ: The human interleukin-2 receptor. Ann Rev Immunol 1985;4:69–95.
114 Durum SK, Schmidt JA, Oppenheim JJ: Interleukin 1. An immunological perspective. Ann Rev Immunol 1985;3:263–287.
115 Schrader JW: The panspecific hemopoietin of activated T lymphocytes (interleukin-3). Ann Rev Immunol 1986;4:205–230.
116 Paul WE, Ohara J: B cell stimulatory factor-1/interleukin 4. Ann Rev Immunol 1987;5:429–459.
117 Kishimoto T, Hirano T: Molecular regulation of B lymphocyte response. Ann Rev Immunol 1988;6:485–512.
118 Vitetta ES, Ohara J, Myers C, et al: Serological, biochemical, and functional identity of B-cell stimulatory factor 1 and B cell differentiating factor for IgG1. J Exp Med 1985;162:1726–1731.
119 Coffman T, Ohara J, Bond MW, et al: B cell stimulatory factor enhances the IgE response of lipopolysaccharide activated B cells. J Immunol 1986;136:4538–4541.
120 Coffman RL, Carty J: A T cell activity that enhances polyclonal IgE production and its inhibition by interferon gamma. J Immunol 1986;136:949–954.
121 Mosmann RR, Cherwinski H, Bond MW, et al: Two types of murine helper T cell clones. I. Definition according to profiles of lymphokines activities and secreted proteins. J Immunol 1986;136:2348–2357.
122 Azuma M, Hirano T, Miyajima H, et al: Regulation of murine IgE production in SJA/9 and in nude mice. Potentiation of IgE production by recombinant interleukin 4. J Immunol 1987;139:2538–2544.
123 Michael JG, Bernstein IL: Thymus dependence of reaginic antibody formation in mice. J Immunol 1973;111:1600–1601.
124 Cher DJ, Mosmann TR: Two types of murine helper T cell clone. J Immunol 1987; 138:3688–3694.
125 Zinkernagel RM, Doherty PC: Restriction of in vitro T cell-mediated cytotoxicity within a syngeneic or semiallogeneic system. Nature 1974;248:701–702.
126 Moore MW, Carbone FR, Bevan MJ: Introduction of soluble protein into class I pathway of antigen processing and presentation. Cell 1988;54:775–785.
127 Robertson M: Tolerance, restriction and the MLs enigma. Nature 1988;332:18–19.
128 Kappler JW, Roehm N, Marrack PC: T cell tolerance by clonal elimination in the thymus. Cell 1987;49:273–280.
129 Kappler JW, Staerz U, White J, et al: Self-tolerance eliminates T cell specific for MLs-modified products of the major histocompatibility complex. Nature 1988;332: 35–40.
130 MacDonald HR, Schneider R, Lees RK, et al: T-cell receptor Vβ use predicts reactivity and tolerance to M1s^{a}-encoded antigens. Nature 1988;332:40–45.

131 Haba S, Nisonoff A: Induction of high titers of anti-IgE by immunization of inbred mice with syngeneic IgE. Proc Natn Acad Sci USA 1987;84:5009–5013.
132 Kisielow P, Teh HS, Bluthmann H, et al: Positive selection of antigen-specific T cells in thymus by restricting MHC molecules. Nature 1988;335:730–733.
133 MacDonald RH, Lees RK, Schneider R, et al: Positive selection of $CD4^+$ thymocytes controlled by MHC class II gene products. Nature 1988;336:471–473.
134 Teh HS, Kisielow P, Scott B, et al: Thymic major histocompatibility complex antigens and the T-cell receptor determine the CD4/CD8 phenotype of T cells. Nature 1988;335:229–233.
135 Palmiter RD, Brinster RL: Transgenic mice. Cell 1985;41:343–345.
136 Jerne NK, Roland J, Cazenave PA: Recurrent idiotypes and internal images. EMBO J 1982;1:243–247.
137 Furusawa S, Okitsu-Negishi S, Yoshino K, et al: Anti-idiotypic antibody as a mirror image of the paratope of the original antibody. Int Archs Allergy Appl Immunol 1987;84:263–270.
138 Furusawa S, Ovary Z: A novel method to investigate the heterocliticity of antibodies. J Immunol Methods 1987;104:275–279.
139 Hammer P, Steiner LA: Specificity and heterocliticity of rabbit antisera to the 1,4-dinitrophenyl determinant. J Immunol 1982;128:343–350.
140 Gershon RK, Kondo K: Cell interaction in the induction of tolerance. The role of thymic lymphocytes. Immunology 1970;18:723–732.
141 Tada T, Taniguchi M, Okumura K: Regulation of homocytotropic antibody formation in the rat. II. Effect of X-irradiation. J Immunol 1971;106:1012–1018.
142 Watanabe N, Kojima S, Shen FW, et al: Suppression of IgE antibody production in SJL mice. II. Expression of Ly-1 antigen on helper and nonspecific suppressor T cells. J Immunol 1977;118:485–488.
143 Hirano T, Ovary Z: Studies on immunity in hybridoma-bearing mice. A. Immune response to antigens. I. Role of subcutaneously injected IgE-producing hybridoma on antibody production of different isotypes against immunizing antigen. Int Arch Allergy Appl Immunol 1984;73:338–341.

Zoltan Ovary, MD, Department of Pathology, New York University Medical School, 550 First Avenue, New York, NY 10016 (USA)

Waksman BH (ed): 1939–1989: Fifty Years Progress in Allergy.
Chem Immunol. Basel, Karger, 1990, vol 49, pp 121–142

Histamine, Mast Cells and Basophils

G.B. West

The first description of tissue mast cells was made in 1877 when Ehrlich [1] observed that some cells in the connective tissue of animals contained granules which changed the color of toluidine blue and certain other dyes as staining proceeded. The number of such cells was greater when chronic inflammation or conditions characterized by increased local nutrition were present. Two years after his first discovery, he found similar cells in the blood, but whereas the blood mast cells (called basophils) originated in the bone marrow and, with other leukocytes, entered the peripheral blood, the more common tissue mast cells were thought to be born, to live and die, in the connective tissues. The similarity of the two kinds of cell lies in their content of water-soluble cytoplasmic granules which have the strong affinity for basic dyes, several of which change color as staining proceeds (that is, metachromasia).

Besides discovering and naming the mast cells (the well-fed cells), Ehrlich also described their morphology, staining properties and distribution but he then left it to others to elucidate the chemical nature of the granules. For the next 60 years, research on this subject remained almost entirely histological, until Swedish workers in 1937 solved one of the riddles of the metachromatic granules of the mast cell.

Jorpes [2] and his colleagues in Stockholm had for long been working on the powerful anticoagulant material first isolated from dog liver and hence called heparin. On finding that heparin stained metachromatically with toluidine blue, these workers searched the tissues for metachromatism as a possible clue to the site of formation of heparin. They were able to show that there was a good correlation of the mast cell count of a particular tissue, and the amount of heparin that was extracted from it. At that time, mast cells were thought to be perivascular in location so as to produce heparin which poured into the bloodstream. However, a later study of the movement of the perivascular mast cells revealed that they migrated from the blood vessels and hence probably produced a secretion for both blood and tissues. Their function was less concerned with blood clotting than with the maintenance and repair of the connective tissues. Since it is now known that they bear receptors for IgE (immunoglobulin E) and they play an essential role in immunological and inflammatory reactions, they are important cells in the body. The proof that histamine was present in these tissue mast cells was produced by Riley and West in 1953, and this is one of the most important steps towards the understanding of the role of histamine in physiology and pathology reported over the last 50 years.

Chemical Releasers of Histamine

By the time the first volume of *Progress in Allergy* appeared in 1939, inhalation of allergen in specifically sensitized guinea pigs (often referred to as experimental asthma) was known to produce clinical symptoms similar to those following inhalation of an aerosol of histamine, though the cell changes were different. In sensitized animals given the allergen, there was a massive eosinophilia in the lungs but no such changes occurred after histamine. Furthermore, after different forms of chemical shock in the dog, the liver enlarged, the blood became incoagulable, and the animal sank into a state of confusion, a syndrome reported by Portier and Richet [3] in 1902 to occur after anaphylactic shock. In fact in 1910 Dale and Laidlaw studied the physiological actions of histamine, reporting on constriction of smooth muscle of the respiratory tract yet dilatation of the capillaries of peripheral organs leading to edema formation, hemoconcentration, and lowering of systemic blood pressure. They had also shown conclusively that the systemic effect of histamine mimicked the signs and symptoms of anaphylactic shock as elicited in sensitized animals by the specific allergen. Later, in 1927, histamine was shown by Lewis to produce the triple response in the human skin – local vasodilatation, local edema and erythema. Despite all these observations, the site of storage of histamine in the body remained undetermined although the amine had been identified in extracts of tissues.

Progress along these lines was slow during the years of the Second World War and for almost 10 years few major advances were made during attempts to identify the physiological role of histamine in human and animal tissues. First class work continued to be reported from the laboratories of Dale, Feldberg, Schild and others, but the great step forward took place in 1949 when a classical paper by MacIntosh and Paton [4] was published describing chemicals which released histamine from stores in the body. These chemicals were basic substances such as diamines, diamidines and diguanides and, as they bore at least some chemical structural resemblance to histamine itself, histamine was considered to be displaced from its normal location in the tissues. These chemicals produced the triple response when injected locally into the skin of the human. They also produced a profound, though significantly delayed, lowering of the blood pressure when injected intravenously into cats, and in dogs their effects resembled to a remarkable degree the syndrome of anaphylactic or peptone shock. It is important to note that one part of the evidence in this study was related to the delay in response of the blood pressure of the animal following injection, for this was the time required for the chemical to release the histamine into the bloodstream and so exert a hypotensive action. This time period was 2–3 times longer than that after injections of histamine itself. Another part of the evidence related to antagonism of the action by the antihistamine drugs of the day (the so-called H_1-receptor antagonists). More and more specific and accurate methods of analysis were later used to assist in the identification of the released histamine, but these did not advance knowledge of its storage site in cells.

Meanwhile, Rocha e Silva [5] was making the important discovery in 1952 that histamine, as well as heparin, was being simultaneously released from the liver of the dog undergoing anaphylactic shock. He even suggested that both substances might come from the same cell, although he thought the liver of the dog was a special case. This evidence was very carefully considered by many workers including Riley and West, who were of the opinion that both substances might originate from the same cell in other locations in other animals and, accordingly, they started work on this topic in 1951.

Histamine in Mast Cells

When a lethal dose of a fluorescent histamine releaser such as stilbamidine or 2-hydroxy-stilbamidine (both diamidines) was rapidly injected intravenously into a rat, they found that some of the diamidine was temporarily trapped in some of the tissue mast cells which then underwent disruption, with the resultant release of histamine. Other histamine releasers, such as tubocurarine and morphine produced greater disruption of tissue mast cells and a greater release of histamine. At that time, there was still much confusion over the accuracy of methods used to extract and estimate histamine in blood and tissues. Qualitative changes were easily detected, using the original technique of the isolated ileum of the guinea pig, but chemical analyses were being developed. Measuring histamine by fluorimetry after condensation with *o*-phthalaldehyde, according to the method of Shore et al. [6] became popular. As stated earlier, Jorpes and his colleagues found a striking positive correlation between numbers of mast cells in a tissue and the amount of heparin extracted from it, so Riley and West [7] started determining the histamine content of those tissues in which the Swedish workers had found heparin.

Riley, who was a consultant radiotherapist at Dundee Royal Infirmary, had always been fascinated by research and had used every spare minute of his working life in reading and then experimenting. His introduction to the mast cell occurred during his student days (around 1935) when one of his teachers remarked that there is always an increase in mast cell numbers in tissues adjacent to experimentally induced cancers in mouse skin. On being asked by Riley what was the function of these cells with granules which stained purple with eosin and toluidine blue, the teacher admitted that he did not know. From that time on, Riley thought that in some way the mast cell might well provide a clue to the mechanism of carcinogenesis. Having reviewed all the available literature on the mast cell changes occurring in precancerous skin, he believed that the local mastocytosis taking place at different stages of skin carcinogenesis merely reflected the increased and continuing fibroblastic activity in the dermis, as in chronic inflammation elsewhere in the body.

In the Surgery Department of the infirmary, Riley weighed samples of all kinds of available tissue and placed them separately in small bottles of trichloroacetic acid to preserve any histamine that might be present. These were numbered and taken to West, a pharmacologist in Dundee, for assay on the isolated ileum of the guinea pig and on the blood pressure of the cat. Meanwhile, small samples of the same tissues were studied by Riley for their relative mast cell content. Later, Riley handed West a sealed envelope containing hopeful expectations of histamine values, based entirely on mast cell counts. A short time later, West contacted Riley to say that he was correct in all cases. Both then knew that they were right about the mast cell being the main location of tissue histamine, although they spent the next 5 years saying it often enough for all to be convinced. In fact, even Dale, at a meeting of the Physiological Society in Cambridge in May 1952 was heard to say: 'Riley, you imply that histamine may be carried in the acid granules of the mast cell. This is very curious, as Code in the States believes that it is associated with eosinophilic cells. You must do more work on this. You may be right but we would like a little more evidence.' Riley had just been the guest speaker with a joint paper which had attracted almost no attention.

At that time, Riley and West had been unable to obtain a mast cell tumor for analysis and so had carried out a comparative study of different organs and tissues from several animal species as well as those from man. A tissue rich in mast cells was always found to be rich in histamine. For example, ox aorta contained much histamine and many mast cells yet that of the pig had none. The capsule of ox liver had nearly 10 times as much histamine as did the parenchyma and there was a correspondingly greater number of mast cells.

However, in December 1952, Riley and West [8] reported on 3 mast cell tumors from dogs, one of which had a very high histamine content, but even more exciting, was a human mast cell tumor containing enormous numbers of immature mast cells, with only a few polymorphs and eosinophils, and a histamine content of nearly 1 mg/g tissue. The best illustration of the link between histamine and mast cells in a normal tissue, however, was fresh ox lung where enormous numbers of mast cells packed with bulky metachromatically stainable granules were present in the pleura which was easily stripped off. The only other material present in this almost avascular tissue was collagen and there were no eosinophils. Its histamine content was found to be over 250 μg/g tissue and it also had a very high heparin value.

Further work showed that low histamine values for some organs of very young animals contrasted with the higher values found in adults. Painting adult mouse skin with a carcinogenic hydrocarbon such as benzpyrene considerably increased both its mast cell population and its histamine content. When the normal skin of different animals was extracted and assayed, a good correlation existed between the mast cell count and the histamine content (table 1) [9]. When low doses of histamine releasers like stilbamidine or tubocurarine were injected into rats and their tissues were analyzed, both the mast cell counts and the histamine contents were reduced (table 2).

As stated earlier, MacIntosh and Paton [4] reported in 1949 that histamine was released from animal tissues by basic compounds, probably by

Table 1. Comparison of the mast cell count (per HP field) and histamine content (μg/g) in the skin of different species [9]

Parameter	Guinea pig	Rabbit	Man	Dog	Cat	Rat	Mouse	Hamster
Mast cells	3	2	7	40	26	30	40	50
Histamine	2	2	7	8	14	22	38	54

Table 2. Effect of histamine liberators on the histamine content (μg/g) and relative mast cell count of rat tissues [9]

	Control		Stilbamidine		Tubocurarine	
	histamine	mast cells	histamine	mast cells	histamine	mast cells
Omentum	20	++	10	+	2	0
Subcutaneous tissue	19	++	10	+	3	0
Mesentery	8	+	6	+	4	+

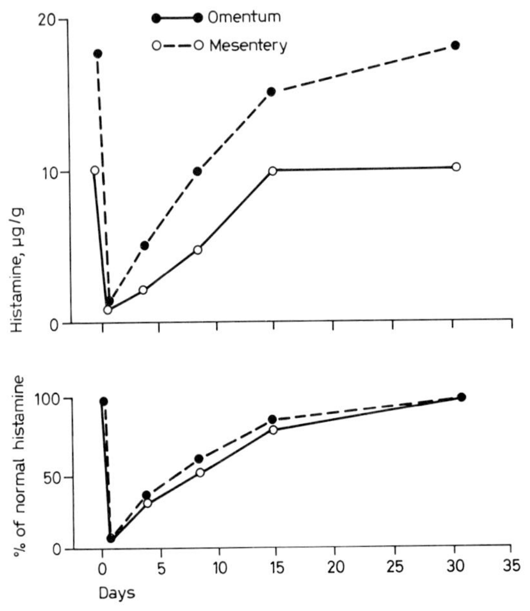

Fig. 1. Effect of 2 intraperitoneal injections of compound 48/80 on the histamine content of rat mesentery and omentum. Note that recovery of the histamine content took about 30 days [11].

displacing the histamine from its normal storage sites in tissues. Then, in 1951, Paton [10] found that a chemical known as compound 48/80, the condensation product of *p*-methoxyphenethylmethylamine with formaldehyde, exerted a depressor action in rats and dogs as a result of the release of histamine into the bloodstream, a rise in portal pressure and a delay in blood coagulation. This depressor response was delayed after the injection and only began some 20 s later, with a slow recovery of blood pressure. It was associated with peripheral vasodilatation and a rise in hematocrit. Compound 48/80 also elicited the triple response of weal and flare from human skin when injected intradermally, resembling that produced by histamine itself. Besides histamine being identified in the plasma, there was also a slow-contracting substance (later termed SRS) released, similar to the material described earlier in experiments on the effect of venoms on different perfused organs, and to that released during anaphylactic shock and peptone shock.

Compound 48/80 was a chemical in short supply in 1951 but when it became freely available a few years later, Riley and West [11] used it in place of the rather toxic basic compounds first tested. After one or two doses, rats acquired tolerance to further daily doses which would once have been fatal and tolerance corresponded to a progressive loss of tissue histamine. In fact, compound 48/80 produced shock in rats only when the histamine was there to be released. Once the tissue was free of histamine, compound 48/80 was no longer toxic (fig. 1) [11]. Compound 48/80 was a selective histamine releaser in the rat, and this was soon confirmed by injecting it directly into rat ears when only the ears lost histamine and the ear mast cells were shattered.

Recovery of peritoneal mast cells after several intraperitoneal doses of compound 48/80 occurred by at least two processes: (1) return of granules to ghost mast cells, and (2) formation of new cells from the adventitia. As the mast cell disappeared after compound 48/80 treatment, there was also progressive swelling and evidence of mobility of other connective tissue cells. This activation of the loose mesenchyme was the result of flooding of the tissue with a protein-rich edema fluid, thereby suggesting a physiological role of histamine in inflammation. As the mast cells of the stomach and intestine remained relatively resistant to compound 48/80 treatment, even after chronic dosage, more than one population of mast cells was considered at that time.

Among the many compounds shown to release histamine from tissue mast cells was a group of polysaccharides of which clinical dextran was typical in the rat. When Riley and West gave dextran to animals in 1951, only the rat responded on first injection with a reaction similar to that found after anaphylactic shock. This so-called anaphylactoid reaction was first reported in 1937 by Selye after he had injected rats with egg-white. Later, in 1964, Harris and West [12] reported that rats from one special colony failed to react either to egg-white or to dextran. This was surprising, as the skin of all adult rats tested contained comparable amounts of histamine and heparin and similar numbers of mast cells. The relative resistance of these rats to dextran (called NR rats to distinguish them from reactive or R rats) was found by selective breeding experiments to be controlled by an autosomal-recessive gene [12]. These rats were later found to be relatively resistant to other forms of shock such as trauma, tourniquet and heat.

A symposium on histamine in honor of Sir Henry Dale's 80th birthday took place at the Ciba Foundation in London in April 1955, and it was then still not completely accepted that much of the body's histamine was stored in tissue mast cells. This was to be expected as Dale's standards of proof of facts had always been rigorous and even the release of histamine in anaphylactic shock was accepted by Dale only after Feldberg had provided unequivocal evidence for it.

While these exciting studies were taking place, a report appeared in the literature stating that certain histamine releasers also released the amine 5-hydroxytryptamine (5-HT) from tissues. This amine was present in the enterochromaffin cells of the gut and in blood platelets and efforts were made to discover if it was also in tissue mast cells. Over half the total 5-HT in adult rats was found to reside in the skin but, when the skin was scraped to separate the subcutaneous layers from the outer epithelium, the major portion of the mast cells and histamine was removed with the scrapings and the majority of the 5-HT remained in the outer epithelium (table 3) [13]. So mast cells were not the only source of 5-HT in rat skin. Pads of the feet contained fewer mast cells and less histamine than did dorsal skin of the feet yet their 5-HT content was higher. In contrast, the skin of most mammals (including that of man) had little or no 5-HT although mast cells and histamine were present (table 4) [13]. The dextran reaction occurred only in rats and so the link was established between 5-HT and the anaphylactoid response. West and his colleagues later found that the antibiotic, polymyxin B, disrupted skin mast cells and reduced the histamine levels in rats without appreciably altering the 5-HT levels, whereas reserpine did the opposite and compound 48/80 lowered both amine levels (fig. 2).

Table 3. Histamine content (μg/g) of rat tissues compared with the 5-HT content (μg/g) and relative mast cell count [13]

Tissue	Histamine	5-HT	Mast cells
Ears	41	1.0	+++
Dorsal skin of feet	41	1.0	+++
Abdominal subcutis	40	0.8	+++
Pads of feet	16	1.5	+
Abdominal epithelium	9	1.6	+
Spleen	2	2.8	0

Table 4. 5-HT content (μg/g) of abdominal skin of different species compared with the relative mast cell count and histamine content (μg/g) [13]

Species	5-HT	Mast cells	Histamine
Guinea pig	0	+	3
Dog	0	++	15
Man	0	+	5
Rabbit	0	+	4
Hamster	0	+++	63
Cat	0.08	++	24
Mouse	0.37	+++	42
Rat	1.34	++	23

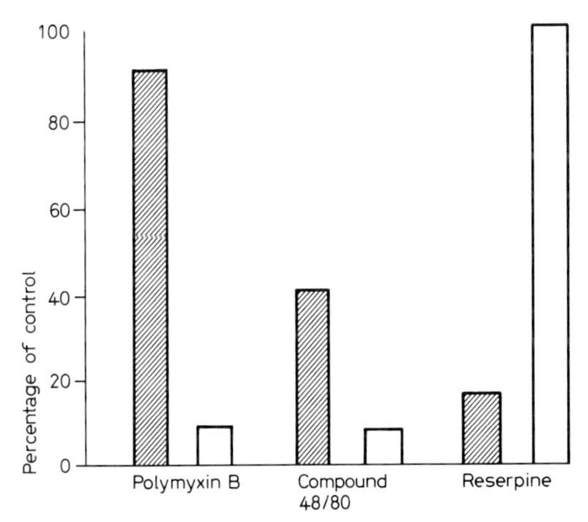

Fig. 2. Effect of repeated intraperitoneal doses of polymyxin B and compound 48/80 and reserpine on the 5-HT (shaded histograms) and the histamine (open histograms) contents of rat skin [13].

Isolated Mast Cells

Since 1970, isolated mast cells have provided a useful and powerful tool for studying stimulus-secretion coupling as well as allergic disorders. They have usually been harvested from the peritoneal cavity of rats using a modified Ringer's solution and they have then been concentrated to over 90% purity. The release of histamine has been shown to occur mainly by exocytosis, with calcium acting as a second messenger. The immunological stimulus has been said to open the calcium gates in the cell membrane, for when the calcium has been removed or has not been present at the time of stimulation, the gates close quickly and the subsequent addition of calcium fails to result in release. It has been found that exogenous phosphatidyl serine (PS) allowed some potentiation of the secretory response, both in magnitude and duration, when calcium was present. The action of PS was highly specific, as no other phospholipid was able to substitute for it. However, differences have been reported in the response of rat isolated mast cells to releasing agents such as antigen (to which the animals had been actively sensitized) and anti-IgE (the immunoglobulin directed against rat IgE molecules). Both these ligands were shown to act by cross-linking receptor proteins on the cell surface but the response patterns were different. Experiments with calcium ionophores like A 23187 [14] and X 537A [15] have confirmed the importance of calcium in these systems, for these agents were capable of forming lipid-soluble complexes with calcium and transferring them directly across the cell membranes. Both their releasing mechanisms were selectively antagonized by benzalkonium compounds [16].

The plant lectin, concanavalin A (Con A), was shown to simulate the anaphylactic reaction in rat isolated mast cells by binding to carbohydrate moieties in the F_c region of the IgE antibodies attached to the cell surfaces. The consequent bridging of antibody receptors then triggered the secretory process. This release, like that induced by anti-IgE molecules was, in 1977, shown to be potentiated by PS and inhibited by disodium cromoglycate (DSCG) [17]. Clinical dextran also simulated the anaphylactic reaction in mast cells in several respects as, for example, the requirement of PS for enhancement of histamine release in the presence of calcium ions (fig. 3). On the other hand, compounds like 48/80 released histamine even in the absence of calcium ions and this release was not enhanced by PS or inhibited by DSCG.

However, release of histamine from rat isolated peritoneal mast cells did not always accurately represent events taking place within the body and

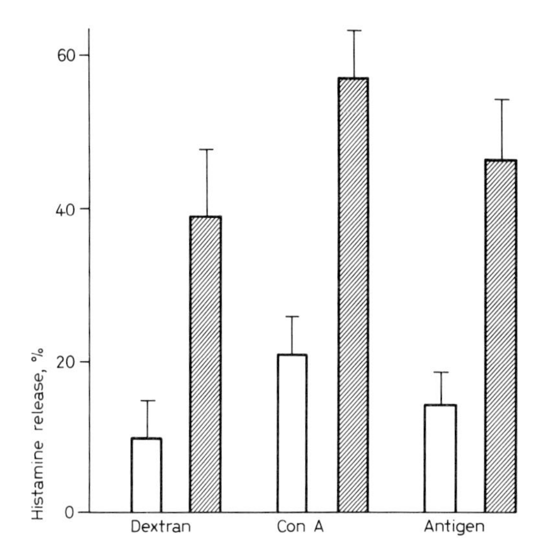

Fig. 3. Histamine release (%) from isolated peritoneal mast cells of reactor rats by dextran (1,000 μg/ml), Con A (10 μg/ml) and antigen (0.1 μg/ml). Shaded histograms represent release in the presence of phosphatidyl serine (10 μg/ml) [13].

Table 5. Amounts of antigen, Con A and dextran required to produce threshold histamine release from reactor rats [13]

Releaser	Isolated mast cells	Skin	Paws	Ratio in vivo/ in vitro
Antigen (BSA)	0.1	10	10	100
Con A	10	2000	1000	150
Dextran	1000	100	100	0.1

conclusions should not therefore be based solely on the results of in vitro experiments (table 5). Whereas amounts of antigen required in the skin and paws of reactor rats greatly exceeded those required in isolated mast cells, the reverse was true for dextran.

In 1984, West [18] analyzed the histamine-releasing activity of 12 agents in vitro and in vivo systems using the two types of genetically different rats (R and NR rats) (table 6). Compounds exerting a non-cytotoxic release initiated by a cross-linking with specific polysaccharide or antibody receptors on the mast cell surface (groups 1, 2 and 3) were used together with those exerting a

Table 6. Amounts of releasers required to produce threshold histamine release from isolated peritoneal mast cells (μg/ml) and threshold anaphylactoid oedema when injected into the paws (μg) of R and NR rats [18]

Group	Releaser	Isolated mast cells			Paws		
		R rats	NR rats	ratio NR/R	R rats	NR rats	ratio NR/R
1	Yeast mannan	200	8,000	40	0	400	40
	Clinical dextran	1,000	50,000	50	50	2,500	50
	Ovomucoid	500	4,000	8	50	2,000	40
	Baker's yeast	1,000	20,000	20	0	2,000	40
	Fresh egg-white	1/32	1/2	16	1/128	1/4	32
2	Dextran 6000	10,000	5,000	0.5	400	800	2
	Dextran 2×10^6	20,000	5,000	0.25	400	600	1.5
3*	Albumen	0.1	0.1	1	10	10	1
	Con A	10	10	1	1,000	2,000	2
4	A23187	5	5	1	10,000	10,000	1
	CTC	25	25	1	20,000	20,000	1
	PA	500	500	1	50,000	50,000	1

*Experiments using sensitized rats.

Table 7. Some properties of mucosal mast cells (MMC) and connective tissue mast cells (CTMC) of the rat [19]

MMC	CTMC
Small, variable shape, few granules	large, uniform shape, dense granules
Soluble granular proteoglycan matrix	less soluble proteoglycan matrix
Chondroitin sulphate in granules	heparin in granules
Low histamine content (2 pg/cell)	high histamine content (30 pg/cell)
Metachromasia blocked by aldehydes	metachromasia preserved by aldehydes
Berberine negative	berberine positive
Stain with alcian blue but no counterstain with safranin	stain with alcian blue and counterstain with safranin
Short life span	long life span
Proliferative response to nematode infection	no proliferative response to nematode infection
IgE in cytosol	no IgE in cytosol
Cromoglycate insensitive	cromoglycate sensitive
Theophylline insensitive	theophylline sensitive

non-cytotoxic release in the absence, as well as in the presence, of calcium ions (1 m*M*) which was not potentiated by PS (group 4). In its release mechanisms, this latter group resembled those involved when basic secretagogues like compound 48/80 were used, such release being unaffected in vivo by insulin or glucose pretreatment.

It is now known that mast cells constitute a heterogenous cell system. A specific type of mucosal mast cell has been identified in rats which required special histochemical procedures for its visualization, and this is in contrast to the familiar mast cell type occurring in connective tissue in the skin, tongue and peritoneal cavity. These differences between subsets may not only relate to structural and biochemical properties but also to functional properties (table 7) [19]. Subpopulations of mastocytes have therefore been grossly divided into these two groups, namely mucosal or atypical mast cells and connective tissue or typical mast cells. This distinction is purely chronological and historical. In the rat intestine, the two cell types were found to be quite separate and confined, respectively, to the lamina propria and submucosa. However, the extent to which these findings may be related to other species, and especially to man, is by no means obvious.

There has been some evidence for the existence of the two types in the nose, skin, lung and intestine of man and these may, respectively, bear some structural and histochemical similarity to the connective tissue and mucosal mast cells of the rat intestine. Subpopulations often seem to be randomly distributed and intermixed within the tissues [20]. As to properties, it was found that rat peritoneal and pleural mast cells were particularly sensitive to compound 48/80 whereas cells from the heart (like those basophils in human blood) were totally unresponsive (fig. 4). DSCG, an inhibitor of anaphylactic histamine release from connective tissue mast cells of the rat, was totally ineffective against human basophils.

Since 1980, mast cells have also been recovered by bronchoalveolar lavage from human subjects and shown to react quantitatively differently from enzymatically dispersed lung cells (obtained by incubation with collagenase) in their response to anti-human IgE [21]. In addition, human mast cells have been isolated from adenoidal tissues of generally healthy children within 2 h after surgery. Isolation of free mast cells from both guinea pig and rat hearts, using collagen dispersion techniques, were made in 1985 [22] but these cells were extremely resistant to a variety of receptor-mediated and ionophoretic stimuli. All these results show that mast cells from different species, and even from various tissues within a given animal, exhibited marked variations in their functional properties.

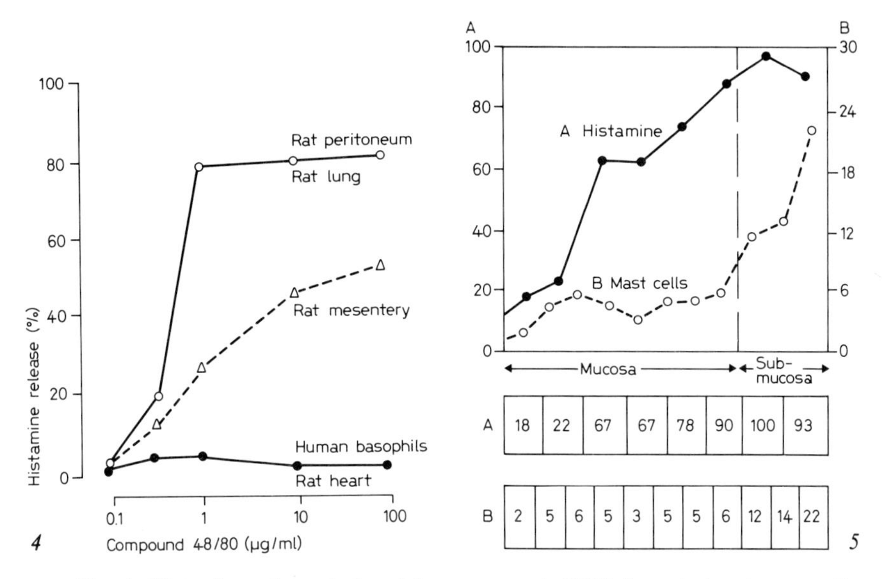

Fig. 4. Histamine release induced by compound 48/80 from isolated mast cells obtained from different sources. Mast cells were obtained by enzymic dissociation of the named tissue or by direct lavage [19].

Fig. 5. Relative histamine values (A, ng/section) and mast cell count (B, per HP field) at different depths in the mucosa and submucosa of hog pylorus. Note that histamine is considered to be bound to cells other than mast cells [23].

In 1956, Riley and West [23] cut serial vertical sections through the mucosa and submucosa of hog pyloric stomach and then stained them or assayed them for histamine (fig. 5). Many differences in mast cell sizes were found, such as large cells in the submucosa and much smaller ones accompanying loose tissue upwards between the tubular pyloric glands. Histamine, measured at 8 levels of depth, was high in the submucosa and the high mast cell count continued up to the mucosa where there was a fall-off in count but little lowering of histamine. As gastric mucin bound histamine and mucin stained metachromatically with toluidine blue, it was considered that mucins in the mucosal cells might well contribute to the storage of histamine in the gut, which may have properties different from those of histamine in mast cells. Ionic binding of histamine to protein carboxyls in the mast cell granule heparin-protein complex, as shown in 1970 [24], might well apply to a mucin complex. Isolated mast cell granules with cation exchanger properties were later found to be totally depleted of their histamine when superfused with isotonic sodium or potassium chloride solution, the release following the same kinetics as those of histamine release from a synthetic carboxylic cation exchanger like IRC-50.

Hanahoe [25] reported in 1984 that anti-IgE and dextran compete for a similar receptor system in isolated peritoneal mast cells from dextran-sensitive rats. He washed them with a pH 3 glycine-HCl buffer for 4 min at 0 °C and then showed that they did not react either to rat anti-IgE or to dextran, even in the presence of optimal amounts of calcium and PS. Presumably, the acid washing removed endogenous antibody which occurs naturally in these rat cells. A similar state may be present in cholinergic histamine release, for acetylcholine only produced a dose-dependent release from human mast cells when the cells were capable of reacting to anti-human IgE. Acetylcholine also produced histamine release from guinea pig mast cells only when the cells were isolated from actively sensitized animals, and a similar situation applied to rats and dogs. Active sensitization may cause mast cells to acquire a particular functional state of susceptibility for stimuli and a readiness for secretion, whereas passive sensitization did not always do this. Cholinergic drugs and nerve stimulation also effected histamine release from submandibular glands of the cat, in the heart of the guinea pig, in the ileum of the rat, and in most whole mammals. It has also been found that carbachol by itself did not always release histamine from mast cells of rats but it potentiated the release induced by substance P (an endogenous basic polypeptide). In this case, histamine release did not occur through interaction with cell-bound IgE but through the action on specific receptors for substance P. The synergistic effect may be of physiological importance in view of the co-existence of peptides with neurohormones at nerve endings. Substance P released from sensory nerves may act on mast cells in and around blood vessels [26], thereby releasing histamine which in turn produces vasodilatation.

Functions of Histamine

Mast cells have for many years been at the center of interest for the study of immediate allergic and anaphylactic reactions. They have a large number of receptors (40,000–150,000) generally with a high affinity for the F_c part of the IgE, the reaginic or anaphylactic antibody. Binding of the specific allergen to the F_{ab} part of two adjacent IgE molecules in the surface of the cell (the so-called bridging) has been shown to elicit complex biochemical events which result in the release of histamine and other stored or de novo synthesized

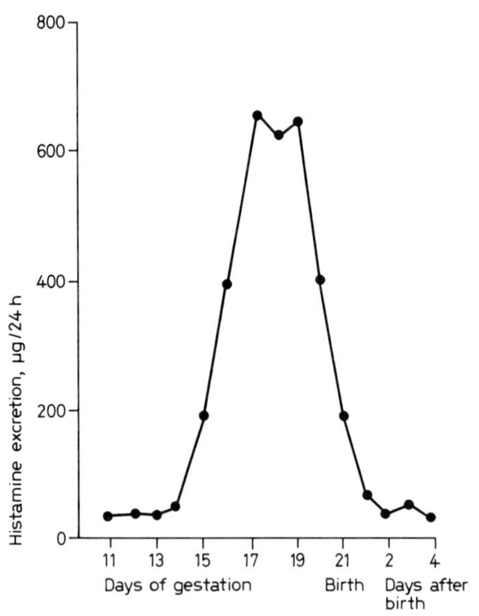

Fig. 6. Effect of age on the rate of histamine excretion (µg/24 h) in rats during pregnancy and up to 4 days post-partum, Note that an abrupt increase occurs from the 15th day of gestation, yet just before term the excretion rapidly declines [28].

mediators. It is now known that two of the biochemical events are an increase in intracellular cyclic AMP and a simultaneous increase in phospholipid methylation, both being inhibited by most anti-allergic agents. Histamine released then stimulated mesenchymal cells to phagocytose and even digest heparin which was released, not for its anticoagulant action, but for its action in speeding up the repair of injured tissue. The primary action of histamine may be to prepare the tissues to receive the heparin. Release of histamine was usually accompanied by release of kinin-forming enzymes producing bradykinin and this, like histamine, is a potent vasodilator, causing the accumulation of leukocytes.

In 1960, Kahlson [27] expressed the view that activation of histidine decarboxylase, the enzyme responsible for the formation of histamine from the essential amino acid, histidine, may be controlling the normal physiological events taking place in maturation processes. For example, he showed that this enzyme activity was fantastically high in rat fetuses during the last 3 days of gestation, so that the excess histamine formed diffused freely into the surrounding fluid and was excreted by the mother (fig. 6). However, later work in 1962 proved that this was not mast cell histidine decarboxylase stimulation in the fetal rat liver because this, like the adult liver, was almost devoid of mast cells. However, between days 13 and 21 of gestation, up to 1 mg of histamine was excreted by just one rat [28]. A dramatic fall, both in histamine-forming capacity and in histamine excretion, occurred at parturition. Fetuses of other mammals including man, however, formed little or no histamine during gestation [28]. Growth of transplantable rat tumors was also unrelated to histamine-forming activity [29]. Kahlson also included regeneration and repair in his hypothesis for a physiological role for histamine but partially hepatectomized rats did not show increased enzyme activity in the regenerating liver, and no increased enzyme activity occurred during the healing process after experimentally produced skin wounds had been made in rats. Consequently, the proposed link between histamine formation and growth was abandoned.

Basophils

The origin of basophils, like that of mast cells, is still not known for certain. Both probably have a common origin, arising from a bone marrow precursor. There has been some evidence to show that antigen interacts with B-lymphocytes to stimulate differentiation of these cells into IgE antibody-secreting plasma cells. IgE almost exclusively bound to both basophils and mast cells but this simple binding to the F_c receptor did not activate the cell to secrete histamine. Binding of the specific allergen to the F_{ab} part of two adjacent IgE molecules in the surface of both basophil and mast cell elicited similar biochemical events. The basophil is the only blood cell to contain appreciable quantities of histamine and this was released in response to antigen challenge. Blood basophil counts were significantly elevated in atopy and so were those of eosinophils. It was also shown that the unresponsiveness of basophils in some allergic patients was associated with impaired IgE-receptor complex activation rather than with a lack of cell-bound IgE [30]. On the other hand, reduced histamine release from basophils of asthmatic patients receiving corticosteroids may explain their protective action, particularly as the steroids reduced blood basophil counts and leukocyte histamine content at a time when eosinophils disappeared and neutrophil counts rose [31].

Although basophils are said to be cousins of mast cells, they have been found to react differently in special circumstances. For example, the histamine-releasing effect of human C_{5a}, a complement peptide from human basophils was substantial, whereas release from isolated mast cells taken from the adenoids of the same donors was only marginal, although both types of cell responded markedly to substance P and polylysine [32]. The difference, of course, may be related to the method of preparation of the cells, for collagenase was used for the adenoid cells to effect purification, whereas the basophils were separated from other blood cells by dextran sedimentation and other differentiation techniques. A similar explanation may be used to identify quantitative differences between human blood basophils, mast cells recovered by bronchoalveolar lavage of human subjects, and the enzymatically dispersed lung cells. These three cell types differed in their response to anti-human IgE and in their response to DSCG. This antagonist was highly selective in inhibiting immunological histamine release from bronchoalveolar cells, less active on dispersed lung cells, but ineffective against basophils [21]. In all cases, activation of adenylate cyclase during the histamine release process was found to be involved.

Future Trends

In my view, the future of mast cell, basophil and histamine investigations remains in the areas of receptors in the tissues and brain, storage sites, uptake mechanisms and metabolism. Biochemical and toxicity studies using selective enzyme inhibitors also appear to be hopeful lines for major advances.

As already stated, isolated mast cells of dextran-resistant rats (the so-called NR rats) were over 50 times less sensitive to dextran than were dextran-sensitive rats (R rats). This ratio did not change, even when active sensitization or passive sensitization with *Nippostrongylus brasiliensis* antigen or with monoclonal myeloma IgE were used as pretreatments, although the response to anti-IgE sera was potentiated in each case. In these special NR rats, therefore, the lack of dextran sensitivity was not associated with lack of IgE bound to the mast cell membrane. Nevertheless, *N. brasiliensis* antisera from NR rats was capable of enhancing dextran-induced histamine release from R cells, thereby indicating that IgE from NR rats can interact with dextran. In other words, NR rats have mast cells which possess a viable dextran receptor, but something like a blocking protein prevents the mast cell membrane from combining with dextran [33]. The electrophoresis patterns of the proteins in peritoneal mast cells of R and NR rats showed that there is a 74,000-dalton protein in NR mast cell membranes which is lacking in R membranes (fig. 7). A similar abnormality may explain why the protein in normal human skin mast cells may be lacking in skin mast cells from atopic

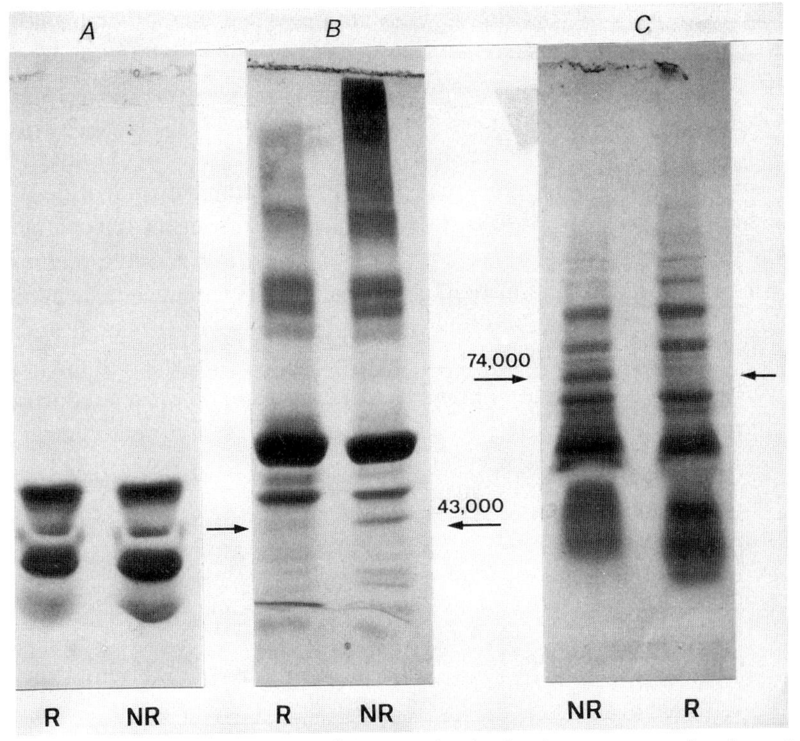

Fig. 7. Electrophoresis patterns of the proteins in the three extracts of peritoneal mast cells from R and NR rats. *A* 3,000 *g* pellet. *B* 55,000 *g* supernatant. *C* 55,000 *g* pellet. Each extract was obtained from equal numbers of mast cells. Note that the 43,000 molecular weight protein is deficient in R cells in *B* and the 74,000 molecular weight protein is lacking in R cells in *C* [33].

patients. However, isolated mast cells from the rat peritoneal cavity were used to identify the abnormality in membrane proteins and there are few mast cells in the peritoneal cavity of man. The degree of purity of the mast cells may also be responsible for the discrepancy, for rat peritoneal fluid has been shown to contain an activator of histamine release from R rat mast cells and this may be absent in fluid from NR rats.

The degranulation process of mast cells treated with potent histamine liberators also needs to be completely identified. Under such treatment, it has been shown that a significant increase in the actin content of the plasma membrane fraction and a corresponding decrease in that of the cytosol fraction was found, indicating that F-actin filaments in the cytoplasm had moved towards the plasma membrane and degranulation followed. The existence of actin filaments in mast cells has enabled three-dimensional images of the subplasma network to be produced and these have shown granules surrounded by a small mesh network. This delicate study has now been started by Tasaka et al. [34] in Japan. The composition of mast cell and basophil plasma membranes has next to be completely identified. Free calcium ions play a vital role in release mechanisms, hence the channels by which these ions pass through both the plasma and the granular membranes need to be traced. Calcium channels have already been shown to be involved, based on the relative activities of calcium entry blockers in allergic and non-allergic release mechanisms. By a strange coincidence, calcium antagonists have recently been found to prevent withdrawal symptoms related to addictive problems of drugs like alcohol and perhaps many other central drug dependencies.

References

1 Ehrlich P: Beiträge zur Kenntnis der Anilinfarbungen und ihrer Verwendung in der mikroskopischen Technik. Mikrosk Anat 1877;13:263–273.
2 Jorpes JC: Heparin in the treatment of thrombosis, ed 2. London, Oxford University Press, 1946.
3 Portier P, Richet C: De l'action anaphylactique de certains venins. C R Soc Biol 1902;54:170–181.
4 MacIntosh FC, Paton WDM: The liberation of histamine by certain organic bases. J Physiol 1949;109:190–219.
5 Rocha e Silva M: Concerning the mechanism of anaphylaxis and allergy. Br Med J 1952;i:779–784.
6 Shore PA, Burkhalter A, Cohn VH: A method for the fluorimetric assay of histamine in tissue. J Pharmacol Exp Ther 1959;127:182–186.
7 Riley JF, West GB: Histamine in tissue mast cells. J Physiol 1952;117:72P.
8 Riley JF, West GB: Mast cells and histamine in normal and pathological tissues. J Physiol 1952;119:69P.
9 Riley JF, West GB: The presence of histamine in tissue mast cells. J Physiol 1953; 120:528–537.
10 Paton WDM: Compound 48/80 A potent histamine liberator. Br J Pharmacol 1951;6: 499–508.
11 Riley JF, West GB: Tissue mast cells. Studies with a histamine liberator of low toxicity (compound 48/80). J Pathol Bact 1955;69:269–282.
12 Harris JM, West GB: A new link between the anaphylactoid reaction in rats and human allergy. Int Archs Allergy Appl Immunol 1964;25:46–57.
13 West GB: Biogenic amines and drug research. Prog Drug Res 1984;28:9–52.
14 Foreman JC, Mongar JL, Gomperts RD: Calcium ionophores and movement of calcium ions following the physiological stimulus to a secretory process. Nature 1973;254:249–251.

15 Pearce FL: Calcium and histamine secretion from mast cells. Prog Med Chem 1982;19:59–109.
16 Reed GW, Kiefer EF: Benzalkonium chloride, a selective inhibitor of histamine release induced by compound 48/80 and other polyamines. J Pharmacol Exp Ther 1979;211:711–715.
17 Read GW, Knoohuizen M, Goth A: Relationship between phosphatidylserine and cromolyn in histamine release. Eur J Pharmacol 1977;42:171–177.
18 West GB: Histamine release from mast cells of the rat. Int Archs Allergy Appl Immunol 1984;74:278–280.
19 Pearce FL: Mast cell heterogeneity the problem of nomenclature. Agents Actions 1988;23:125–128.
20 Befus D, Goodacre R, Dyke N, et al: Mast cell heterogeneity in man. Histologic studies in the intestine. Int Archs Allergy Appl Immunol 1985;76:232–236.
21 Leung KBP, Flint KC, Brostoff J, et al: Some properties of mast cells obtained by human bronchoalveolar lavage. Agents Actions 1986;18:110–112.
22 Ali H, Pearce FL: Isolation and properties of cardiac and other mast cells from the rat and guinea-pig. Agents Actions 1985;16:138–140.
23 Riley JF, West GB: A binding site for histamine in hog pyloric mucosa. Experientia 1956;12:153–154.
24 Uvnas B, Aborg CH, Bergendorff A: Storage of histamine in mast cells. Evidence for an ionic binding of histamine to protein carboxyls in the granular heparin-protein complex. Acta Physiol Scand 1970;336(suppl):1–26.
25 Hanahoe THP: Mechanism of histamine release from rat isolated peritoneal mast cells by dextran. The role of immunoglobulin E. Agents Actions 1984;14:468–474.
26 Foreman J, Jordan C: Histamine release and vascular changes induced by neuropeptides. Agents Actions 1983;13:105–116.
27 Kahlson G: Activation of histidine decarboxylase as a common factor in various kinds of normal and malignant rapid tissue growth. J Physiol 1961;160:12–19.
28 Kameswaran L, Pennefather JN, West GB: Possible role of histamine in rat pregnancy. J Physiol 1962;164:138–149.
29 Buttle GAH, Eperon J, Kameswaran L, et al: Histamine formation and tumour growth. Br J Cancer 1962;16:131–140.
30 Stahl Skov P, Norn S, Weeks B, et al: Impaired basophil histamine release from allergic patients. Agents Actions 1987;20:303–306.
31 Assem ESK: Inhibitors of histamine release from basophil leucocytes of asthmatic patients treated with corticosteroids. Agents Actions 1985;16:256–259.
32 Jurgensen H, Braan U, Pult P, et al: Human C_{5a} induces a substantial histamine release in human basophils but not in tissue mast cells. Int Archs Allergy Appl Immunol 1988;85:487–488.
33 Ludowyke RL, West GB, Harris JR, et al: A possible link between the ability of rat isolated peritoneal mast cells to release histamine when challenged with dextran and the proteins in their plasma membranes. Int Archs Allergy Appl Immunol 1983;68: 188–191.
34 Tasaka K, Akagi M, Miyoshi K, et al: The role of micro filaments in the exocytosis of rat peritoneal mast cells. Int Archs Allergy Appl Immunol, in press.

G.B. West, DSc, 22 Burgh Heath Road, Epsom, Surrey KT17 4LS (UK)

Waksman BH (ed): 1939–1989: Fifty Years Progress in Allergy.
Chem Immunol. Basel, Karger, 1990, vol 49, pp 142–172

New Insights into the Mechanisms of Histamine Release from Rat Peritoneal Mast Cells

Bertil Diamant

Department of Pharmacology, University of Copenhagen Medical School, Copenhagen, Denmark

Our understanding of the cellular reactions involved in histamine release has extended over the last 10 years to involve many new cellular regulatory mechanisms. The heterogeneity not only of different histamine secretory cell types but also among cells of a single type at different locations in the body has become apparent. A totally new regulatory mechanism involving the phosphoinositol cycle, which is considered responsible for perturbations of calcium homeostasis in the cell, is now regarded by many as the ultimate trigger of the secretory process for granule-stored histamine. This regulatory cycle has placed the once (10 years ago) popular unifying theory of cAMP as a regulator of histamine release somewhat in the background. A rise and subsequent decline in the concept of the importance of methylating reactions early in secretion has also been evident for over 10 years and the possible importance of early phosphorylation of key components of the cell membranes has been established. Furthermore, the regulatory influence of sialic acid in the plasma membrane of the secretory cell on secretion has been clearly established, as have new methods to indicate the increase and movement of free calcium within the cell early in the secretory process. To gain further insight into the molecular reactions underlying the secretory process, various methods have been introduced to increase the permeability of isolated rat mast cells and to measure the reactions occurring in such permealized cells.

When Paul Kallos asked me to write a chapter for *Progress in Allergy* on new insights into the mechanisms of histamine release, he did so for several reasons. First, I have had the privilege to have Paul as a dear and close friend and a constantly encouraging mentor for the greater part of my life. Secondly, as early as 1978, Waclaw Kazimierczak and I were asked to write a similar chapter for volume 25 of *Progress in Allergy* [100]. Third, Paul knew that I had been out of the histamine field for at least 5 of the last 10 years and therefore thought that I might be able to look at progress without too much bias. Finally, he evidently considered that over the last 30 years some contributions to the understanding of the process of histamine release had come from my work together with that of various colleagues.

Heterogeneity of Histamine-Secreting Cells

The biochemical processes which occur when different secretory agonists become associated with histamine-storing cells and which lead to the appearance of the characteristic intracellular morphological changes usually accompanied by externalization of granular material, indicative of histamine secretion, are still incompletely understood.

There are several reasons for this. In spite of great efforts at purification, mast cells and basophils obtained from different organs and sources are often contaminated to a disturbing degree by nonsecretory cells. Therefore, supposedly specific pharmacological or biochemical analytical results may give false clues, which are easily misinterpreted. It has long been recognized that quantitative as well as qualitative differences exist in the reactivity of a single cell type from different species towards different secretory agonists [61, 144, 147], and also among cells of the same type from different inbred strains of the same species [80–82, 176]. Even mast cells isolated from different sites in a single rat strain may differ in reactivity [153]. Enerbäck was the first to call attention to the histological heterogeneity of mast cells [56–58]; he noted differences in fixation between rat mast cells from the connective tissue and the intestine. Qualitative differences are also known for the influence of inhibitors (e.g. beta-receptor agonists) on histamine secretion [6, 112]. Such differences are also recognized among human histamine-secreting cells, for example, between the reactivity of human basophils and human lung mast cells [86, 177]. Even within a single organ (human lung) different subtypes of mast cells have been demonstrated with changes of their frequency distribution under pathological conditions [173, 174]. In recent years, these problems have been studied extensively by the groups around Pierce [134, 136], Foreman [1], Church [12, 28] and van Overveld [172]. For example, human mast cells isolated from the intestine were shown to possess secretory characteristics similar to human lung but different from human skin mast cells [115, 146]. The complement peptide C5a effectively induces histamine release from human basophils as well as human skin mast cells, but not human lung mast cells [155]. These examples emphasize that there are major differences in the reactivity of related cells and that generalization from one cell type to another should be avoided. Warnings about these matters appear frequently in the literature and should be heeded more than appears to be the case today. In a critical editorial on mast cells and the actions of flavonoids on secretion, Foreman [62] states that: 'No longer can any single mast cell population be used as a model system for development of studying drugs unless it is that the same mast cell population will ultimately be the target for the drug or unless it is established that the model is comparable with the target.'

On the other hand, Fox et al. [70] recently claimed that: 'We have found no pharmacologic difference between the human lung and intestinal mucosal mast cell such as exists in the rodent models. This underscores the limitations of applying the data from animal mast cells to the human mast cells, and at the same time demonstrates that human basophils and mast cells are different.'

Whether this is a valid conclusion remains to be seen. It is also claimed that the reactivity of human skin mast cells corresponds well with rat mast cells isolated from the peritoneal cavity [27]. Compound 48/80 and ATP release histamine from rat mast cells but not from human basophils [42]. Nevertheless, when injected intracutaneously in the human skin, both agents induce wheal and flare reactions which are antagonized by pretreatment with antihistamines, indicating the engagement of mast cells in the skin [89].

The rat mast cell is the only normally occurring histamine-containing cell that can easily be obtained with a purity of more than 95%. It has been successfully used during the last decades for the study of histological, biochemical and pharmacological cellular changes underlying secretory processes of histamine. If rat peritoneal mast cells may serve as a model for what happens during the allergic reaction in human skin, a lot would be gained in view of their widespread experimental use and the extensive knowledge now available on their behavior. On the other hand, the observed heterogeneity among histamine-secreting cells in different species and bodily localization does not by itself invalidate the assumption that certain steps in the release process may be common for most normal cells and for most releasing agents, in that they depend on certain common factors needed to couple stimulation to secretion. The differences observed may be due to variations between cells in membrane composition and permeability properties, key enzymes and specific receptors; these may all influence the reactivity of the cells to secretory agents as well as the effectiveness of alleged inhibitory agents.

Primary and Secondary Secretory Cofactors

Several reviews have been published during the last 10 years focusing on morphological, biological, biochemical, pharmacological, and pathophysiological aspects of histamine release [62, 71, 116, 118, 135]. Although comprehensive, the final conclusions are often inconclusive, which is quite understandable in view of the complexity of the present state of the art.

A general agreement seems to exist, however, that what is referred to as noncytotoxic or selective release of histamine depends on two primary coupling factors, namely calcium and metabolic energy in the form of ATP. Far less appreciated are the findings, based on studies of Uvnäs and coworkers over the last decades after original observations by Lagunoff and Benditt [106] and Uvnäs [165, 166], that the final step of the release process, after stimulation of mast cells by immunological or nonimmunological agents, is a simple ion exchange reaction between histamine bound to the granule matrix of the mast cells and cations normally excluded from the cell. By using naturally occurring granule matrix isolated from rat mast cells or artificially produced from protamine and heparin, the kinetics of such an exchange have been found to be a highly efficient mechanism for the release of histamine from the granules [15, 166, 167, 171]. It is important to emphasize that the granule matrix need not be expelled from the cells (degranulation) in order to release histamine by such an ion exchange mechanism, although this is seen for most agents, unless special conditions are instituted [3, 19, 103, 107, 108, 150].

Calcium and ATP are designated as primary secretory coupling factors simply because it has repeatedly been demonstrated that if one or the other becomes unavailable, the histamine secretory cell becomes highly resistant to stimulation, irrespective of the releasing agent to which it is exposed. Not only is histamine not released, but characteristic morphological changes, as seen in the light or electron microscope and which are a prerequisite for release, fail to appear. The following question must therefore be answered in order to understand the mechanism of secretion [100]: What calcium- and energy-requiring biochemical processes are triggered in the histamine-secreting cell, when various agonists associate themselves with the cell via receptors or by other means, leading to morphological changes which allow cations access to the granular matrix of the cell and thereby to exchange with its histamine?

The divergent opinions that have become apparent on this issue during the last decade seem to concern the qualitative and quantitative importance of what may be considered as secondary coupling factors. In this group may be included enzymes (phospholipases, protein kinases and cyclases), phospholipids and metabolites (arachidonic acid, phosphoinositol, diacyl glycerol and inositol-triphosphate), calcium-binding proteins (calmodulin) and cofactors of phosphorylation (cAMP and cGMP). The function of such secondary cofactors for secretion should perhaps be looked upon as regulators or modulators for the availability or utilization of the primary ones. An attempt will be made in the following pages to maintain this attitude.

Histamine-Retaining Rat Mast Cells Granules

Since the granules of the mast cells have a central role in the release process, it is pertinent to discuss their behavior in an isolated milieu. By sonication of rat mast cells suspended in either sucrose (0.34 *M*) with human serum albumin (1 mg/ml) [2] or in Ficoll (6.4% w/v)/Hypaque (10.5% w/v)/ human serum albumin (1 mg/ml) [77] at a suitable frequency, the cells break and granular material can be separated from the cell residues. After wash in a balanced salt solution (BSS) that will exchange all histamine that is bound to membrane-free granular material with monovalent cations, 30–50% of the original histamine of the cells can be recovered from granules which retain their histamine, as well as an intact perigranular membrane after isolation [77, 145]. In our experiments, this fraction showed a higher spontaneous release of histamine than intact rat mast cells, amounting to 20–30% of the total histamine content after incubation at 37 °C for 10 min in BSS containing NaCl (131 m*M*), KCl (2.4 m*M*), Tris buffer (10 m*M*, pH 7.0) and 0.5 mg/ml of human serum albumin. In contrast to what is found with intact mast cells, a slight but significant release of histamine above the spontaneous level was noted for ATP in the presence of 3 m*M* magnesium. Calcium could not substitute for magnesium and ATP by itself was without effect. The ionophore A23187 (5 μ*M*) efficiently released histamine provided divalent cations were present.

In contrast to intact rat mast cells, magnesium was as efficient as calcium in supporting the release from the granules, whereas strontium (10 m*M*) or barium (3 m*M*) were without effect. Also in contrast to intact rat mast cells, antimycin, DFP and N-ethyl-malemide did not inhibit histamine release induced by A23187 in the presence of calcium. Neither antigen nor compound 48/80 [77], fluoride [132] or thapsigargine [133] released histamine from histamine-retaining granules, irrespective of the presence of calcium or magnesium. Both the cytotoxic compound decylamine and the monovalent cationic ionophore X537A were highly efficient in releasing most of the histamine from the granules [77]. In contrast to decylamine, X537A was dependent on the presence of sodium or potassium in the incubation medium to release histamine from granules in good agreement with its function as a carrier of monovalent cations across membranes. Also in contrast to decylamine, the action of X537A decreased at 0 °C [45]. The action of X537A should be considered in relation to its effect on intact rat mast cells where, at optimal concentrations, it causes the release of histamine without dependence on calcium or metabolic energy [68, 101].

These studies on histamine-retaining granules clearly distinguish the mechanism of histamine release from intact rat mast cells from that of granules in that most agents are only active on intact cells. X537A is an exception and seems to be able to deliver monovalent cations directly to the granular matrix for exchange with histamine equally well in intact rat mast cells as in histamine-retaining granules. In contrast, the ionophore A23187 is able to transport some but not all divalent cations to isolated histamine-containing granules for exchange. However, it is important to keep in mind that histamine release induced by A23187 from intact rat mast cells is absolutely dependent on calcium and that calcium cannot be replaced by magnesium [68].

The results with histamine-retaining mast cell granules seem to exclude the possibility that agents which depend on calcium and cellular energy for histamine release from intact rat mast cells may have a direct effect by themselves on the granular matrix. The results conform very well to the nature of the granule matrix, as demonstrated regarding the binding kinetics of mono- and divalent cations to membrane-free granule material [15]. The granule matrix can bind any compound that has a sufficiently high affinity for the binding sites, as has also been shown for biogenic amines other than histamine [13, 14].

In the intact rat mast cell, the normal electron-dense compact appearance of the granules, as seen in the electron microscope, changes when histamine release is initiated. They become swollen, less dense and have the appearance of a perigranular space. Characteristically, membrane-free granule material appears outside the cell and there are intracellular vacuoles, which may each contain several granules with broken perigranular membranes, which communicate not only with each other but also with the exterior [19]. This morphological process has been named sequential exocytosis [150] and results in communication between the granule matrix and extracellular milieu. It has been suggested that the granules of amine-secreting cells are localized in microtubules which traverse the cell and connect with the outer membrane. This has also been claimed to be the case for histamine-containing granules in rat mast cells [128, 129] and might help explain the phenomenon of vacuole formation. However, no signs of microtubules can be demonstrated in nonsecreting rat mast cells, so their existence is not finally proved. Vacuoles may theoretically also be formed during the release process by an active engagement of the granules themselves. Mast cell protease is a major component of the granules [10, 106] and normally has a low activity, due to an acid intragranular pH of around 6.0 [109]. Protease might become activated after contact between the plasma and perigranular membrane has been established, allowing communication with the more alkaline pH of the extracellular space. Since this will happen initially with those granules that are localized most superficially in the cell, the activated protease may facilitate a further breakdown of the membranes of adjacent granules, which could lead to communication and vacuole formation spreading to the interior of the cell.

In 1967 it was established that isolated rat mast cells hydrolyze ATP added to the incubation medium by an enzyme activated by calcium or magnesium. The enzyme was described as an ecto-ATPase [38, 39]. A similar enzyme has also been localized by morphological methods to the membranes of isolated mast cell granules. Its activity seemed, however, not to differ between secreting and nonsecreting cells [23]. The function of this enzyme is unknown but it cannot be overlooked that it might be involved in an increase in the permeability of the granular membranes between adjacent granules, thereby facilitating influx of water and cations and consequent vacuolization and release of histamine by ion exchange. Such an increase in the permeability of the plasma membrane has been established in isolated rat mast cells incubated with micromolar concentrations of ATP leading to marked swelling of the cells [20] and a pronounced uptake of sodium and release of potassium [35]. It is also possible that the ATPase found on the granules is identical to a phosphatidylinositol kinase associated with mast cell granules, which transforms phosphatidylinositol (PI) to phsophatidylinositol-4-phosphate (PIP) in the presence of ATP and Mg^{2+} or Mn^{2+} [104]. This kinase is present on the exterior of the granule membrane where it has access to endogenous ATP and other regulatory cofactors. It was recently reported to become activated after exposure of mast cells to histamine-releasing agents like concanavalin A (Con A) and compound 48/80 [105].

An enhanced turnover of phosphatidylinositol with accompanying changes of protein phosphorylation as well as calcium transport may be a key to the control of secretory processes in general [18]. It has been speculated that these PI products participate in the mast cell not only as a trigger of secretion but also in repair processes after secretion has terminated [105]. Irrespective of the significance of these ideas, the observations made at the granule level suggest the possibility of a more active engagement of key components and enzymes in the granules in the secretory process than generally anticipated.

Sources of Calcium and Metabolic Adenosine Triphosphate for Histamine Release

Adenosine Triphosphate (ATP)

Anaphylactic histamine release is one of the first secretory processes for which an energy requirement became apparent. Parrot [130] demonstrated that when minced, desensitized guinea pig lung tissue was incubated in an anoxic milieu, histamine release did not occur upon antigen challenge. This was soon confirmed by the use of various inhibitors of oxidative phosphorylation [22, 119, 123]. The energy requirement can be met under anerobic conditions through a breakdown of glucose to lactate [36]. After methods had been devised for the isolation of peritoneal mast cells [169], these phenomena were confirmed in several laboratories studying energy requirements for histamine release induced by different agents in pure rat mast cells.

This was later followed by direct measurements of lactate production from hexoses under anerobic conditions [51] and of ATP content in rat mast cells using enzymatic fluorometric methods [37, 141]. The kinetics of the decrease in cellular ATP in the presence of metabolic inhibitors were established, as well as the production of ATP through glycolytic metabolism [48, 137]. The degree of histamine release was shown to be directly related to the cellular content of ATP and, furthermore, ATP was demonstrated to be utilized in the release process [96, 97, 140, 142]. Energy utilization appeared quantitatively similar for different energy-requiring releasing agents, a process that is best registered under glycolytic conditions when oxidative phosphorylation is inhibited [95, 97, 142].

A decrease of the ATP content by 25% does not interfere with histamine release when induced by either ionophore A23187 or compound 48/80, whereas anaphylactic histamine release was found to be sensitive to such a decrease of cellular ATP which was ascribed to nonspecific actions of the metabolic inhibitor used [95, 96, 139]. The utilization of ATP by the release process has been calculated to be 25–30% of the total available amount [48, 97, 142]. It was estimated that the ATP turnover in rat mast cells is high with renewal of all ATP at least twice per minute [40]. Therefore, the energy utilized in the release process would be restored within 10 s provided the cellular reactions and morphological changes that develop do not by themselves influence energy production. This cannot be excluded for certain releasing agents. Indications of an uncoupling effect on oxidative phosphorylation have been noted for anaphylactic histamine release from rat mast cells [48].

Extracellular ATP inhibits glycolytic reactions in rat mast cells and, since the presence of increasing concentrations of calcium counteracts inhibition of lactate production as well as histamine release, it was concluded that both effects depend on free ATP in the incubation medium [143]. Furthermore, inhibition of glycolysis in rat mast cells was not secondary to the release of histamine, since inhibition of glycolysis was optimal in the absence of calcium when, as a consequence, no histamine release took place. In contrast, compound 48/80 under similar conditions did not inhibit glycolysis; this is the most likely reason for the finding that, when oxidative phosphorylation was blocked by antimycin A, glucose counteracted the inhibition of histamine release by compound 48/80 but not by ATP [138]. It is important to emphasize that histamine release induced by extracellular ATP is not only dependent on calcium in the incubation medium but also on an adequate metabolic energy production by the cells [46]. Besides histamine-releasing properties and metabolic effects, extracellular ATP exerts a number of other actions on isolated peritoneal rat mast cells, some of which are highly specific among histamine-releasing agents (see below). It is thus important to distinguish between effects exerted by extracellular ATP on rat mast cells from those exerted by intracellular ATP in the form of metabolic energy which functions as a primary secretory cofactor.

Since ATP production in rat mast cells under anerobic conditions in the presence of glucose is somewhat higher than under aerobic conditions in the absence of glucose and since the ATP content of rat mast cells is stable for at least 4 h under aerobic conditions even in the absence of glucose, there seems to be no risk that under normal conditions cellular ATP would limit the histamine-releasing reactions [30, 40]. On the basis of our own experiments, it is highly recommended, when working with unknown substances with alleged inhibitory actions, to establish if they exert such an effect by inhibiting ATP production in the cells. With bioluminescence methods using firefly luciferase, ATP determinations are now fast, inexpensive and sensitive. Only 10,000 cells are needed for an adequate reading.

Calcium

The functional importance of calcium has been much more intensively studied than metabolic energy in investigations of the mechanisms behind stimulus-secretion coupling. For the histamine release process, the first indication of the importance of calcium was the finding of Mongar and Schild [120] in 1958 that antigen-induced histamine release from lungs of sensitized guinea pigs was enhanced in the presence of calcium. In 1973, Foreman et al. [68] showed that ionophore A23187 induced histamine release from isolated rat mast cells provided that calcium was present in the incubation medium. Also in 1973, Kanno et al. [98] showed that microinjections of calcium into single rat mast cells induced morphological changes consistent with histamine release (degranulation). In 1978, Theoharides and Douglas [163] demonstrated that calcium introduced into rat mast cells with the help of liposomes triggered histamine release without the presence of any additional stimulator.

Histamine-releasing agents may induce release by activating calcium-mobilizing surface receptors. Besides the IgE receptor, which has been extensively studied, little is known about the structure of any other surface receptor involved in histamine release. Histamine-releasing peptides seem to bind to surface receptors [69]. An example is neurotensin [110], which by itself seems to be a partial agonist competitively antagonizing receptor binding and consequently the histamine-releasing activity of substance P [66]. Polymyxin and compound 48/80 may share a common receptor or part of one, as indicated by cross-desensitizing experiments between the two agents [122]. Ionophores seem to bypass surface receptors and can transport ions directly to the cell interior. Other lipophilic histamine-releasing agents like thapsigargine [127] may enhance the permeability of the membranes for calcium by other mechanisms so far not well described. It also seems unlikely that sodium fluoride would activate the mast cells through a surface receptor [131].

Over the years there has been a mutual exchange of ideas regarding receptor-mediated and other reactions which underlie secretory histamine release and release of many hormones and neurogenic transmitters [179]. Various membrane-bound enzymes have been identified (e.g. phospholipase

C and adenyl cyclase) which, when activated, are responsible for an energy-requiring production of cofactors for phosphorylation of different protein structures by various protein kinases. These reactions appear to be involved in the triggering of secretory processes including histamine release, as will be discussed below. Calcium, which has a central role both in the activation of these enzymes and in the further function of the cofactors, can be utilized from both internal and external sources.

In isolated rat mast cells it is possible to differentiate between different secretory agonists on the basis of their sensitivity to calcium.

(1) Certain phospholipids like phosphatidylserine (PS) have been found to enhance the sensitivity of IgE-mediated histamine secretion to calcium in isolated rat mast cells [121]. Dextran is a very poor histamine releaser in the absence of PS [9, 73]. It has been suggested that this effect of PS may be ascribed to an interaction between PS and calcium with formation of calcium-PS bridges between the granule and plasma membranes, which may enhance exocytosis [121]. Such a unifying concept is, however, difficult to accept in view of observations that histamine release induced by chymotrypsin and extracellular ATP was not affected by PS, and that PS inhibited the action of such diverse agents as compound 48/80, the ionophore A23187 and decylamine [73, 75].

Sialic acid also seems to regulate the calcium permeability of the plasma membrane. When rat mast cells were treated with a sialidase to decrease the content of sialic acid in the plasma membrane, histamine release induced by antigen or A23187 was enhanced [7, 9]. This effect has also been shown in human basophils where it was established that treatment with sialidase made the cells supersensitive to calcium after IgE stimulation [93]. In addition, inhibition of histamine release was noted when the amount of sialic acid was enhanced in the cells by pretreatment with various gangliosides, and this inhibition could be overcome by increasing the concentration of calcium in the medium [94].

The finer details behind the influence of sialic acid in membranes on the permeability of calcium are not known.

(2) After dilution of rat mast cells incubated in a calcium-containing medium (1:100) to a medium devoid of calcium, the cells respond to histamine-releasing agents like antigen [74], the ionophore A23187 [49] or Con A (but not to dextran) [178], provided the cells are exposed to these agents immediately upon dilution. When the releasing agents are added at varying times after dilution, the secretory response gradually decreases to the low values observed in control cells which lack calcium during the preincuba-

tion period. In our laboratory the half-life of the secretory response was found to be about 20 s, indicating that a superficial pool of calcium, loosely bound and rapidly dissociating, can be utilized in the release process. A finding of interest is that antigen-induced histamine release was significantly higher when calcium-treated cells were diluted (1:100) to antigen in a calcium-free, as compared with a calcium-containing, medium, indicating that calcium might affect the binding or configuration of the antigen [74]. In similar experiments only 70% of the histamine released after calcium-treated cells had been diluted to ionophore A23187 in a calcium-containing medium was found when the medium was free of calcium [49]. This may be ascribed to a comparatively fast dissociation of this calcium pool in relation to the slow histamine release induced by the ionophore and possibly also to a different mechanism of calcium uptake induced by A23187 as compared with that involved after IgE stimulation.

(3) Agents like compound 48/80 [170], MCD-peptide [5], polymyxin B, somatostatin, polylysin, and protamine sulfate [125] release considerable amounts of histamine from rat mast cells which have never been in contact with calcium after isolation. Not even a short-term presence of EDTA prior to the releasing agent influences the release [32, 43, 67].

Other histamine-releasing agents like antigen, Con A, the ionophore A23187, thapsigargine, sodium fluoride, and ATP have been found by us to be comparatively poor releasers in the absence of extracellular calcium, which usually exerts an optimal effect at 0.5–2 m*M*. The implication of such a difference between various agonists is important but at present difficult to explain adequately.

If mast cells are treated with millimolar concentrations of EDTA [52] for 3 h or with 0.25 μ*M* of the ionophore A23187 for 1 min [49], the secretory response to compound 48/80 totally disappears, indicating that the first group of agents can utilize a second pool of calcium for histamine release to occur. It was suggested that this pool, which does not seem to disappear spontaneously from the cells even after prolonged incubation, may be localized deep in the cell. The endoplasmatic reticulum, which constitutes an intracellular tubule system extending from the plasma membrane, may well be the site for this pool of calcium, in analogy with the intracellular calcium released by inositol triphosphate after activation of the phosphatidylinositol cycle [18]. Interestingly, after being emptied, this pool can functionally be refilled in the rat mast cell [41]. In ionophore-treated cells incubated with 10 μ*M* calcium for up to 15 min, the secretory response towards compound 48/80 was gradually restored. It should be noted that 10 μ*M* calcium in the incubation medium is normally never sufficient to stimulate histamine release above the level seen in its complete absence, irrespective of the histamine-releasing agent studied. Furthermore, calcium could be substituted with strontium (10 μ*M*), indicating that the secretory response which relies on this intracellular pool may not be strictly calcium-specific.

(4) It is also possible to differentiate between various secretagogues on the basis of their sensitivity to addition of calcium after stimulation. When rat mast cells or human basophils are exposed to IgE-stimulating agents (antigen, anti-IgE, Con A) and calcium is added to the same incubation medium at various times thereafter, the secretory response gradually disappears, indicating that the process involves a short-lasting increase in the permeability of the cell membrane to calcium [74, 111, 113]. The time course for this decay of sensitivity to calcium is markedly prolonged in the presence of phosphatidylserine [64]. A similar decay for the triggering effect of calcium on secretion has been noted for mast cells stimulated with dextran [8, 64]. In contrast, calcium was found to induce histamine release whenever it was added to rat mast cells preexposed to the ionophore A23187 [49, 63], thapsigargine [133], or fluoride [131]. These findings point to differences in the mechanism by which calcium is made available to the cells. Some agents evidently transport calcium themselves into the cells. Others seem to induce a state of increased permeability to calcium which, depending on the agonist, may decay quickly or remain as long as the stimulator is present in the incubation medium. These variations most likely reflect differences in biochemical reactions and secondary cofactors activated by different agonists which, however, are all directed to achieve an enhanced entrance of calcium into the cell necessary to trigger secretion.

(5) With rat mast cells it is often difficult to dissociate a stimulatory, nonsecretory step from a second calcium-triggered, secretory step if the cells are diluted or washed between the two steps. This is, however, possible to demonstrate for the ionophore A23187 [49]; thapsigargine [131] and sodium fluoride [133]. When rat mast cells were first exposed to these agents in the absence of calcium, followed by dilution (1:100) to normally ineffective concentrations of the agonist, histamine release was initiated when calcium was introduced to the incubation medium. In contrast to findings with human granulocytes [42], however, the sensitivity of rat mast cells to calcium gradually disappeared, indicating dissociation of the agonists from the cells. The half-life of the sensitivity of ionophore-treated cells to calcium was about 20 s, whereas the same values amounted to 10–15 min for the other two agents with no change in reactivity during the first 5 min after dilution of the cells.

(6) Hypertonicity blocks histamine release induced by all agents tested so far (compound 48/80, ionophore A23187, thapsigargine, sodium fluoride, and extracellular ATP) in spite of the presence of calcium [50]. Hypertonicity inhibited not only histamine release but also swelling of rat mast cells induced by extracellular ATP as well as ionophore X-537A, as evaluated by the use of a Coulter channel analyzer. In addition, when a two-step reaction was used, it was found, for both sodium fluoride and thapsigargine, that hypertonicity inhibited the second calcium-triggered secretory step and not the first, activating step. Furthermore, when the medium was made normotonic during the second step, histamine release occurred to the same degree as in controls never exposed to hypertonicity. These results indicate that following stimulation by different secretory agonists, the influx of calcium into the mast cells occurs by simple passive diffusion.

(7) Due to the high binding affinity of the granule matrix for calcium it is generally accepted today that binding of Ca^{45} to rat mast cells in connection with histamine release does not represent calcium involved in the release process. This point is supported by finding the same values for Ca^{45} binding when the isotope was added to mast cells after histamine release had ended as before stimulation had been induced [78, 159]. When the isotope was added at extended periods of time after termination of histamine release, Ca^{45} binding gradually disappeared; 30 min after histamine release had been initiated by antigen or compound 48/80, the plasma membrane had returned to its normally impermeable state for calcium [78]. A rapid recovery of partially degranulated cells was suggested by Fawcett [60] in 1954 and this has been verified for both rat and human mast cells in a number of morphological and biochemical studies during recent years [16, 26, 54, 91, 157]. Local degranulation can be obtained when a secretory agent like compound 48/80 is delivered to the surface of a mast cell by use of a micropipette [47]. The process is short-lasting (seconds) and does not spread spontaneously but can be repeated several times with the same cell by engaging different parts of the surface. These observations have been extended to include the visual demonstration of a postdegranulating repair process [162]. The time course for decay of the secretory response after IgE-stimulating agents referred to above is much faster than restoration of the permeability properties of the plasma membrane and are therefore most likely not interrelated.

(8) Changes in the level of free cytoplasmic calcium in rat mast cells have been evaluated using calcium fluorescing probes like quin-2 and fura-2. When histamine release was determined in parallel samples, it was found

that maximum fluorescence of quin-2 occurred 20 s after antigen challenge and that it then gradually declined. Similar results, which seemed to precede histamine release, were obtained with compound 48/80 and the ionophore A23187 [177]. The quantitative relationships between the two parallel phenomena studied are unclear. The presence of EGTA decreased the calcium-induced fluorescence by 50–70% without affecting the histamine release. Theophylline and methyltransferase inhibitors not only inhibited antigen-induced fluorescence of the calcium probe to the same extent but histamine release as well. Of great interest, but difficult to interpret, was the finding that histamine release was induced by the phorbol ester TPA without any changes in the fluorescence of the calcium probe. Phorbol esters are generally considered to activate protein phosphorylation by a direct stimulation of protein kinase C [4] in the presence of calcium. The action of phorbol esters on rat mast cells seems, however, to be highly complex, including both inhibitory and enhancing effects [79]. It could not be excluded that TPA may also act on energy metabolism as well as influencing intracellular stores of calcium utilized for histamine release by compound 48/80 in the absence of extracellular calcium.

By using fura-2 to monitor calcium fluxes and the light microscope to monitor morphological changes in single rat mast cells, it was found that stimulation with antigen or compound 48/80 caused a transient rise in intracellular calcium above 5 μM which did not depend on extracellular calcium. This did not lead to degranulation. It sometimes occurred spontaneously, in particular in the presence of PS [124]. This method can elegantly establish transient changes in free calcium in individual mast cells. There are, however, certain limitations. First, it is not possible to establish histamine release objectively by this method. Histamine release may occur to a great extent without concomitant degranulation (see above). Even when degranulation occurs, the percentage of histamine or serotonin released exceeds the percentage of granules extruded; this is easily understandable in view of the development of intracellular granule-containing vacuoles [21, 168]. Thus degranulation, as observed in the light microscope, cannot be considered an adequate indicator of histamine release. Nor can changes in capacitance values, when related to the degree of degranulation, be accepted as a quantitative measure of histamine release. Finally, these calcium probes have not helped to explain the function of calcium in the release process.

(9) So far, little is known regarding the involvement or even the existence of calcium channels in histamine secretory cells. From experiments regarding the inhibitory action of different classes of channel-blocking agents on

histamine release from rat mast cells and human basophils there seems to be general agreement that these compounds are effective only at concentrations at least 100 times higher than needed to block voltage-regulated calcium channels in cardiac or smooth muscle [59, 92]. Counting against the involvement of specific calcium channels is the fact that these agents also inhibit histamine release induced by the ionophore A23187, which is generally considered to bypass receptor-mediated calcium channels with a direct transfer of calcium to the cytosol. Furthermore, their inhibitory effect on histamine release is not efficiently counteracted by increasing the concentration of calcium. A specific action on calcium channels in secretory cells has been questioned and their inhibitory action on secretion has been ascribed to a nonspecific membrane-stabilizing effect [148]. Using a special patch-clamp technique in which the patch under the micropipette tip was not disrupted but instead permeabilized, thus preventing the diffusion of large molecules out of the cell, it was found that calcium channel blockers inhibited conductance (interpreted as the opening of calcium channels) without affecting degranulation after stimulation with antigen. It was therefore concluded that secretory cells such as rat mast cells do not use ion channels in stimulus-secretion coupling [114]. Again the conclusion may be correct but cannot be fully appreciated since histamine release was not measured objectively.

Functional Role of Calcium and Metabolic Energy in the Histamine Secretory Process

For obvious reasons it is impossible to visualize a single mechanism valid for all agents that can induce histamine release. In view of the differences pointed out earlier, relating to species and localization of mast cells, the following will relate only to findings in isolated rat peritoneal mast cells. It would not be meaningful to make generalizations to other histamine-secreting cells, not even to rat mast cells in different locations. Two observations will form the basis for further discussion:

(1) Ionophore A23187 transports Ca^{45} into rat peritoneal mast cells even in the absence of metabolic energy production. The fact that no histamine is released under these conditions indicates that transfer of extracellular calcium to the intracellular compartment per se does not induce histamine release [78]. A similar conclusion was reached on the basis of experiments using the patch-clamp together with fura-2 [124]. Obviously, energy is needed somewhere after calcium has entered the cell in order to institute

proper morphological changes which allow cations to reach the granule matrix and exchange for histamine. This by itself does not eliminate the possibility that agents which act by binding to the plasma membrane may require energy for those processes which lead to a state (often transient) of enhanced permeability for calcium.

(2) As already mentioned, hypertonicity inhibits histamine release during the calcium-induced secretory step after stimulation with thapsigargine or sodium fluoride, and release is instituted immediately upon restoration of normotonic conditions. This suggests that calcium may pass through the membrane by passive diffusion in accordance with its concentration gradient.

It is of interest to relate these two findings to various current theories regarding the function of energy and calcium in the release process of histamine.

Contraction Theory

Actin filaments connecting intracellular granules with the plasma membrane were first identified by Røhlich [149] in 1971. Their existence was recently confirmed by Tasaka [162] who, elaborating on suggestions by Røhlich, presented a hypothetical contraction mechanism for the apposition of the granules to the plasma membrane involving the combined action of calcium and metabolic ATP. This intriguing model does so far not explain the biochemical background for the fusion of the two membranes and the subsequently occurring enhanced permeability to intragranular sites.[1]

Phosphorylation Theory

Receptor functions are frequently regulated by protein phosphorylation. Activation as well as desensitization of receptors may be the result of phosphorylation [88]. By analogy with the autophosphorylation of the insulin receptor after binding of insulin, which is a prerequisite for some of the functions of the hormone, it has been shown that the IgE receptor also becomes phosphorylated after ligand binding [84]. This process may account for an enhanced permeability allowing passive diffusion of calcium into the cell. Such a state of enhanced permeability may be terminated by either dephosphorylation or possibly prolonged phosphorylation of the same or additional sites. For obvious reasons this suggested turn-off mechanism is

[1] Of special interest in this context is the recent demonstration that antigen stimulation of rat basophilic leukemia cells enhanced the phosphorylation of a membrane bound protein with a mol. wt of 18.000 which corresponded to myosin light chain. (Teshina R, et al, Molecular Immunology 1989;26:641–648).

only applicable to agents which show a decay in sensitivity to calcium when added at a later stage than the agonist. However, the significance of phosphorylation of the IgE receptor remains to be established since, somewhat surprisingly, during histamine release by ionophore A23187, the same receptor became phosphorylated [85]. There are several kinases that may be candidates for inducing phosphorylation. As demonstrated for the insulin receptor, the IgE antibody, upon binding of an antigen, may be regulated by a cAMP-dependent protein kinase and/or by protein kinase C. Ca-calmodulin may also play a role in activating phosphokinases.

When rat mast cells are exposed to compound 48/80 or ionophore A23187, several protein bands become phosphorylated [156]. The inhibitory action of chromoglycate has also been associated with phosphorylation of a specific mast cell protein [164].

The pattern of endogenous phosphorylation differs between agonists that can utilize intracellular calcium for release and those that cannot [151]. Among the former, when mast cells were exposed to compound 48/80, substance P or histones, there was a common pattern involving increased phosphorylation of one 35,000 MW protein and dephosphorylation of one 15,000 MW protein. In contrast, when mast cells were exposed to a releasing agent like anti-IgE, which depends on extracellular calcium for histamine release, two additional proteins of larger size were phosphorylated (68,000 and 56,000 MW). Although the location of these phosphorylated proteins in rat mast cells has not been established, the findings are intriguing and constitute the first biochemical demonstration that in the process of IgE-mediated histamine release specific proteins are phosphorylated that can be distinguished from non-IgE-inducing secretory agonists. Whether, as suggested, this indicates an involvement of different protein kinases or different isoforms of the same kinase is not known.

Calmodulin. Calmodulin has been identified by immunological methods in rat mast cells in association with the granule membranes [24, 55]. Since it is known to activate myosin light chain kinase, it has been thought to be involved in phosphorylation and contraction of actomyosin in microfilaments [24]. It may thus have a function in the contractile theory. However, calmodulin has many other activities which could all be involved in secretion at various stages, e.g. effects on cAMP and arachidonic acid metabolism, control of microtubule assembly, and calcium-magnesium ATPase activity. In view of all these possible functional sites, further speculations on the role of calmodulin are not fruitful at the present time.

Cyclic AMP. An inhibitory role of cAMP has been discussed since 1968, when Lichtenstein and Margolis [112] demonstrated that methylxanthines as well as catecholamines inhibited antigen-induced histamine release from human basophils. The finding that antigen-induced histamine release from sensitized rat mast cells was inhibited by dibutyryl-cAMP as well as theophylline strengthened the concept that IgE-mediated histamine release was in general under the control of a cAMP-dependent protein kinase [65]. No corresponding inhibition was noted when histamine release was induced by A23187. Regarding rat mast cells, this concept was challenged, since data in the literature suggested little correlation between the cAMP content of rat mast cells after exposure to these agents and effects on histamine release [44]. The role of cAMP has been further complicated by the finding that in rat mast cells, stimulation of IgE by anti-IgE and Con A is accompanied by a short monophasic rise in the cellular content of cAMP [160]. Recent data indicate that histamine release is not directly related to increase in cAMP levels in rat mast cells activated by Con A, anti-IgE, antigen, prostaglandin D2, and isoprenalin [161]. Thus, the significance of cAMP as a regulator remains controversial.

Protein Kinase C. Protein kinase C is another enzyme responsible for protein phosphorylation. It is activated by intramembranous 1,2-diacylglycerol (DG) which is supposed to form a membrane-associated macrocomplex with a cytosolic soluble kinase; the complex in turn phosphorylates adjacent proteins. Cytosolic calcium and phosphatidylserine may be needed for the formation of the complex [17, 126, 154]. DG and inositol-1,4,5,-triphosphate (IP_3) are generated from receptor-coupled activation of phospholipase C which hydrolizes inositol lipids. IP_3 may furnish the calcium necessary for activation of protein kinase C through release from intracellular stores [158]. The complexity of the protein kinase C system, the multiple modes for activation and signal transduction have recently become apparent. Protein kinase C is no longer a single entity but already has seven different identified isoforms [53].

In rat mast cells histamine release is accompanied by a significant activation of phosphatidylinositol metabolism and diacylglycerol phosphorylation [29, 102, 152]; this supports the concept of protein kinase C involvement in the secretory mechanism. Another approach has been to use phorbol esters, which act as direct agonists on protein kinase C. The phorbol ester TPA, as well as diacylglycerol, exerts a synergistic action with other releasing agents [83, 99]. Protein phosphorylation after exposure of rat mast cells to

TPA has the same spectrum as observed after histamine release induced by Con A [99].

Several observations thus favor the concept that protein phosphorylation by protein kinase C may be involved in the mechanism of histamine release induced by at least some agents. The finding that TPA by itself releases histamine in the absence of calcium, and that TPA at higher concentrations inhibits release, was recently the background to a careful reevaluation of the specificity of the mechanics behind the various effects of TPA in relation to histamine release [79]. Although not conclusive, the results indicated an involvement of protein kinase C in both the enhancing and the inhibitory effects of TPA on histamine release from rat mast cells, at least when induced by antigen or low concentrations of ionophore A23187. However, no such synergism has been observed by TPA on histamine release induced by compound 48/80 [79, 83]. It must be emphasized that biochemical observations in intact histamine-secreting cells seldom allow final conclusions, no matter what relation the observed changes in biochemical parameters bear to the release process. This is especially obvious with histamine secretion from isolated rat mast cells, where the release often occurs with a high velocity.

To simplify matters, attempts are now made to control the composition of the cytosol in permeabilized rat mast cells. The finding that stable analogs of GTP may induce secretion in such cells in the presence of calcium [72] suggests that G proteins (which play a role in both activation of adenyl cyclase and phospholipase C and which may exist in both an activating and an inhibiting form) are involved. A new histamine release activating G protein has been suggested. When this protein was activated by GTP, further phosphorylation did not seem to be an obligatory step for secretion to occur. There are reasons to question the significance of these models since the action of the 'permealizing' agents (Sendai virus, high-voltage discharge, ATP^{+4}, streptolysin-O and patch pipettes) on the mast cells per se have usually not been sufficiently considered. For example, streptolysin-O has cytotoxic actions in the sense that lactic dehydrogenase leaks out of the cells [87]. When ATP was the agent used to permeabilize the cells, the many other actions exerted by this compound on rat mast cells seem to have been completely overlooked. Besides histamine release (in the presence of extracellular calcium) and metabolic effects mentioned above, free ATP (ATP^{+4}) causes pronounced fluxes of monovalent cations across the plasma membrane with a net uptake of sodium, a loss of potassium and a concomitant swelling of cells [20, 34]. Swelling and ion fluxes are not influenced by ouabain [35] or inhibition of metabolic energy production [31] but are effectively counteracted by using a hypertonic incubation medium [50]. Furthermore, exposure of rat mast cells to micromolar concentrations of exogenous ATP in the absence of divalent cations over a period of time renders the cells insensitive to the histamine-releasing action of compound 48/80 [32, 33], antigen [76] and ionophore A23187 [49].

In view of these observations and a general failure to investigate similar actions of other permeabilizing agents, the value of the conclusions drawn from studies on permeabilized cells are limited in the understanding of the mechanism of histamine release from normal rat mast cells.

Phospholipid Methylation. In bacteria, methylation of glutamic acid residues in specific proteins has been shown to control the calcium gating mechanism which, in turn, is considered to control bacterial movement [25]. Phosphorylation mediated by hydrolysis of ATP is involved in the biosynthesis of S-adenosylmethionine, the most effective methyl donor for biosynthetic reactions, including methylation of phosphatidylethanolamine to phosphatidylcholine. Methylation of this kind has been observed in rat peritoneal mast cells during IgE-receptor-mediated secretion of histamine [90] and has been proposed as an initial event responsible for enhanced permeability of the plasma membrane to calcium. This hypothesis has been questioned on the grounds that only a small percentage of radioactivity from tritium-labelled methionine was incorporated in the phospholipids which could not be excluded as artefact [175]. Recent reinvestigation of the matter has completely failed to support the concept [11]; therefore, at the present time, methylation as an obligatory event in IgE-mediated secretion remains an open question.

Conclusion and Perspectives

Progress during the last decade leaves us far from the goal of understanding the finer details of histamine release from histamine-containing cells. It is generally accepted that several mechanisms exist by which histamine release can be induced. Although there may be common basic elements in different releasing systems, major differences, insufficiently defined from a biochemical point of view, also exist.

Methods in biochemistry at the cellular and subcellular level have developed much further during this time than new insights into cellular physiopathological processes. The many new possibilities are far more difficult to evaluate or prove in intact cells than in cell-free systems. The striking heterogeneity of mast cells, both with regard to species and organ localization, makes generalization of biological and biochemical results impossible when working with histamine release mechanisms in different cells.

It is hoped that newer cell purification procedures, the possibility of separating biological and biochemical events from each other in time, and general knowledge on how one event in the secretory mechanism may influence another, will all improve to a point so that a decade from now, scientists may speak with greater certainty on the finer details of how histamine is released.

References

1 Ali H, Leung KBP, Pearce FL, et al: Comparison of the histamine-releasing action of substance P on mast cells and basophils from different species and tissues. Int Arch Allergy Appl Immunol 1986;79:413.
2 Anderson P, Röhlich P, Slorach SL, et al: Morphology and storage properties of rat mast cell granules isolated by different methods. Acta Physiol Scand 1974;91:143.
3 Anderson P, Slorach SA, Uvnäs B: Sequential exocytosis of storage granules during antigen-induced histamine release from sensitized rat mast cells in vitro. An electron microscopic study. Acta Physiol Scand 1973;88:359.
4 Anderson WB, Estival A, Tapiovaara H, et al: Altered subcellular distribution of protein kinase C (a phorbol ester receptor). Possible role in tumor promotion and the regulation of cell growth: relationship to changes in adenylate cyclase activity; in Cooper DMF, Seamon KB (eds): Advances in Cyclic Nucleotide Protein Phosphorylation, ed 19. New York, Raven Press 1985, p 287.
5 Assem ESK, Atkinson G: Histamine release by MCDP (401), as peptide from the venom of the honey bee. Br J Pharmacol 1973;48:337P.
6 Assem ESK, Schild HO: Antagonism by beta-adrenoceptor blocking agents by the anaphylactic effect of isoprenaline. Br J Pharmacol 1971;42:620.
7 Bach MK, Brashler JR: On the nature of the presumed receptor for IgE on mast cells. I. The effect of sialidase and phospholipase C treatment on the capacity of rat peritoneal cells to participate in IgE-mediated, antigen-induced histamine release in vitro. J Immunol 1973;110:1599.
8 Baxter JH: Role of Ca^{++} in mast cell activation, desensitization, and histamine release by dextran. J Immunol 1973;111:1470.
9 Baxter JH, Adamik R: Differences in requirements and actions of various histamine-releasing agents. Biochem Pharmacol 1978;27:497.
10 Benditt EP, Arase M: An enzyme in mast cells with properties like chymotrypsin. J Exp Med 1959;110:451.
11 Benyon RC, Church MK, Holgate ST: IgE-dependent activation of mast cells is not associated with enhanced phospholipid methylation. Biochem Pharmacol 1986;35:2535.

12 Benyon RC, Lowman MA, Church MK: Human skin mast cells: their dispersion, purification and secretory characterization. J Immunol 1987;138:861.
13 Bergendorff A: Intracellular distribution of amines taken up by rat mast cells in vitro. Acta Physiol Scand 1975;95:133.
14 Bergendorff A, Uvnäs B: Uptake of biogenic amines by mast cell granules. Acta Pharmacol Toxicol 1967;25:32.
15 Bergendorff A, Uvnäs B: Storage properties of rat mast cell granules in vitro. Acta Physiol Scand 1973;78:213.
16 Berlin G, Enerbäck L: Mast cell secretion: Rapid sealing of exocytotic cavities demonstrated by cytofluorometry. Int Arch Allergy Appl Immunol 1984;73:256.
17 Berridge MJ: Intracellular signalling through inositol triphosphate and diacylglycerol. Biol Chem Hoppe-Seyler 1986;367:447.
18 Berridge MJ: Inositol triphosphate. A new second messenger; in Schou JS, Geisler A, Norn S (eds): Drug Receptors and Dynamic Processes in Cells, ed 22, p 90. Alfred Benson Symposium. Copenhagen, Munksgaard, 1986.
19 Bloom GD, Hägermark O: A study on morphological changes and histamine release induced by compound 48/80 in rat peritoneal mast cells. Exp Cell Res 1965;40:637.
20 Bloom G, Diamant B, Hägermark O, et al: The effect of adenosine-5′-triphosphate (ATP) on structure and amine content of rat peritoneal mast cells. Exp Cell Res 1970;62:61.
21 Carlsson SA, Ritzén M: Mast cells and 5-HT. Intracellular release of 5-hydroxytryptamine (5-HT) from storage granules during anaphylaxis or treatment with compound 48/80. Acta Physiol Scand 1969;77:449.
22 Chakravarty N: Observations on histamine release and formation of a lipid-soluble smooth muscle stimulating principle ('SRS') by antigen-antibody reaction and compound 48/80; thesis. Stockholm, 1959, p 449.
23 Chakravarty N, Holm Nielsen E: $Ca2^{+}$-$Mg2^{+}$-activated adenosine triphosphatase in plasma and granule membranes in non-secreting and secreting mast cells. An electron-microscopic histochemical study. Exp Cell Res 1980;130:175.
24 Chakravarty N, Holm Nielsen E: Calmodulin in mast cells and its role in histamine secretion. Agents Actions 1985;16:122.
25 Chelsky D, Dahlqvist FW: Methyl-accepting chemotaxis proteins of Escherichia coli methylated at three sites in a single tryptic fragment. Biochemistry 1981;20:977.
26 Chock SP, Schmauder-Chock EA: Evidence of de novo membrane generation in the mechanism of mast cell secretory granule activation. Biochem Biophys Res Commun 1985;132:134.
27 Church MK, Lowman MA, Rees PH, et al: Mast cells, neuropeptides and inflammation. Agents Actions 1989;27:8.
28 Church MK, Pao GJ-K, Holgate ST: Characterization of histamine secretion from human dispersed human lung mast cells: effect of anti-IgE, calcium ionophore A23187, compound 48/80 and basic polypeptides. J Immunol 1982;129:2116.
29 Cockcroft S, Gomperts BD: Evidence for a role of phosphatidylinositol turnover in stimulus-secretion coupling. Studies with rat peritoneal mast cells. Biochem J 1979;178:681.
30 Coffey RG, Middleton E: Release of histamine from rat mast cells by lysosomal cationic protein. Possible involvement of adenylate cyclase and adenosine triphosphate in pharmacological regulation. Int Arch Allergy Appl Immunol 1973;45:593.
31 Dahlquist R: Relationship of uptake of sodium and 45calcium to ATP-induced histamine release from rat mast cells. Acta Pharmacol Toxicol 1974;35:11.

32 Dahlquist R, Diamant B, Elwin K: Interaction of adenosine-5′-triphosphate (ATP) with histamine release induced by compound 48/80. Acta Physiol Scand 1973;87: 145.
33 Dahlquist R, Diamant B, Elwin K: Effect of divalent cations on the interaction of adenosine-5′-triphosphate (ATP) with histamine release induced by compound 48/80. Acta Physiol Scand 1973;87:158.
34 Dahlquist R, Diamant B, Krüger PG: ATP-induced uptake of sodium and swelling of mast cells and its relation to histamine release. Acta Pharmacol Toxicol 1972;31 (suppl 1):79.
35 Dahlquist R, Diamant B, Krüger PG: Increased permeability of the rat mast cell membrane to sodium and potassium caused by extracellular ATP and its relationship to histamine release. Int Arch Allergy Appl Immunol 1974;46:655.
36 Diamant B: The influence of anoxia and glucose on histamine liberation caused by a principle in *Ascaris suis*. Acta Physiol Scand 1960;50 (suppl 175):34.
37 Diamant B: The effects of compound 48/80 and distilled water on the adenosine triphosphate content of isolated rat mast cells. Acta Physiol Scand 1967;71:283.
38 Diamant B: The mechanism of histamine release from rat mast cells induced by adenosine triphosphate. Acta Pharmacol Toxicol 1967;25:33.
39 Diamant B: The influence of adenosine triphosphate on isolated rat peritoneal mast cells. Int Arch Allergy Appl Immunol 1968;36:3.
40 Diamant B: Energy production in rat mast cells and its role for histamine release. Int Arch Allergy Appl Immunol 1975;49:155.
41 Diamant B: Influence of calcium ions and metabolic energy (ATP) on histamine release; in Johansson SGO, Strandberg K, Uvnäs B (eds): Molecular and Biological Aspects of the Acute Allergic Reaction. New York, Plenum, 1976, p 255.
42 Diamant B: Intracellular mechanisms leading to mediator release; in Oehling A, et al. (eds): Advances in Allergology and Immunology. Oxford, Pergamon Press, 180, p 163.
43 Diamant B, Grosman N, Skov PS, et al: Effect of divalent cations and metabolic energy on the anaphylactic histamine release from rat peritoneal mast cells. Int Arch Allergy Appl Immunol 1974;47:412.
44 Diamant B, Kazimierczak W, Patkar SA: Does cyclic AMP play any role in histamine release from rat mast cells. Allergy 1978;33:50.
45 Diamant B, Kazimierczak W, Patkar SA: Mechanism of histamine release induced by the ionophore X537A from isolated rat mast cells. III. Actions of X537A on isolated histamine-retaining granules, on a heparin-protamin complex saturated with histamine and on transport of histamine into an organic phase. Int Arch Allergy Appl Immunol 1978;56:179.
46 Diamant B, Krüger PG: Histamine release from isolated rat peritoneal mast cells induced by adenosine-5′-triphosphate. Acta Physiol Scand 1967;71:291.
47 Diamant B, Krüger PG, Uvnäs B: Local degranulation of individual rat peritoneal mast cells induced by compound 48/80. Acta Physiol Scand 1970;79:1.
48 Diamant B, Norn S, Felding P, et al: ATP level and CO2 production of mast cells in anaphylaxis. Int Arch Allergy Appl Immunol 1974;47:809.
49 Diamant B, Patkar SA: Stimulation and inhibition of histamine release from isolated rat mast cells. Dual effect of the ionophore A23187. Int Arch Allergy Appl Immunol 1975;49:183.
50 Diamant B, Patkar SA: The influence of hypertonicity on histamine release from isolated rat mast cells. Agents Actions 1980;10:140.

51 Diamant B, Peterson C: The metabolism of monosaccharides in isolated rat mast cells and its influence on histamine release induced by adenosine-5′-triphosphate. Acta Physiol Scand 1971;83:324.
52 Douglas WW, Ueda Y: Mast cells secretion (histamine release) induced by 48/80. Calcium-dependent exocytosis inhibiting strongly by cytochalasin only when glycolysis is rate-limiting. J Physiol 1973;234:97P.
53 Dreher ML, Hanley MR: Multiple modes of protein kinase C regulation and their significance in signalling. Trends Pharmacol Sci 1988;9:114.
54 Dvorak AM, Schleimer RP, Lichtenstein LM: Human mast cells synthesize new granules during recovery from degranulation. In vitro studies with mast cells purified from human lungs. Blood 1983;71:76.
55 Egsmose C, Bock E, Møllgard K, et al: Immunocytochemical demonstration of calmodulin in cells secreting by exocytosis. Experientia 1985;41:1340.
56 Enerbäck L: Mast cells in the rat gastrointestinal mucosa. II. Dye-binding and metachromatic properties. Acta Pathol Microbiol Immunol Scand 1966;66:303.
57 Enerbäck L: Mast cells in the gastrointestinal tract. I. Effects of fixation. Acta Pathol Microbiol Immunol Scand 1966;66:289.
58 Enerbäck L: Mast cells in the gastrointestinal tract. III. Reactivity towards compound 48/80. Acta Pathol Microbiol Immunol Scand 1966;66:313.
59 Ennis M, Ind PW, Pearce FL, et al: Calcium antagonists and histamine secretion from rat peritoneal mast cells. Agents Actions 1983;13:144.
60 Fawcett DW: Cytological and pharmacological observations on the release of histamine by mast cells. J Exp Med 1954;100:217.
61 Feldberg W, Mongar JL: Comparison of histamine release by compound 48/80 and octylamine in perfused tissue. Br J Pharmacol 1954;9:197.
62 Foreman JC: Mast cells and the actions of flavonoids. J Allergy Clin Immunol 1984; 73:769.
63 Foreman JC, Garland LG: Desensitization in the process of histamine secretion induced by antigen and dextran. J Physiol 1974;239:381.
64 Foreman JC, Garland LG, Mongar JL: The role of calcium in secretory processes. Model studies in mast cells. Calcium in biological systems, ed 30. London, Cambridge University Press, 1976, p 193.
65 Foreman JC, Hallet MB, Mongar JL: The relationship between histamine release and ^{45}calcium uptake by mast cells. J Physiol 1977;271:193.
66 Foreman JC, Jordan CC, Piotrowski W: Interaction of neurotensin with the substance P receptor mediating histamine release from rat mast cells and the flare in human skin. Br J Pharmacol 1982;77:531.
67 Foreman JC, Mongar JL: The role of the alkaline earth ions in anaphylactic histamine secretion. J Physiol 1972;224:753.
68 Foreman JC, Mongar JL, Gomperts BD: Calcium ionophores and movement of calcium ions following the physiological stimulus to a secretory process. Nature 1973;245:249.
69 Foreman J, Jordan C: Histamine release and vascular changes induced by neuropeptides. Agents Actions 1983;13:105.
70 Fox CC, Wolf EJ, Kagey-Sobotka A, et al: Comparison of human lung and intestinal mast cells. J Allergy Clin Immunol 1988;81:89.
71 Gomperts BD: Calcium and cellular activation; in Chapman D (ed): Biological membranes, ed 5. London, Academic Press, 1984, p 289.

72 Gomperts BD, Cockcroft S, Howell TW, et al: The dual effector system for exocytosis in mast cells. Obligatory requirement for both Ca and GTP. Biosci Rep 1987; 7:368.
73 Goth A, Adams HR, Knoohuizen M: Phosphatidylserin. Selective enhancer of histamine release. Science 1971;173:1034.
74 Grosman N, Diamant B: Studies on the role of calcium in the anaphylactic histamine release from isolated rat mast cells. Acta Pharmacol Toxicol 1974;35:284.
75 Grosman N, Diamant B: The influence of phosphatidyl serine on the release of histamine from isolated rat mast cells induced by different agents. Agents Actions 1975;5:296.
76 Grosman N, Diamant B: Effect of adenosine-5′-triphosphate (ATP) on rat mast cells. Influence on anaphylactic and compound 48/80 induced histamine release. Agents Actions 1975;5:108.
77 Grosman N, Diamant B: Studies on histamine-retaining granules obtained from isolated rat mast cells. Agents Actions 1976;6:394.
78 Grosman N, Diamant B: Binding of 45calcium to isolated rat mast cells in connection with histamine release. Agents Actions 1978;8:338.
79 Grosman N, Nielsen KA: The influence of tetradecanoyl-phorbol-acetate (TPA) on histamine release from isolated rat mast cells. Agents Actions 1988;24:40.
80 Hanahoe TAP, Tanner T, West GB: Resistance of rats to the potentiating action of phosphatidyl serine on dextran response. J Pharm Pharmacol 1973;25:429.
81 Harris JM, Kalmus H, West GB: Genetic control of the anaphylactic reaction in rats. Genet Res 1963;4:346.
82 Harris JM, West GB: Rats resistant to the dextran anaphylactoid reaction. Br J Pharmacol 1963;20:550.
83 Heiman AS, Crews FT: Characterization of the effects of phorbol esters on rat mast cell secretion. J Immunol 1985;134:548.
84 Hempstead BL, Kulczycki A Jr, Parker CW: Phosphorylation of the IgE receptor from ionophore A23187 stimulated intact rat mast cells. Biochem Biophys Res Commun 1981;98:815.
85 Hempstead BL, Parker CW, Kulczycki A Jr: Selective phosphorylation of the IgE receptor in antigen-stimulated rat mast cells. Proc Natl Acad Sci USA 1983;80:3050.
86 Hook WA, Shiffman E, Aswanilaunar S, et al: Histamine release by chemotactic formyl-methionine-containing peptides. J Immunol 1976;117:594.
87 Howell TW, Gomperts BD: Rat mast cells permeabilised with streptolysin O secrete histamine in response to Ca at concentrations buffered in the micromolar range. Biochim Biophys Acta 1987;927:177.
88 Huganir RL, Greengard P: Regulation of receptor function by protein phosphorylation. Trends Pharmacol Sci 1987;8:472.
89 Hägermark Ö, Diamant B, Dahlquist R: Release of histamine from human skin induced by intracutaneous injection of adenosine-5′-triphosphate. Int Arch Allergy Appl Immunol 1974;47:167.
90 Ishizaka T, Hirata F, Ishizaka K, et al: Stimulation of phospholipid methylation, Ca-influx, and histamine release by bridging of IgE receptors on rat mast cells. Proc Natl Acad Sci USA 1980;77:1903.
91 Jamur MC, Vugman I: Rat peritoneal mast cell degranulation and acid phosphatase and trimetaphosphate activity induced after stimulation by 48/80. A fluorescence, ultrastructural and cytochemical study. Cell Molec Biol 1988;34:231.

92 Jensen CB, Skov PS, Noren S: Inhibitory effect of calcium antagonists on histamine release from human leucocytes. Allergy 1983;38:233.
93 Jensen C, Dahl BT, Norn S, et al: Enhancement of histamine release from human basophils pretreated with different sialidases. Agents Actions 1986;18:499.
94 Jensen C, Svensen UG, Thastrup O, et al: Complexicity of the influence of gangliosides on histamine release from human basophils and rat mast cells. Agents Actions 1987;21:79.
95 Johansen T: Utilization of adenosine triphosphate in rat mast cells during histamine release induced by the ionophore A23187. Br J Pharmacol 1979;65:103.
96 Johansen T, Chakravarty N: Dependence of histamine release from rat mast cells on adenosine triphosphate. Arch Pharmacol 1972;275:457.
97 Johansen T, Chakravarty N: The utilization of adenosine triphosphate in rat mast cells during histamine release. Arch Pharmacol 1975;288:243.
98 Kanno T, Cochrane DE, Douglas WW: Exocytosis (secretory granule extrusion) induced by injection of calcium into mast cells. Can J Physiol 1973;51:1001.
99 Katakami Y, Kaibuchi K, Sawamura M, et al: Synergistic action of protein kinase C and calcium for histamine release from rat peritoneal mast cells. Biochem Biophys Res Commun 1984;121:573.
100 Kazimierczak W, Diamant B: Mechanisms of histamine release in anaphylactic and anaphylactoid reactions. Prog Allergy. Basel, Karger, 1978, vol 24, p 295.
101 Kazimierczak W, Patkar SA, Diamant B: The mechanism of histamine release induced by the ionophore X537A from isolated rat mast cells. I. Significance of monovalent cations, calcium, metabolic energy and temperature. Acta Physiol Scand 1978;102:265.
102 Kennerly DA, Sullivan TJ, et al: Activation of phospholipid metabolism during mediator release from stimulated rat mast cells. J Immunol 1979;122:152.
103 Krüger PG: The histamine release process and concomitant structural changes in rat peritoneal mast cells. In vitro study on effects of compound 48/80 and the dependence of the process on cell preparation, temperature and calcium. Int Archs Allergy Appl Immunol 1976;51:608.
104 Kurosawa M, Parker CW: A phosphatidylinositol kinase in rat mast cell granules. J Immunol 1986;136:616.
105 Kurosawa M, Parker CW: Changes in polyphosphoinositide metabolism during mediator release from stimulated rat mast cells. Bioochem Pharmacol 1989;38:431.
106 Lagunoff B, Benditt EP: Proteolytic enzymes of mast cells. Ann N Y Acad Sci 1963; 103:185.
107 Lagunoff D: The mechanism of histamine release from mast cells. Biochem Pharmacol 1972;21:1889.
108 Lagunoff D: Membrane fusion during mast cell secretion. J Cell Biol 1973;57:252.
109 Lagunoff D, Rickard A: Evidence for control of mast cell granule protease in situ by low pH. Exp Cell Res 1983;144:353.
110 Lazarus L, Perrin M, Brown M: Mast cell binding of neurotensin: I. Iodination of neurotensin and characterisation of neurotensin in mast cell receptor sites. J Biol Chem 1977:252:7174.
111 Lichtenstein LM: The immediate allergic response. In vitro separation of antigen activation, decay and histamine release. J Immunol 1971;107:1122.
112 Lichtenstein LM. Margolis S: Histamine release in vitro: Inhibition by catecholamines and methylxanthines. Science 1968;161:902.

113 Lichtenstein LM, Osler AG: Studies on the mechanism of hypersensitivity phenomena. IX. Histamine release from human leucocytes by Ragweed pollen antigen. J Exp Med 1964;120:507.
114 Lindau M, Fernandez JM: IgE-mediated degranulation of mast cells does not require opening of ion channels. Nature 1986;319:150.
115 Lowman MA, Rees PH, Benyon RC, et al: Human mast cell heterogeneity: Histamine release from mast cells dispersed from skin, lung, adenoids, tonsils and intestinal mucosa in response to IgE-dependent and non-immunological stimuli. J Allergy Clin Immunol 1988;81:590.
116 Ludowyke R, Lagunoff D: Drug inhibition of mast cell secretion; in Junker E (ed): Progress in Drug Research, ed 29. Basel, Birkhauser, 1985, p 277.
117 Marone G, Kagey-Sobotka A, Lichtenstein LM: Effects of archidonic acid and its metabolites on antigen-induced histamine release from human basophils in vitro. J Immunol 1979;123:1669.
118 Martin TM, Lagunoff D: Mast cell secretion; in Canon M (ed): Cell Biology of the Secretory Process. Basel, Karger, 1984, p 481.
119 Mongar JL, Schild HO: Inhibition of the anaphylactic reaction. J Physiol 1957;135: 301.
120 Mongar JL, Schild HO: The effect of calcium and pH on the anaphylactic reaction. J Physiol 1958;140:272.
121 Mongar JL, Svec P: The effect of phospholipids on anaphylactic histamine release. Br J Pharmacol 1972;46:741.
122 Morrison DC, Henson PM, Roser JF, et al: Two independent recognition sites for the initiation of histamine release from mast cells. Leukocyte Membrane Determinants Regulating Immune Reactivity. New York, Academic Press, 1976, p 551.
123 Moussatché H, Provoust-Danon A: Influence of oxidative phosphorylation inhibitors on the histamine release in the anaphylactic reaction in vitro. Experientia 1958; 14:414.
124 Neher E, Almers W: Fast calcium transients in rat peritoneal mast cells are not sufficient to trigger exocytosis. EMBO J 1986;5:51.
125 Nemeth EF, Douglas WW: Differential inhibitory effects of the arachidonic acid analog ETYA on rat mast cell exocytosis evoked by secretagogues utilizing cellular and extracellular calcium. Eur J Pharmacol 1980;67:439.
126 Nishizuka Y: Studies and perspectives of protein kinase C. Science 1986;233:305.
127 Norup E, Smitt UW, Brøgger Christensen S: The potencies of thapsigargin and analogues as activators of rat peritoneal mast cells. Planta Med 1985;251–255:305.
128 Padawer J: Microtubules in rat peritoneal fluid mast cells. J Cell Biol 1967;35:180A.
129 Padawer J: The reaction of rat mast cells to polylysine. J Cell Biol 1970;47:352.
130 Parrot JL: Sur la réaction cellulaire de l'anaphylaxie. Son caractère aerobique. C R Séanc Soc Biol 1942;136:361.
131 Patkar SA, Kazimierczak W, Diamant B: Sodium fluoride. A stimulus for a calcium-triggered secretory process. Int Arch Allergy Appl Immunol 1977;55:193.
132 Patkar SA, Kazimierczak W, Diamant BH: Histamine release by calcium from sodium fluoride-activated rat mast cells. Further evidence for a secretory process. Int Archs Allergy Appl Immunol 1978;57:146.
133 Patkar SA, Rasmussen U, Diamant B: On the mechanism of histamine release induced by thapsigargin from *Thapsia garganica* L. Agents Actions 1979;9:53.

134 Pearce F, Befus AD, Bienenstock J: Mucosal mast cells. III. Effect of quercetin and other flavonoids on antigen-induced histamine secretion from rat intestinal mast cells. J Allergy Clin Immunol 1984;73:819.
135 Pearce FL: Calcium and histamine secretion from mast cells; in Ellis GB, West GB: Progress in Medicinal Chemistry, ed 19. Amsterdam, Elsevier Biomedical Press, 1982, p 59.
136 Pearce FL: Mast cell heterogeneity. Trends Pharmacol Sci 1983;4:165.
137 Peterson C: Histamine release induced by compound 48/80 from isolated rat mast cells. Dependence on endogenous ATP. Acta Pharmacol Toxicol 1974;34:356.
138 Peterson C: Inhibitory action of antimycin A on histamine release from isolated rat mast cells. Acta Pharmacol Toxicol 1974;34:347.
139 Peterson C: Role of energy metabolism in histamine release. A study on isolated rat mast cells. Acta Physiol Scand 1974;413 (suppl):1.
140 Peterson C, Diamant B: Endogenous ATP in isolated rat mast cells and its relation to histamine release induced by compound 48/80. Acta Pharmacol Toxicol 1972;31 (suppl 1):80.
141 Peterson C, Diamant B: Utilization of endogenous ATP during histamine release from isolated rat mast cells. Acta Physiol Scand 1973;(suppl 396):121.
142 Peterson C, Diamant B: Increased utilization of endogenous ATP in isolated rat mast cells during histamine release induced by compound 48/80. Acta Pharmacol Toxicol 1974;34:337.
143 Peterson C, Diamant B: Inhibitory action of exogenous adenosine-5′-triphosphate on glycolysis in isolated rat mast cells: significance for histamine release. Acta Pharmacol Toxicol 1974;34:325.
144 Poyser RH, West GB: Changes in vascular permeability produced in rats by dextran, ovomucoid and yeast cell wall polysaccharides. Br J Pharmacol 1952;25:602.
145 Raphael GD, Henderson WR, Kaliner M: Isolation of membrane-bound rat mast cell granules. Exp Cell Res 1978;115:428.
146 Rees PH, Hillier K, Church MK: The secretory characteristics of mast cells isolated from the human large intestinal mucosa and muscle. Immunology 1988;65:437.
147 Riley JF, West GB: The presence of histamine in tissue mast cells. J Physiol 1953; 120:528.
148 Rubin RP: Actions of calcium antagonists on secretory cells; in Weiss GB (ed): New Perspectives on Calcium Antagonists. Bethesda, American Physiology Society, 1981, p 1.
149 Röhlich P: Membrane-associated actin filaments in the cortical cytoplasm of the rat mast cell. Expl Cell Res 1975;93:293.
150 Röhlich P, Anderson P, Uvnäs B: Electron microscope observations on compound 48/80-induced degranulation in rat mast cells. Evidence for sequential exocytosis of storage granules. J Cell Biol 1971;51:546.
151 Sagi-Eisenberg R, Foreman JC, Raval PJ, et al: Protein and diacylglycerol phosphorylation in the stimulus-secretion coupling of rat mast cells. Immunology 1987;61: 203.
152 Schellenberg RR: Enhanced phospholipid metabolism in rat mast cells stimulated to release histamine. Immunology 1980;41:123.
153 Shanahan F, Denburg JA, Fox J, et al: Mast cell heterogeneity. Effects of neuroenteric peptides on histamine release. J Immunol 1985;135:1331.
154 Shatzman RC, Turner RS, Kuo JF: Phospholipid-sensitive Ca-dependent protein

phosphorylation; in Cheung WY (ed): Calcium and Cell Function. New York, Academic Press, 1984.

155 Shulman ES, Post TJ, Henson PM, et al: Differential effects of the complement peptides, C5a and C5a des arg on human basophil and lung mast cell histamine release. J Clin Invest 1988;81:918.

156 Sieghart W, Theoharides TC, Alper SL, et al: Calcium-dependent protein phosphorylation during secretion by exocytosis in the mast cell. Nature 1978;275:329.

157 Slutsky B, Jarvis D, Bibb P, et al: Viability and recovery from degranulation of isolated rat peritoneal mast cells. Exp Cell Res 1987;168:63.

158 Streb H, Bayerdorffer E, Haase W, et al: Effect of inositol-1,4,5-triphosphate on isolated subcellular fractions of rat pancreas. J Membr Biol 1984;81:241.

159 Sugiyama K: Significance of ATP-splitting activity of rat peritoneal mast cells in the histamine release induced by exogenous ATP. Jpn J Pharmacol 1971;21:531.

160 Sullivan TJ, Parker KL, Kulczycki A Jr, et al: Modulation of cyclin AMP in purified rat mast cells. III. Studies on the effects of concanavalin A and anti-IgE on cyclic AMP concentrations during histamine release. J Immunol 1976;117:713.

161 Takei M, Matumoto T, Itoh T, et al: Role of cyclic AMP during histamine release. Histamine release is not directly related to increase in cyclic AMP levels in rat mast cells activated by concanavalin A, anti-IgE, antigen, prostaglandin D2 and isoproterenol. Hoppe-Seyler's Biol Chem 1988;369:765.

162 Tasaka K: Degranulation: exocytosis and its repair. Abstract: XIII Int Congr Allergol Clin Immunol, Montreaux, 1988;p 765.

163 Theoharides TC, Douglas WW: Secretion in mast cells induced by calcium entrapped with phospholipid vesicles. Science 1978;201:1143.

164 Theoharides TC, Sieghart W, Greengard P, et al: Antiallergic drug cromolyn may inhibit histamine secretion by regulating phosphorylation of a mast cell protein. Science 1980;207:80.

165 Uvnäs B: Release process in mast cells and their activation by injury. Ann N Y Acad Sci 1964;116:880.

166 Uvnäs B: Mast cells and histamine release. Indian J Pharmacol 1969;1:123–132.

167 Uvnäs B: Mast cells and inflammation; in Bertelli, Houck (eds): Inflammation: Biochemistry and Drug Interactions. Amsterdam, Excerpta Medica, 1969, p 221.

168 Uvnäs B: Quantitative correlation between degranulation and histamine release in mast cells; in Austen F, Becker E (eds): Biochemistry of the Acute Allergic Reaction. Oxford, Blackwell 1971, p 175.

169 Uvnäs B, Thon I-L: Isolation of 'biologically' intact mast cells. Exp Cell Res 1959; 18:512.

170 Uvnäs B, Thon I-L: Evidence for enzymatic histamine release from isolated rat mast cells. Exp Cell Res 1961;23:45.

171 Uvnäs B, Aborg C-H, Bergendorff A: Storage of histamine in mast cells. Evidence for an ionic binding of histamine to protein carboxyls in the granule heparin-protein complex. Acta Physiol Scand 1970;78:1.

172 van Overveld FJ: Some aspects of mast cell subtypes from human lung tissue. A comparison between normal individuals and patients with obstructive lung disease; thesis, Utrecht, 1988, p 1.

173 van Overveld FJ, Houben LAMJ, Bruijnzeel PLB, et al: Mast cell subtypes from human lung tissue: their identification, separation, and functional characteristics. Agents Actions 1988;23:227.

174 van Overveld FJ, Houben LAMJ, Terpstra GK, et al: Patients with chronic bronchitis differ in their mast cell subtypes as compared with normal subjects. Agents Actions 1989;227.
175 Vance DE, Kruijff B: The possible functional significance of phosphatidyl ethanol amine methylation. Nature 1980;288:277.
176 West GB: Calcium channels and histamine release from mast cells. Int Arch Allergy Appl Immunol 1987;84:101.
177 White JR, Ishizaka T, Ishizaka K, et al: Direct demonstration of increased intracellular concentration of free calcium as measured by quin-2 in stimulated rat peritoneal mast cell. Proc Natl Acad Sci USA 1984;81:3978.
178 White JR, Pearce FL: Role of membrane bound calcium in histamine secretion from rat peritoneal mast cells. 1981;11:324.
179 Winkler H: Occurrence and mechanism of exocytosis in adrenal medulla and sympathetic nerve; in Trendelenburg U, Weiner N (eds): Handbook of Experimental Pharmacology. Catecholamines I, ed 90. Berlin, Springer, 1988, p 43.

Bertil Diamant, MD, Department of Pharmacology, University of Copenhagen Medical School, Julinane Mairies Vej 20, DK2100 Copenhagen Ø (Denmark)

Waksman BH (ed): 1939–1989: Fifty Years Progress in Allergy.
Chem Immunol. Basel, Karger, 1990, vol 49, pp 173–205

Lipid Mediators of the Allergic Reaction

Floyd H. Chilton, Lawrence M. Lichtenstein

Department of Medicine, Division of Clinical Immunology,
The Johns Hopkins University School of Medicine, Baltimore, Md., USA

Clemens Freiherr von Pirquet first coined the term 'allergy' in the early 1900s and defined it as any altered reaction, helpful or harmful, of the immune system. Since that time, we have embraced the term 'inflammation' to indicate a protective response which tends to enhance resistance. By contrast, allergy has come to imply an autotoxic response which leads to increased susceptibility to a specific substance. Yet, research in the last 30 years has demonstrated the complex biological responses associated with inflammation, and allergy involves a similar interrelated series of cellular and biochemical events. A pivotal step in this series of reactions is the release of preformed and newly formed products from cells involved in the allergy/inflammatory (A/I) response. These released products then alter the function and biochemistry of surrounding cells and tissues. A major part of the ensuing biological response, as well as much of the pathogenesis which is attributed to allergy, is thought to be dependent on the effects these products have on adjacent cells within the inflammatory region.

In reviewing these processes, it is apparent that the cells involved in the allergic/inflammatory reactions must recognize, as well as produce, a complex network of mediators in order to control the events associated with inflammation. Two classes of lipid mediators have been discovered in the last 50 years, which almost always appear during inflammatory cell activation in vitro and are often associated with allergy/inflammatory reactions in vivo. One class consists of the endogenously formed oxygen-carrying metabolites of arachidonic acid; these products include the prostaglandins, thromboxanes, leukotrienes and HETEs. The second class is a group of endogenously formed phospholipid molecular species commonly referred to as platelet-activating factor. Because of the capacity of these products to mimic signs and symptoms of inflammation and allergy, it is generally accepted that they play crucial roles in a variety of allergic diseases. This chapter provides a brief discussion of the history of these two classes of lipid mediators in the allergic response. In particular, we focus on the biochemistry of these lipid classes in cells thought to be involved in this response. Several in-depth review articles provide a more complete perspective on the biology of these compounds than is presented here.

Early History of Eicosanoids

The initial experiments which recognized the importance of arachidonic acid can be traced back some 60 years when two biochemists, George and Mildred Burr, found that certain polyunsaturated 'essential' fatty acids were important dietary components necessary for growth and development [1, 2]. The first discovery of prostaglandins came about the same time, when a gynecologist, Raphael Kurzrok, and a pharmacologist, Charles Lieb, found that there was often a violent contraction of the uterus after installation of semen [3]. This substance in some cases also relaxed strips of human uterine tissue in vitro. A few years later, Goldblatt [4] and Von Euler [5] described a factor in prostate gland tissue which caused smooth muscle contraction and changes in blood pressure. Von Euler found that these factors were lipid soluble and termed them 'prostaglandins' [5]. Some 30 years would elapse before the association and ramifications of these early findings would be understood. Several key observations by Von Euler, Bergstrom, and Van Drop and their colleagues, led to the description of many biological activities of prostaglandins, the identification of the structures of prostaglandin E_1, E_2 and $F_2\alpha$ and the recognition that prostaglandins were formed from arachidonic acid [6–9]. From the work of these early pioneers until today, the eicosanoid field has exploded with new discoveries. Major areas where progress has been made include the identification of a wide variety of arachidonic acid metabolites, the description of additional biochemical pathways which utilize arachidonic acid and its metabolites and a better understanding of the role of arachidonic acid and its metabolites in mammalian biology.

Initiation of the Eicosanoid Biosynthesis

For some time, it has been recognized that one of the initial events of the allergic response is the activation of mast cells and basophils by the cross-

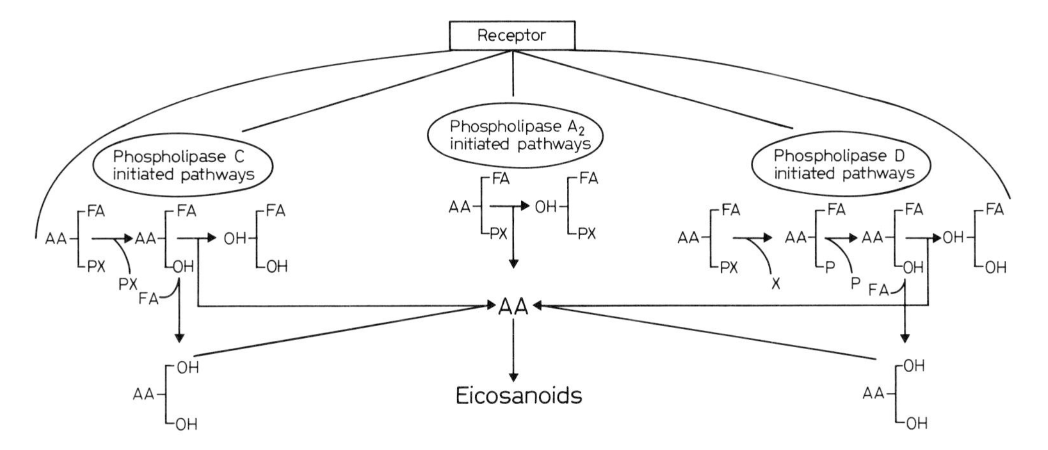

Fig. 1. Some potential pathways by which arachidonic acid may be released from A/I cells. Shown are the initial phospholipase activities which may eventually lead to arachidonic acid mobilization. All pathways may not be functional in all A/I cells. Moreover, there may be other pathways not listed which are functional in some A/I cells.

linking of IgE on Fc receptors with specific antigen. This event leads to a complex series of biochemical events which includes phospholipid turnover, calcium influx, changes in cyclic nucleotides and arachidonic acid metabolism [10], often followed by the directional migration and accumulation of circulating inflammatory cells such as neutrophils, eosinophils, monocytes and lymphocytes to the site of the insult. Each of the cell types involved in this response produce and secrete a unique set of these eicosanoids; the quantities and nature of the metabolites depend on which enzymes and precursor pools are available to the cell.

Eicosanoids are formed in cells by a series of enzymatic steps all working in concert. It was recognized early on that polyunsaturated fatty acids, and in particular arachidonic acid, are necessary precursors for the eicosanoids [8, 9]. Endogenous stores of arachidonate are located at the sn-2 position of phosphoglycerides in mammalian cells [11, 12]. It has generally been accepted that the production of eicosanoids in A/I cells is initiated by the activation of phospholipase(s) (fig. 1) which release arachidonate from endogenous phospholipids during cell stimulation [13, 15]. Several phospholipase mechanisms were initially proposed in the platelet to account for the release of arachidonic acid from membrane phospholipids. Studies in the late 1970s by

Bills et al. [16] suggest that arachidonic acid mobilized directly by phospholipase A_2 provides the bulk of the arachidonic acid which forms eicosanoids. Subsequent studies suggest that phospholipase C acts on inositol-linked phospholipids resulting in the formation of diacylglycerols and cyclic inositol phosphates (when inositol phospholipids are substrates) [17]. Both diacylglycerides and cyclic inositol phosphates are thought to be important in regulating protein kinase C and cellular calcium concentrations, respectively, without directly entering the arachidonic acid cascade. In addition, arachidonic acid can subsequently be mobilized from these diacylglycerides by the action of diglyceride lipase(s) and/or monoacylglycerol lipases [16, 19]. Still others implied that a significant portion of arachidonic acid is released by a specific phospholipase A_2 which utilizes phosphatidic acid (derived from phospholipase C and diglyceride kinase as its major arachidonate-containing substrate) [20]. All these mechanisms have been proposed to regulate arachidonic acid release in many different A/I cells. In addition, diradylglycerides have recently been shown to be mobilized from choline containing phospholipids via phospholipase C and/or phospholipase D [21–24]. Little is known about their role as a source of arachidonic acid during A/I cell activation. It is clear from the present literature in this area that many unanswered questions remain concerning the source and enzymes responsible for mobilizing arachidonic acid which forms eicosanoids in response to any stimuli.

In addition to the enzyme systems which release arachidonate from phosphoglycerides, there are equally complicated systems which facilitate its reincorporation into complex lipids. In resting cells, the levels of free arachidonic acid are extremely low [16]. Several enzyme systems are responsible for maintaining low levels of free arachidonic acid within the cell. For example, it has been recognized for some time that free arachidonic acid is esterified into lyso phospholipids through CoA-dependent acyl transferase(s) [12, 25]. However, arachidonic acid must first be converted to arachidonoyl-CoA by an acyl-CoA synthetase. A specific acyl-CoA synthetase for arachidonic acid has been described in several cells [26–28]. It has recently become clear that other CoA dependent and independent enzyme systems may be responsible for the acylation of arachidonate into individual phosphoglycerides [29–36]. In addition, the capacity of many of these acylation systems to incorporate arachidonic acid into complex lipids is often increased during cell activation [37, 38]. The aforementioned studies suggest that the level of cellular arachidonic acid available for eicosanoid biosynthesis during A/I cell activation represents a complex balance between that which is released and that which is reincorporated into complex lipids.

Cyclooxygenase Pathway

Once arachidonic acid is released from cellular phosphoglycerides, it can be enzymatically converted to a variety of eicosanoids including prostaglandins, thromboxanes, leukotrienes and HETEs (fig. 2). Much of the original work in this field focused on the biosynthesis of these compounds by an enzyme known as cyclooxygenase. Initial studies describing this activity began in the mid-1960s when Bergstrom and his colleagues in Sweden [8] and Van Drop and his colleagues in The Netherlands [9] reported that radiolabeled arachidonic acid was incorporated into prostaglandins. It was subsequently discovered that this biosynthetic pathway proceeded via intermediate endoperoxide structures (PGG_2/PGH_2). Much of the early work focused on the mechanism of the enzyme system that catalyzes the formation of endoperoxides (PGG/PGH) from arachidonic acid [39–41]. It was later discovered that a number of different products with a wide range of biological activities were formed (enzymatically and non-enzymatically) from these

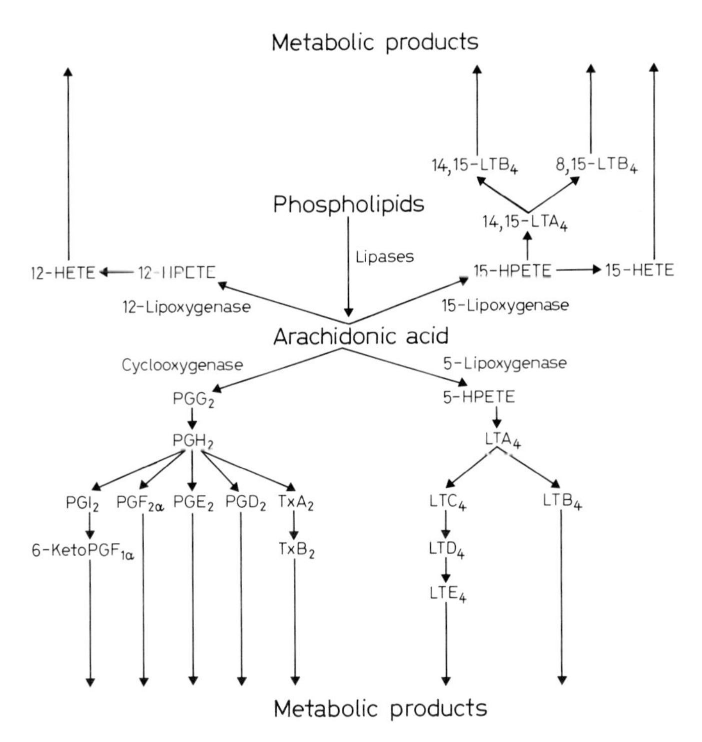

Fig. 2. Major pathways of arachidonic acid metabolism in A/I cells.

endoperoxides. These products included prostaglandins E, F, and D as well as thromboxanes and prostacylin.

Much of the original work with cyclooxygenase was done on tissues such as seminal vesicles, lungs and the renal medulla [42]. The cyclooxygenase catalyzes the transfer of molecular oxygen to arachidonic acid to form an unstable 15-hydroperoxy-9,11-intermediate, PGG_2. A separate peroxidase transforms PGG_2 into another unstable 15-hydroxy-9,11-endoperoxide intermediate, PGH_2 [43]. Both activities appear to be contained within a single protein [44–46]. The enzyme has been purified and reported to have a molecular weight of approximately 70,000 [47, 48]. It has been characterized as a membrane-bound enzyme which requires one heme group bound per enzyme monomer [47]. In addition, the enzyme undergoes auto-inactivation during prostaglandin formation. Studies utilizing monoclonal antibodies have localized this enzyme to the endoplasmic reticulum and nuclear membrane [49].

Once PGH_2 is formed, it can be metabolized to a variety of cyclooxygenase products by A/I cells. The nature of the final product depends on tissues where the endoperoxides are formed, For example, PGH_2 formed in or obtained by vascular endothelial cells will be transformed into prostaglandin I_2 by the prostaglandin synthase [50]. This reaction involves the isomerization of the 9,11-endoperoxide of PGH_2 into a 6,9-epoxide and an 11 hydroxyl group [51]. PGI_2 is spontaneously transformed into 6-keto-$PGF_1\alpha$ under physiological conditions. In contrast, PGH_2 formed in platelets will be converted into another unstable product, thromboxane A_2 by the thromboxane synthase located in the dense tubular system [43]. 12-*L*-Hydroxy-5,8,10-heptadecatrienoic acid (HHT) and malondialdehyde are formed concomitantly with TxA_2 from PGH_2. TxA_2 is rapidly converted into TxB_2 in a matter of seconds by the addition of water. PGD synthase catalyzes the isomerization of PGH_2 to PGD_2 [43, 54]. This involves the transformation of the 9,11-endoperoxy moiety of the molecule to a 9-hydroxyl and 11-keto group [52, 53]. This enzyme seems to be especially prevalent in the mast cell and brain tissue. PGH_2 can also be converted to PGE_2 by the isomerization of the 9,11-endoperoxide moiety to a 9-keto and 11-hydroxyl group [43]. Macrophages/monocytes appear to be one of the major A/I cells which utilize PGH_2 in this manner [43]. Finally, $PGF_2\alpha$ may be converted directly from PGH_2 by the reductive hydrolysis of the 9,11-endoperoxide moiety of the molecule [54] or formed from PGE_2 and PGD_2 by reductions of the 9-keto group or the 11-keto group, respectively [55–57]. Each of these pathways for $PGF_2\alpha$ synthesis have been proposed in cells participating in the A/I response. The enzymes involved in prostaglandin, thromboxane or prostacylin production, including the PGD synthetase, PGE synthetase, PGF synthetase, PGI synthetase and thromboxane synthetase have all been characterized in a number of cells and tissues. Several reviews provide a more in-depth review of the enzymology of these than is provided here [42, 43].

Nonsteroidal anti-inflammatory agents such as aspirin and indomethacin block the cyclooxygenase pathway [58]. Aspirin appears to inhibit the oxygenase activity of cyclooxygenase by acetylating a serine residue within the protein and preventing access of AA to the heme-containing active site. In addition, these agents also inhibit prostaglandin, thromboxane and prostacylin synthesis from the endoperoxides PGH_2 and PGG_2 in varying degrees.

Lipoxygenase Pathway

Although much of the initial work in eicosanoids focused on the cyclooxygenase pathway, an alternative pathway (lipoxygenase) has been described in a number of cells. The lipoxygenase pathway (fig. 2) gives rise to a family of biologically active molecules (leukotrienes and HETEs) with multiple actions that implicate them in the pathophysiology of a number of allergic diseases [59, 60]. The first intermediate synthesized by the 5-lipoxygenase pathway in the production of any of these eicosanoids is 5-hydroperoxyeicosatetraenoic acid. The reduction product of this intermediate (5-HETE) was initially isolated from polymorphonuclear leukocytes which had been incubated with arachidonic acid [61, 62]. Alternative to reduction of 5-hydroperoxyeicosatetraenoic acid, this intermediate can be converted to another epoxide intermediate [63], the 5(S)-trans-oxido-7,9-trans-11,14-cis-eicosatetraenoic acid (LTA_4). This intermediate has a very short half-life in cellular preparations. At this point, the lipoxygenase pathway diverges to form leukotriene B_4 by the enzymatic addition of a hydroxyl group at carbon 12 and opening of the 5,6-epoxide moiety of LTA_4 via the LTA_4 hydrolase [64]. Alternatively, the addition of the thiol from glutathione to carbon-6 of the LTA_4 epoxide is catalyzed by a glutathione S-transferase to form leukotriene C_4 [65, 66].

The relative ratios of LTB_4 and LTC_4 in various A/I cells depends on the availability of enzyme cofactors and substrates within the cell. For example, LTA_4 is metabolized quite differently when examined in a number of human A/I cells. Human neutrophils synthesize large quantities of LTB_4 via the epoxide hydrolase upon stimulation with little or no synthesis of LTC_4 [67].

In contrast, human basophils and eosinophils produce predominantly LTC_4 with little LTB_4 [68, 69]. Human mast cells, monocytes and macrophages seem to have the capacity to produce LTC_4 and LTB_4 in varying quantities [70–73]. However, it is important to point out that there can be dramatic differences in the amount and relative ratios of leukotriene B_4 and leukotriene C_4 produced in the same cell from different species.

Initial studies on the purified 5-lipoxygenase suggested that this was a cytosolic enzyme which was dependent on calcium and ATP for its activation [74]. In the last 5 years, several important findings have revealed the complex nature of this enzyme activity. This enzyme has been purified from several A/I cells and reported to have a molecular weight of between 70,000 and 80,000 [75–77]. In addition, it is now recognized that a single monomeric protein catalyzes both the synthesis of 5-HPETE and the conversion of 5-HPETE to leukotriene A_4 [76–79]. One of the most interesting aspects of the enzyme is that it appears to translocate from the cytosol to cellular membranes during cell activation [80]. However, very little is known about the mechanism and the ramifications of this translocation. It is possible that this provides a mechanism by which the 5-lipoxygenase can obtain arachidonic acid from phospholipid precursors.

Leukotriene C_4 and Its Metabolites

Work on lipoxygenase products began some 50 years ago when Feldberg and Kellaway [81] discovered that cobra venom released a substance from animal lungs which caused smooth muscle contraction in vitro. Two years later, Kellaway and Trethewie [82] described a substance with similar characteristics produced in the lungs of sensitized guinea pigs following antigen challenge. This activity appeared to be different from histamine based on the kinetics of the contractile response and was therefore given the name 'slow reacting substance of anaphylaxis' (SRS-A). Subsequently, Brocklehurst [83] provided evidence that this substance could be differentiated from histamine by demonstrating its activity in the presence of an antihistamine. Much of the early work by Brocklehurst [84] and colleagues implicated SRS-A as a major participant in the allergic response.

Characterization of the structure of SRS-A began in the 1960s and would not be completed until 10 years later. Early progress was slow due primarily to the small quantities which could be obtained from biological systems and the extremely labile nature of the compound. This began to change in the mid-1970s when it was discovered that the challenge of several cells in vitro with calcium ionophore A23187 led to the production of large quantities of

SRS-A [85, 86], and the addition of cysteine to these cell preparations greatly potentiated the amount of SRS-A produced during cell activation [87]. In addition to providing a method for producing larger amounts of SRS-A, this latter finding provided evidence that cysteine was part of the structure of SRS-A. Several important findings led up to the final characterization of the structure including the facts that: (1) SRS-A contained a sulfur moiety [88]; (2) SRS-A was derived from arachidonic acid [89]; (3) SRS-A contained a conjugated triene system as evidenced by its characteristic UV triplet absorption spectra (absorption maximum at 280 with shoulders at 270 and 292). Finally in 1979, Murphy et al. [90] reported that SRS-A was a cysteine-containing derivative of 5-hydroxy-7,9,11,14-eicosatetraenoic acid with an amino acid attached in a thioether linkage at carbon 6. The compound was called 'leukotriene C' based on the presence of a conjugated triene found in this molecule derived from a leukocyte-type cell. Further proof of the structure of this molecule and identification of the amino acid portion of the molecule was obtained that same year [91]. The following year, the cysteinyl-glycyl form of the molecule (LTD) was independently characterized by two groups [92, 93]. Subsequent studies revealed that the glycine portion of cysteinyl-glycyl residue of LTD_4 was also cleaved in biological systems to yield LTE_4 [94, 95]. With these discoveries, a family of molecules was characterized which fully accounted for the biological activity attributed to SRS-A. Several pathways have been described for the inactivation of the sulfidopeptide leukotrienes. As previously mentioned, several cell types contain F-glutamyltranspeptidase and dipeptidase activities which will convert LTC_4 to LTD_4 and LTE_4, respectively, by the peptide cleavage of the tripeptide moiety attached at carbon 6 of the arachidonate backbone. In addition, a peroxidase-hypochlorous reaction has been shown to transform LTC_4, LTD_4 and LTE_4 to sulfoxides in activated neutrophils and eosinophils [96, 97]. In all these cases, the metabolic products are generally less potent with respect to their biological activities than the parent compound.

Leukotriene B_4 and Its Metabolites

Just prior to the structure elucidation of the sulfidopeptide leukotrienes, Borgeat and Samuelsson [64] recognized the existence of another molecule which was derived from incubating arachidonic acid with polymorphonuclear leukocytes. This molecule contained three conjugated double bonds and a hydroxyl group at carbon 5. In contrast to LTC_4, it also contained a hydroxyl group at carbon 12 and no sulfidopeptide moiety within the molecule. The structure of this compound was determined to be 5S, 12R-dihydroxy-6,14-cis-8,10-trans-eicosatetraenoic acid and was called 'LTB_4'. In contrast to many lipid mediators, the structure of this compound was elucidated before any activity was attributed to it. However, it did not take long before this molecule was recognized as a powerful stimulus for the chemokinesis, chemotaxis, aggregation and degranulation of neutrophils [98]. More recently, it has been demonstrated that LTB_4, much like LTC_4, is rapidly converted to less active metabolites 20-OH LTB_4 and 20-COOH LTB_4 by the neutrophil [99]. The ω-oxidation of LTB_4 to these products appears to be catalyzed by a NADPH-dependent, cytochrome P-450 enzyme [100, 101].

Other Lipoxygenase Pathways

Other lipoxygenase pathways have been described in a variety of cells. The platelet contains a 12-lipoxygenase [102] which converts arachidonic acid to 12-hydroperoxy-5,8,10,14-eicosatetraenoic acid (12-HPETE). 12-HPETE is subsequently reduced to 12-HETE. The 12-lipoxygenase appears to be loosely associated with internal membranes within the platelet [103]. The platelet enzyme requires no cofactors or exogenous calcium for its activity [104]. Similarly, neutrophils also contain a 15-lipoxygenase activity which converts arachidonic acid to 15-HPETE [105]. This enzyme isolated from reticulocytes has a molecular weight of 78,000 with one non-heme ion atom per molecule [106]. Similar to 5-HPETE and 12-HPETE, 15-HPETE can be reduced to 15-HETE. Cells which contain both 5-lipoxygenase and 15-lipoxygenase can convert arachidonate to the double lipoxygenase product [107] 5,15-dihydroxy-6,13-trans-8,11-cis-eicosatetraenoic acid (5, 15-di-HETE). Finally, 15-HPETE may be dehydrated [108, 109] to form a 14,15-epoxy-5,8,10,12-eicosatetraenoic acid (14, 15-LTA_4) which can subsequently be hydrolyzed into two isomers of 14,15-dihydroxy-5,8,10,12-eicosatetraenoic acid (14, 15-LTB_4) and two isomers of 8,15-dihydroxy-5,9,11,13-eicosatetraenoic acid (8,15-LTB_4).

Transcellular Metabolism of Eicosanoids

Metabolic intermediates released from one cell can be used by an acceptor cell for the production of eicosanoids. This allows a cell to produce a given eicosanoid while being devoid of some of the enzyme activities necessary for its entire synthesis. In general, this process involves the transfer of an oxygenated metabolite of arachidonic acid. Several models of this can

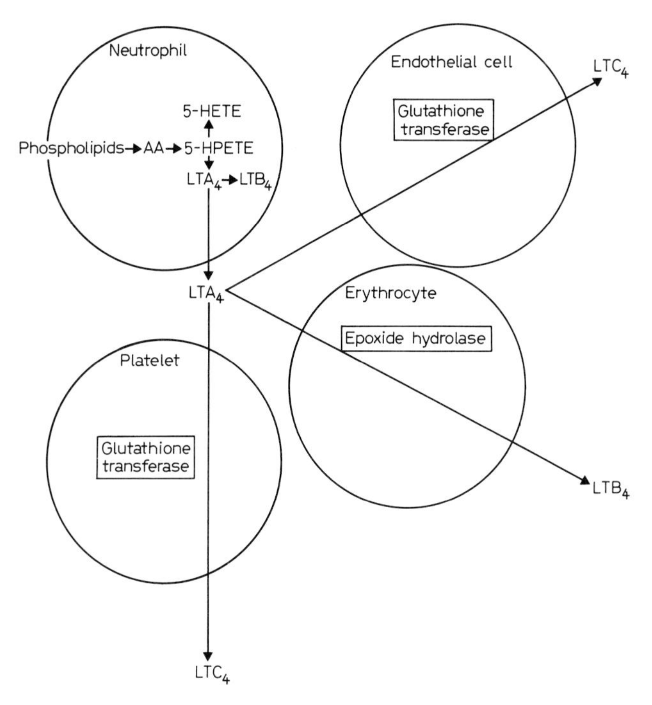

Fig. 3. Transcellular metabolism of LTA_4 by a few A/I cells.

be found in prostaglandin synthesis. One of the early examples of cell-cell interactions were studies by Marcus et al. [110] which demonstrated that stimulated platelets provided endoperoxides for vascular endothelial cells in the production of prostacylin. Other examples of this include the conversion of released prostacyclin and its nonenzymatic product 6-keto-$PGF_1\alpha$ into 6-keto-PGE_1 by the 9-hydroxyprostaglandin dehydrogenase in the platelet [111], and the reduction of released PGD_2 to 9α ,11β-PGF_2 by a variety of cells [112].

In addition to prostaglandin synthesis, several lipoxygenase intermediates have been shown to participate in cell-cell interactions. One of the most interesting of these is the 5,6-epoxy derivative of arachidonic acid, LTA_4 (fig. 3). As previously described, this intermediate can either be hydrolyzed by the epoxide hydrolase to form LTB_4 or it can be conjugated with glutathione to form LTC_4. Several studies have now shown that this intermediate can be released from the stimulated neutrophil and it appears to be stabilized by albumin. Once released, this 5-lipoxygenase product can be converted to LTB_4 by erythrocytes [113, 114] and to LTC_4 by vascular endothelium [115] and platelets [116]. In both these later cases, the acceptor cell does not have the capacity to synthesize LTA_4. These findings point out the potential of cells seemingly devoid of 5-lipoxygenase activity to play an important role in this leukotriene biosynthesis.

Other studies have reported the transcellular metabolism of monohydroxy derivatives of the 5- and 12-lipoxygenase pathways. For example, both 5-HETE and 12-HETE are converted to (5S,12S)-dihydroxy-6,10-trans-10,14-cis-eicosatetraenoic acid when neutrophils and platelets are incubated with 12(S)-HETE and 5(S)HETE, respectively [117, 118]. This product can then be hydroxylated at carbon 20 to form 5(S),12(S),20-triHETE [119]. In addition, 12-HETE can also undergo ω-hydroxylation by neutrophils to form 12(S),20-diHETE. The biologic activity or roles played by these compounds are still poorly understood [120].

Platelet-Activating Factor

Platelet-activating factor (PAF) is a unique phosphoglyceride which possesses many potent biological activities including the activation of platelets and neutrophils, stimulation of glycogenolysis and antihypertensive effects [121–132]. In addition, this versatile molecule has been shown to induce several responses relevant to airway disease including bronchoconstriction, increases in vascular permeability, increases in bronchial hyperreactivity and infiltration of inflammatory cells [132]. PAF was first recognized in the early 1970s in studies which demonstrated that specific antigen would release a substance from sensitized rabbit leukocytes that would aggregate and degranulate platelets [121–123]. Based on this biological activity, this substance was termed PAF in 1972 [124]. Much of the early work in PAF centered around defining the chemical structure(s) of the substance. From 1972 until 1979, several findings, including the fact that it was extracted into organic solvents and could be inactivated by phospholipases A_2, C and D, suggested that this molecule was a phospholipid [133, 134]. In 1979, two groups (Demopoulus et al. [135] and Benveniste et al. [136]), synthesized a molecule, 1-0-alkyl-2-acetyl-sn-glycero-3-phosphocholine (fig. 4) and showed that it had biological activity indistinguishable from naturally occurring PAF. That same year, Blank et al. [131] independently reported the structure of an antihypertensive substance produced by the kidney to be 1-0-alkyl-2-acetyl-sn-glycero-3 phosphocholine. A year later, the

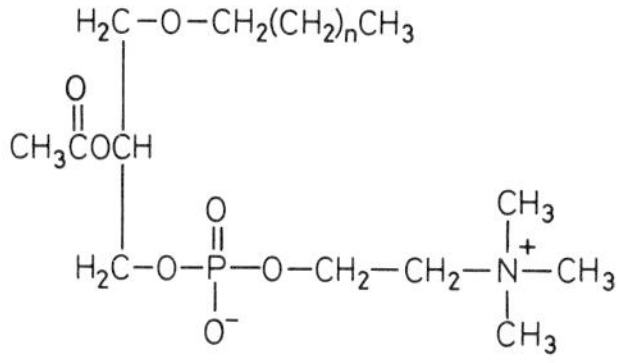

```
          H2C-O-CH2(CH2)nCH3
       O   |
       ||  |
    CH3COCH
           |     O                  CH3
           |     ||                 |+
          H2C-O-P--O-CH2--CH2--N--CH3
                 |                  |
                 O-                 CH3
```

Fig. 4. Structure of PAF. n = Predominantly 14 and 16.

structure of PAF from antigen-stimulated leukocytes was also identified as 1-0-alkyl-2-acetyl-sn-glycero-3-phosphocholine using mass spectrometry following chemical degradation [137].

Structure of Platelet-Activating Factor

Since these early studies, we have now recognized that PAF is a group of 1-alkyl-2-acetyl-sn-glycero-3-phosphocholine molecular species containing various fatty alcohol chains (16:0, 18:0, 18:1, etc.) at the sn-1 position of the molecule. PAF is remarkable for the fact that it resembles a membrane phospholipid in a number of ways and yet contains hidden within its structure are unique requirements necessary for its activity. For example, it was early recognized that the acetyl moiety at the sn-2 position was crucial to the activity of the molecule [138]. The molecule has little activity if this portion of the molecule is removed or if fatty acyl chains with more than 3 carbons are placed at this position. An ether linkage with a long hydrocarbon chain ($\geq$16 carbons) at the sn-1 position is required for optimal activity [135]. An acyl moiety at this sn-1 position renders the molecules severalfold less active [135]. A choline head group at the sn-3 position is necessary for optimal activity although ethanolamine or mono- and dimethyl-ethanolamine display some activity [138, 139]. In addition to the need for certain functional groups at the sn-1, sn-2 and sn-3 positions, the enantiomer of PAF at its one chiral center (sn-2) has very little activity relative to naturally occurring PAF [140, 141]. These findings have suggested stringent structural requirements necessary for potent biological activity.

Biosynthesis of Platelet-Activating Factor

Similar to the leukotrienes, PAF biosynthesis in A/I cells is thought to be initiated by the activation of phospholipase(s) during cell stimulation. In the 'deacylation-acetylation' pathway of biosynthesis, PAF is derived from 1-0-alkyl-2-acyl-sn-glycero-3-phosphocholine cellular pools through the action of

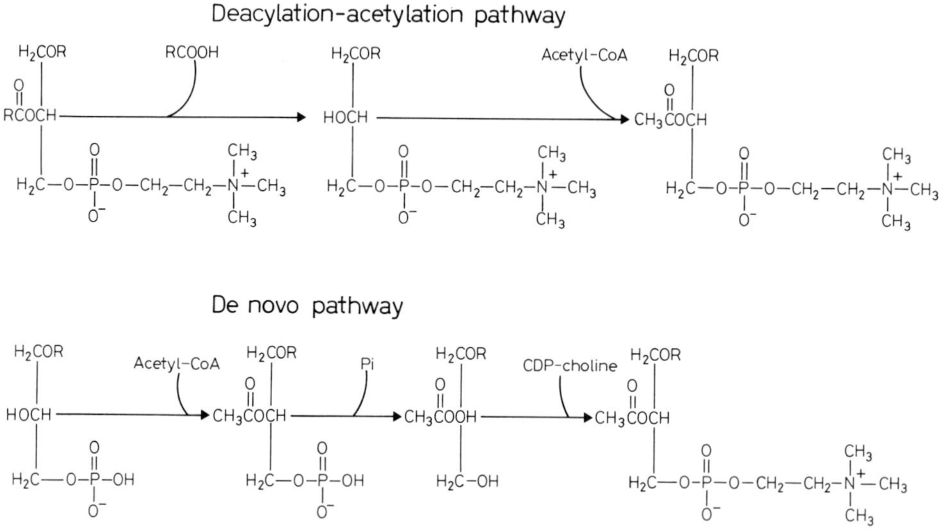

Deacylation-acetylation pathway

De novo pathway

Fig. 5. Pathways of PAF biosynthesis.

a phospholipase A_2 (fig. 5). A number of A/I cells including neutrophils, macrophages and mast cells contain large naturally occurring 1-alkyl-2-acyl-sn-glycero-3-phosphocholine pools [142–145]. In addition, these pools are mobilized in A/I cells during immunologic and nonimmunologic activation [146–150]. Once 1-0-alkyl-2-lyso-sn-glycero-3-phosphocholine is formed in these cells, it is rapidly acetylated by an acetyl-CoA requiring enzyme called acetyltransferase [151]. Both phospholipase A_2 and acetyltransferase are activated during A/I cell stimulation [152–155]. This fact, along with studies which have demonstrated that potential pathway intermediates exogenously provided to A/I cells are ultimately incorporated into PAF during cell stimulation, have suggested that the deacylation-acetylation pathway is the primary pathway by which PAF is formed in A/I cells [146, 149, 152].

A second pathway de novo (fig. 5) for the synthesis of PAF has been proposed for the A/I cell as well as isolated tissues [156]. The final enzymatic step of this pathway involves a specific dithiothreitol-insensitive choline phosphotransferase which transfers phosphocholine from CDP-choline to 1-alkyl-2-acetyl glycerol [157]. A number of tissues have been reported to contain this enzyme activity [156–159]. Although the complete de novo synthesis pathway has been demonstrated in some tissues, there are still major questions relating to if and how 1-0-alkyl-2-acetyl glycerol, which receives the phosphocholine via the choline phosphotransferase, is formed in A/I cells. In addition, many questions remain as to whether enzymes in this pathway are activated during cell stimulation.

Catabolism of Platelet-Activating Factor

In view of the potent biological activity of PAF, it is clear that levels of PAF must be tightly controlled. This requires efficient pathways which catabolize large quantities of PAF often produced during cell activation. One of the primary ways in which this is accomplished is by the enzyme acetyl hydrolase found in serum [160–162]. This enzyme rapidly converts PAF (1-0-alkyl-2-acetyl-sn-glycero-3-phosphocholine) to lyso PAF (1-alkyl-2-lyso-sn-glycero-3-phosphocoline). It was originally characterized by Farr [160, 161] and his co-workers; more recently Stafforini et al. [162] found this enzyme to be associated with both low- and high-density lipoproteins in the serum.

In A/I cells, PAF is rapidly catabolized to lyso-PAF by cytosolic acetyl hydrolase. This acetyl hydrolase catalyzes the rate-limiting step for the catabolism of PAF by A/I cells [163]. In A/I cells, the subsequent fate of lyso PAF is rapid reacylation with unsaturated fatty acids at the sn-2 position [164–166]. Determination of the nature of the acyl groups at the sn-2 position has revealed that arachidonate is the predominant fatty acyl chain incorporated into lyso PAF in human neutrophils, rabbit macrophages, human and murine mast cells, human and rabbit platelets and the isolated perfused lung [167–170]. These findings have suggested that both PAF and arachidonic acid can be immobilized during cell activation with this one enzymatic step.

Sources of Platelet-Activating Factor

Initial studies with PAF characterized its release from rabbit basophils in response to antigen. Since these studies, PAF has been shown to be produced by a number of A/I cells in response to a variety of stimuli (table 1). Each of these cells produces differing amounts and PAF molecular species, reflecting the precursor pools available to each cell. Mast cells from a variety of sources have recently been shown to produce another type of PAF species. We have found that human lung mast cells produce predominantly 1-acyl-2-acetyl-sn-glycero-3-phosphocholine in addition to 1-alkyl-2-acetyl-sn-glycero-3-phosphocholine in response to anti-IgE. The biological relevance of this molecule has yet to be described. Neutrophils and macrophages/monocytes from a variety of sources also produce relatively large concentrations of PAF in

Table 1. Cellular sources of PAF[1]

Cells	References
Neutrophil	171–177
Macrophage	178–180
Basophil	124, 173, 174, 181–185
Mast cell	181, 186–188
Eosinophil	154, 155
Monocyte	173, 176, 177
Endothelial cell	189–191
Platelet	192–197

[1]This list does not include all reports of PAF production but many of the early references which identified PAF in each of the aforementioned cell types.

response to stimulation with A23187, F'met leu Phe, and opsonized zymosan [171–180]. In contrast to large amounts of PAF produced by these cells, other cells such as endothelial cells and platelets appear to produce smaller quantities of PAF [189–191, 198, 199]. From species to species, an individual cell type may vary a great deal in the amount of PAF it produces. An example of this is found in the basophil. It is generally accepted that rabbit basophils make relatively large quantities of PAF, whereas human basophils have been reported to produce little if any PAF [124, 173, 174, 181–185]. PAF is also found in vivo by IgE-related stimuli suggesting that mast cells and basophils are responsible for PAF found in these situations [200–204]. Alternatively, the mast cell or basophil may only act as intermediates which eventually stimulate PAF release from another cell.

Since PAF has been highly touted as an extracellular mediator, it has generally been assumed that, like prostaglandins and leukotrienes, it is completely released from the cell as it is produced. However, there have been several studies which suggest that PAF remains cell associated after it is synthesized by a variety of cells [205]. This has raised questions as to the potential role of PAF as an intracellular mediator. However, we have recently provided evidence that PAF is released from the neutrophil under conditions where buffer containing albumin is superfused over cells in a matrix [206]. This study suggests that PAF can be released if it is partitioned away from the cell with its catabolic capacity for PAF. It is clear from these studies that the factors which lead to the release of synthesized PAF from any given cell are not known at this time and are likely to be very complex.

Biochemical Interaction between Platelet-Activating Factor and Arachidonic Acid

It has become increasingly apparent that the biochemistry of arachidonic acid and PAF intersects at a number of points. Suspicions that this was the case were raised in early observations which showed that both PAF and arachidonic acid metabolites are often synthesized together and act together to produce a biological response. We and others have recently proposed that PAF and arachidonic acid share common biochemical pathways in a number of A/I cells (fig. 6). In this pathway (PAF cycle) both precursors of PAF (lyso PAF) and eicosanoids (arachidonic acid) are shuttled in and out of a common molecule 1-0-alkyl-2-arachidonoyl-sn-glycero-3-phosphocholine. Two observations in the neutrophil have strongly supported this hypothesis:

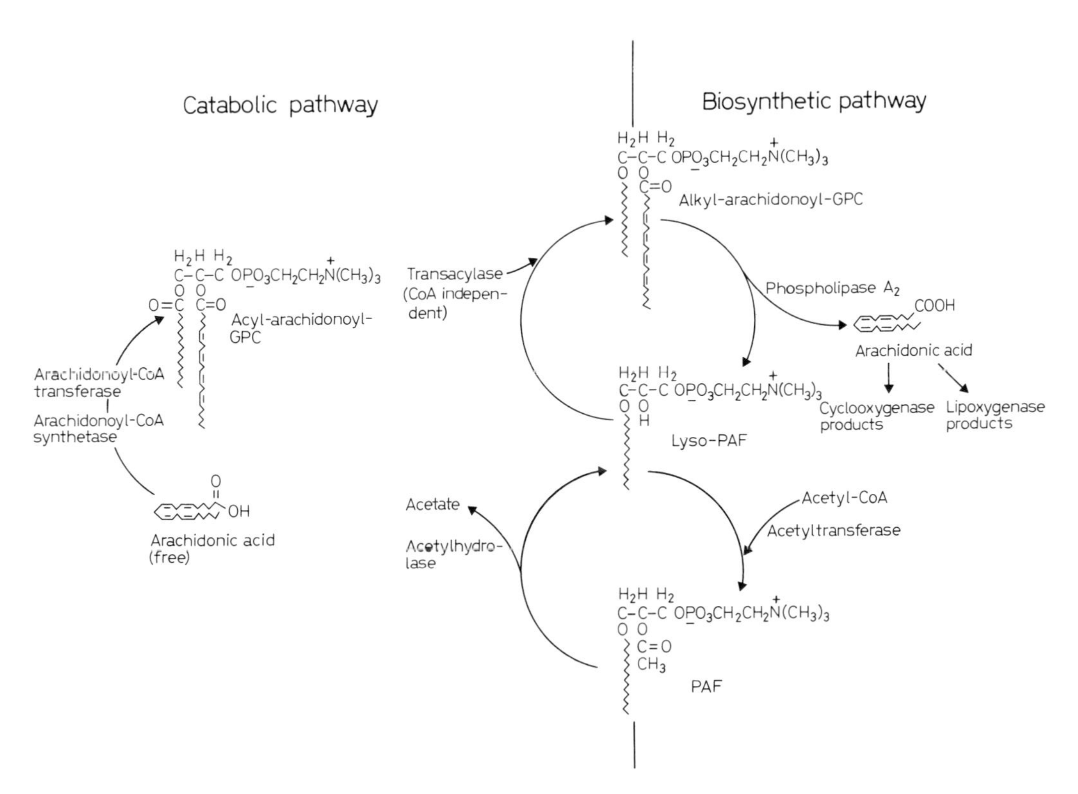

Fig. 6. PAF cycle. Common pathways for arachidonic acid and PAF metabolism in choline-linked phosphoglycerides. Chilton and Murphy, J Biol Chem 261:7771–7777.

(1) PAF is catabolized by removing the acetate moiety at the sn-2 position and replacing it specifically with arachidonic acid [164–167]. This pathway provides an efficient means by which the metabolic precursors of (e.g. lyso PAF and arachidonic acid) two biologically active classes of molecules (e.g. PAF and eicosanoids) are metabolized and stored. It now appears that the key enzyme responsible for the selective transfer of arachidonate to lyso PAF is a CoA-independent transacylase which accepts arachidonate from 1-acyl-2-arachidonoyl-sn-glycero-3-phosphocholine [37, 38, 207, 208]; (2) Both precursors of PAF and eicosanoids, lyso PAF and arachidonic acid, can be mobilized from a common precursor molecule, 1-0-alkyl-2-arachidonoyl-sn-glycero-3-phosphocholine by the action of a phospholipase A_2 during cell activation.

Several lines of circumstantial and direct evidence support this hypothesis. First, A/I by and large contain large substrate pools of this potential common precursor, 1-alkyl-2-arachidonoyl-sn-glycero-3-phosphocholine [142–145]. In fact, the bulk of the arachidonate found in phosphatidylcholine of the neutrophil [142, 143, 209] are located in 1-alkyl-linked pools. In addition, these are major pools of arachidonate which are lost during cell activation [209]. Furthermore, it has been shown that a major portion of lyso PAF mobilized from 1-0-alkyl-2-arachidonoyl GPC during cell activation forms PAF [146]. Other evidence for the obligatory nature of arachidonic acid in PAF production has been provided by studies which have demonstrated that both PAF and leukotriene biosynthesis are inhibited in neutrophils depleted of arachidonate [210]. In addition, PAF and leukotriene B_4 biosynthesis can be partially restored in arachidonic acid-deficient cells by incubating cells with arachidonic acid. The common regulation of PAF and leukotriene B_4 biosynthesis has been further suggested by a recent study which demonstrated that inhibitors of protein kinase C inhibit both PAF and leukotriene B_4 biosynthesis [211]. Furthermore, this process can be reversed for both mediators by the addition of the protein kinase C activator, PMA. Finally, we have recently attempted to identify the phosphoglyceride subclasses which provide arachidonic acid to the leukotrienes by measuring the specific activity of arachidonate in phosphoglyceride subclasses as well as leukotrienes produced during cell activation. These experiments strongly suggest that 1-alkyl-2-arachidonoyl-GPC is a major source of the arachidonic acid used for leukotriene biosynthesis [212]. All of these experiments support the hypothesis that PAF and leukotriene biosynthesis may be tightly coupled in A/I cells.

A number of more casual relationships between PAF and eicosanoids has also been established. For example, it has been shown that PAF can

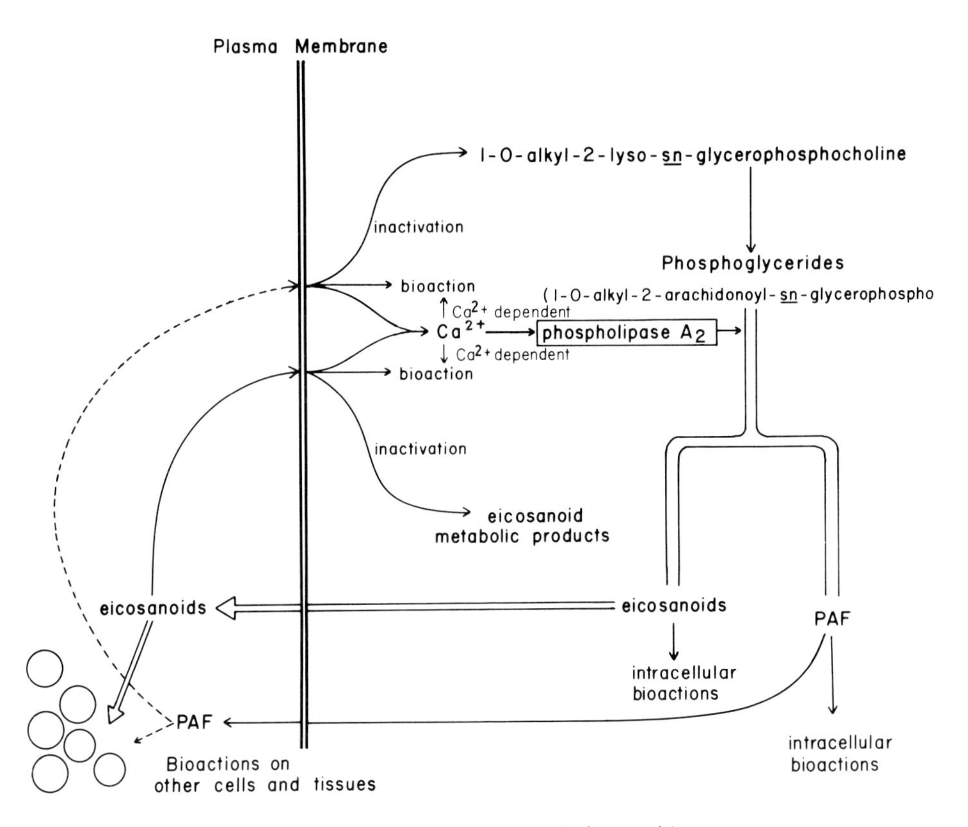

Fig. 7. Potential interactions between PAF and eicosanoids.

induce the synthesis of leukotrienes, prostaglandins and thromboxanes in a number of systems [213–222]. Furthermore, blockers of arachidonic acid metabolism have been shown to inhibit a number of the biological responses of cells to PAF, suggesting that a portion of bioaction of PAF may be due to arachidonic acid metabolism. Conversely, lipoxygenase products have been shown to regulate PAF biosynthesis in several cells [223]. In addition, PAF has been shown to induce its own biosynthesis [224, 225]. It is clear from the aforementioned studies that a complex system of regulatory and feedback mechanisms exists for PAF and arachidonic acid metabolites in the organs and cells that produce them (fig. 7). Lipid mediators released from one cell may act on another cell or in feedback on itself to produce a second lipid mediator. The secondary mediator may then affect a response attributed to the initial mediator.

Summary and Future Directions

In reviewing the 20 or so molecules which encompass these two classes of lipid mediators (e.g. arachidonic acid and PAF), several factors make it difficult to assess the relative importance of any one mediator in the allergic response: (1) There are several biological activities characteristic of the allergic response including effects on smooth muscle contractility, vascular permeability, hyperresponsiveness, and activation of inflammatory cells in which several of these lipid mediators are considered to play a role. A number of factors contribute to the perception of rank importance of these mediators in the allergic response. These include the isolated cells and tissues in which they are studied (e.g. pulmonary vs. arteriole), the animal species in which they are studied, the models in which they are studied (e.g. in vivo and in vitro), the availability of good antagonist for the mediator, and years since the mediator was discovered. (2) There are many cells which participate in the allergic response. Each of these cells produces a unique combination of lipid mediators depending on enzymes and substrates within the cell. Furthermore, the amount and ratio of lipid mediators may vary considerably, depending on the stimulus and cell source studied. (3) Metabolic intermediates released from one cell can be used by an acceptor cell for the production of lipid mediators. This allows cells devoid of some of the enzymes necessary for mediator synthesis to participate in the allergic response. There have been examples of this with PGH_2 and LTA_4. (4) A lipid mediator released from one cell may act on another cell or feedback on itself to produce a second lipid mediator. In some cases, this secondary product may mediate the biological response attributed to the initial stimulus. For example, several studies indicate that PAF induces arachidonic acid metabolism in a number of cells. Furthermore, blockers of arachidonic acid metabolism often inhibit the biological response of cells to PAF indicating some portion of the bioaction of PAF may be regulated by the intracellular production of arachidonic acid. Conversely, arachidonic acid metabolites have also been shown to stimulate PAF biosynthesis in some cells. (5) PAF and arachidonic acid may share common biochemical pathways in some A/I cells. For example, PAF and leukotrienes may be synthesized from a common phospholipid precursor in some A/I cells upon activation. Conversely, PAF and arachidonic acid may be immobilized in some A/I cells by common metabolic pathways.

The role of an individual mediator class in the allergic response may be compared to the role of a branch in a large oak tree. A complex network of branches make up this tree we call the allergic response. Work in the last 50 years has provided a basic understanding of the individual biochemical and biological branches of arachidonic acid and PAF. However, a number of factors suggest that it will be difficult to anticipate the shape of the tree or the role of any single branch within the tree without understanding the interactions between the branches. A challenge of the future in understanding the role of lipid mediators in the allergic reaction will be to better define the interactions between the individual molecules that make up the general class of lipid mediator.

References

1 Burr GO, Burr MM: On the nature and role of the fatty acids essential in nutrition. J Biol Chem 1930;86:587.
2 Burr GO, Burr MM: A new deficiency disease produced by the rigid exclusion of fat from the diet. J Biol Chem 1929;82:345.
3 Kurzrok R, Lieb CC: Biochemical studies of human semen. II. The action of semen on the human uterus. Proc Soc Exp Biol Med 1930;28:268.
4 Goldblatt MW: Properties of human seminal fluid. J Physiol 1935;84:208.
5 von Euler US: On the specific vasodilating and plain muscle stimulating substance from accessory genital glands in man and certain animals (prostaglandin and vesiglandin). J Physiol 1936;88:213.
6 Bergstrom S, Dressler F, Krabisch L, et al: Isolation and structure of a smooth muscle stimulating factor in normal sheep and pig lungs. Arkiv Kemi 1962;20:63.
7 Bergstrom S, Dressler F, Krabisch L, et al: The isolation of two further prostaglandins from sheep prostate glands. Arkiv Kemi 1962;19:563.
8 Bergstrom S, Danielsson H, et al: The enzymatic formation of prostaglandin E_2 from arachidonic acid prostaglandins and related factors 32. Biochim Biophys Acta 1964; 90:207.
9 van Drop DA, Beerthvis RK, Nugteren DH, et al: The biosynthesis of prostaglandins. Biochim Biophys Acta 1964;90:204.
10 Kagey-Sobotka A, MacDonald SM, Fox C, et al: In vitro and in vivo studies of human basophils and mast cells: Mediators, secretagogues, and pharmacology; in Marone, Lichtenstein, Condorelli et al (eds): Human Inflammatory Disease. Publisher B.C. Decker, Toronto/Philadelphia 1988, pp 39–47.
11 Lands WEM, Samuelsson B: Phospholipid precursors of prostaglandins. Biochim Biophys Acta 1968;164:426.
12 Hill EE, Husbands DR, Lands WEM: The selective incorporation of 14C-glycerol into different species of phosphatidic acid, phosphatidylethanolamine and phosphatidylcholine. J Biol Chem 1968;243:4440–4447.
13 Irvine RF: How is the level of free arachidonic acid controlled in mammalian cells. Biochem J 1983;204:3.
14 Van den Bosch H: Phospholipases; in Hawthorne JH, Ansell GB (eds), Phospholipids. Amsterdam, Elsevier Biomedical, 1982, p 317.

15 Dennis EA: Phospholipases; in Boyer P, The Enzymes. New York, Academic Press, 1983, p 307.
16 Bills TK, Smith JB, Silver MJ: Selective release of arachidonic acid from the phospholipids of human platelets in response to thrombin. J Clin Invest 1977;60:1.
17 Rittenhouse-Simmons S: Production of diglyceride from phosphatidylinositol in activated human platelets. J Clin Invest 1979;63:580.
18 Bell RL, Kennerly DA, Stanford U, et al: Diglyceride lipase. A pathway for arachidonate release from human platelets. Proc Natl Acad Sci USA 1979;76:3238.
19 Prescott SM, Majerus PW: Characterization of 1.2 diacylglycerol hydrolysis in human platelets. Demonstration of an arachidonoyl-monoacylglycerol intermediate. J Biol Chem 1983;258:764.
20 Lapetina EG, Billah MM, Cuatrecasas P: The initial action of thrombin on platelets: conversion of phosphatydlinositol to phosphatidic acid preceding the production of arachidonic acid. J Biol Chem 1981;256:5037.
21 Chilton FH, Ellis JM, Olson SC, et al: 1-0-alkyl-2-arachidonoyl sn-glycero-3-phosphocholine. A common source of platelet activating factor and arachidonate in human polymorphonuclear leukocytes. J Biol Chem 1984;259:12014.
22 Daniel LW, Waite M, Wykle RL: A novel mechanism of diglyceride formation. 12-0-tetradecanoylphorbol-13-acetate stimulates cyclic breakdown and resynthesis of phosphatidylcholine. J Biol Chem 1986;261:1928.
23 Pai J-K, Siegel MI, Egan RW, et al: Phospholipase-D catalyzes phospholipid metabolism in chemotactic peptide-stimulated HL-60 granulocytes. J Biol Chem 1988;263:12472–12477.
24 Kennerly DA: Diacylglycerol metabolism in mast cells. Analysis of lipid metabolic pathways using molecular species analysis of intermediates. J Biol Chem 1987;262: 16305–16310.
25 Lands WEM, Crawford CG: in Martonosi A. (ed): The Enzymes of Biological Membranes. New York, Plenum Press, 1976, vol 2, pp 3–25.
26 Wilson DB, Prescott SM, Majerus PW: Discovery of an arachidonoyl Co enzyme A synthetase in human platelets. J Biol Chem 1982;257:3510–3516.
27 Neufeld EJ, Bross TE, Majerus PW: A mutant $HSDM_1$ C_1 fibrosarcoma line selected for defective eicosanoid precursor uptake lacks arachidonate-specific acyl-CoA synthetase. J Biol Chem 1984;259:1986–1992.
28 Laposata M, Reich EL, Majerus PW: Arachidonoyl CoA synthetase: separation from non-specific acyl-CoA synthetase and distribution in various cells and tissues. J Biol Chem 1985;260:11016–11022.
29 Kameyama Y, Yosioka S, Imai A, et al: Possible involvement of 1-acyl-glycerophosphorylinositol acyl transferase in arachidonate enrichment of phosphatidylinositol in human platelets. Biochim Biophys Acta 1983;752:244–250.
30 Trotter J, Flesch I, Schmidt B, et al: Acyltransferase-catalyzed cleavage of arachidonic acid from phospholipids and transfer to lysophosphatides in lymphocytes and macrophages. J Biol Chem 1982;257:1816–1821.
31 Kramer RM, Pritzker CR, Deykin D: Co enzyme A-mediated arachidonic acid transacylation in human platelets. J Biol Chem 1984;259:2403–2408.
32 Kramer RM, Deykin D: Arachidonoyl transacylase in human platelets: CoA-independent transfer of arachidonate from phosphatidylcholine to lyso plasmenylethanolamine. J Biol Chem 1983;258:13806–13811.
33 Sugiura T, Waku K: CoA-independent transfer of arachidonic acid from 1,2-diacyl-

sn-glycero-3-phosphocholine to 1-0-alkyl-sn-glycero-3-phosphocholine (lyso) platelet-activating factor by macrophage microsomes. Biochem Biophys Res Commun 1985;127:384–389.
34 Chilton FH, Murphy RC: Remodeling of arachidonate-containing phosphoglycerides within the human neutrophil. J Biol Chem 1986;261:7771–7777.
35 Chilton FH, Hadley JS, Murphy RC: Incorporation of arachidonic acid into 1-acyl-2-lyso-sn-glycero-3-phosphocholine of the human neutrophil. Biochim Biophys Acta 1987;917:48–53.
36 Sigiura T, Masuzawa Y, Nakagawa Y, et al: Transacylation of lyso platelet-activating factor and other lysophospholipids by macrophage microsomes: distinct donor and acceptor selectivities. J Biol Chem 1987;262:1199–1205.
37 Walsh CE, DeChatelet LR, Chilton FH, et al: Mechanism of arachidonic acid release in human polymorphonuclear leukocytes. Biochim Biophys Acta 1983;750:32–40.
38 Tou J-S: Platelet activating factor promotes arachidonate incorporation into phosphatidylinositol and phosphatidylcholine in neutrophils. Biochem Biophys Res Commun 1985;127:1045–1049.
39 Samuelsson B: On the incorporation of oxygen in the conversion of 8,11,14-eicosatrienoic acid to prostaglandin E_1. J Am Chem Soc 1965;87:3011–3013.
40 Hamberg M, Samuelsson B: Detection and isolation of an endoperoxide intermediate in prostaglandin biosynthesis. Proc Natl Acad Sci USA 1973;70:899–903.
41 Nugteren DH, Hazelhof E: Isolation and properties of intermediates in prostaglandin biosynthesis. Biochim Biophys Acta 1973;326:448–461.
42 Samuelsson B, Goldyne M, Granstrom E, et al: Prostaglandins and thromboxanes. Ann Rev Biochem 1978;47:997–1029.
43 Needleman P, Turk J, Jakschik BA, et al: Arachidonic acid metabolism. Ann Rev Biochem 1986;55:69–102.
44 Miyamoto T, Ogino N, Yamamoto S, et al: Purification of prostaglandin endoperoxide synthetase from bovine vesicular gland microsomes. J Biol Chem 1986;251: 2629–2636.
45 Onki S, Ogino N, Yamamoto S, et al: Prostaglandin hydroperoxidase, an integral part of prostaglandin synthetase from bovine vesicular gland microsomes. J Biol Chem 1979;254:829–836.
46 Van der Ouderaa FJ, Buytennek M, Nugteren DH, et al: Purification and characterization of prostaglandin endoperoxide synthetase from sheep vesicular glands. Biochim Biophys Acta 1977;487:315–331.
47 Roth GS, Machuga ET, Strittmatter P: The heme-binding properties of prostaglandin synthetase from sheep vesicular glands. J Biol Chem 1981;256:10018–10022.
48 Pagels WR, Sachs RJ, Marnett LJ, et al: Immunochemical evidence for the involvement of prostaglandin H synthase in hydroperoxide-dependent oxidations by ram seminal vesicle microsomes. J Biol Chem 1983;258:6517–6523.
49 Rollins TE, Smith WL: Subcellular localization of prostaglandin-forming cyclooxygenase in swiss mouse 3T3 fibroblast by electron microscopic immunocytochemistry. J Biol Chem 1980;255:4872–4875.
50 Johnson RA, Morton DR, Kinner SH, et al: The chemical structure of prostaglandin X (prostacyclin). Prostaglandins 1976;12:915–928.
51 Pacc-Asciak C, Gryglewski R: The prostacyclins; in Pace-Asciak C, Granstrom E (eds): Prostaglandins and Related Substances. 1983, pp 95–126 (Elsevier, Amsterdam).

52 Shimizu T, Mizuno M, Amano T, et al: Prostaglandin D_2, a neuromodulator. Proc Natl Acad Sci USA 1979;76:6231.
53 Lewis RA, Soter NA, Diamond PT, et al: Prostaglandin D_2 generation after activation of rat and human mast cells with anti-IgE. J Immunol 1982;129:1627–1631.
54 Wlodawer P, Kindahl H, Hamberg M: Biosynthesis of prostaglandin $F_2\alpha$ from arachidonic acid and prostaglandin endoperoxides in the uterus. Biochim Biophys Acta 1976;431:603–614.
55 Hamberg M, Israelsson U: Metabolism of prostaglandin E_2 in guinea pig liver. I. Identification of seven metabolites. J Biol Chem 1970;245:5107–5114.
56 Wong PY-K: Purification and partial characterization of prostaglandin D_2 11-keto reductase in rabbit liver. Biochiom Biophys Acta 1981;659:169–178.
57 Reingold DF, Kawasaki A, Needleman P: A novel prostaglandin 11-keto reductase found in rabbit liver. Biochim Biophys Acta 1981;659:179–188.
58 Vane JR: Inhibition of prostaglandin synthesis as the mechanism of action for aspirin-like drugs. Nature New Biol 171;231:232–235.
59 Lewis RA, Austen KF: The biologically active leukotrienes. Biosynthesis, metabolism, receptors, functions, and pharmacology. J Clin Invest 1984;73:889–897.
60 Samuelsson B: Leukotrienes: mediators of immediate hypersensitivity reactions and inflammation. Science 1981;220:568–575.
61 Borgeat P, Hamberg M, Samuelsson B: Transformation of arachidonic acid and homo-γ-linolenic acid by rabbit polymorphonuclear leukocytes: monohydroxyacids from novel lipoxygenases. J Biol Chem 1976;251:7816–7820.
62 Borgeat P, Samuelsson B: Arachidonic acid metabolism in polymorphonuclear leukocytes. Effect of ionophore A23187. Proc Natl Acad Sci USA 1979;76:2148–2152.
63 Borgeat P, Samuelsson B: Arachidonic acid metabolism in polymorphonuclear leukocytes. Unstable intermediate in formation of dihydroxy acids. Proc Natl Acad Sci USA 1979;76:3213–3217.
64 Borgeat P, Samuelsson B: Transformation of arachidonic acid by rabbit polymorphonuclear leukocytes: formation of a novel dihydroxyeicosatetraenoic acid. J Biol Chem 1979;254:2643–2646.
65 Murphy RC, Hammarstrom S, Samuelsson B: Leukotriene C. A slow reacting substance from murine mastocytoma cells. Proc Natl Acad Sci USA 1979;76:4275–4279.
66 Hammarstrom S, Murphy RC, Samuelsson B, et al: Structure of leukotriene C. Identification of the amino acid part. Biochem Biophys Res Commun 1979;91: 1266–1279.
67 Radmark O, Malmsten C, Samuelsson B, et al: Leukotriene A: Isolation from human polymorphonuclear leukocytes. J Biol Chem 1980;255:11828–11831.
68 Warner JA, Freeland HS, MacGlashan DW, et al: Purified human basophils do not generate LTB_4. Biochem Pharmacol 1987;36:3195–3199.
69 Bruynzeel PL, Kok PT, Hamelink MC, et al: Exclusive leukotriene C_4 synthesis by purified human eosinophils indirectly by opsonized zymosan. FEBS Lett 1985;189: 350–354.
70 Peters SP, MacGlashan DW, Schulman ES, et al: Arachidonic metabolism in purified human lung mast cells. J Immunol 1984;132:1972–1979.
71 Pawloski NA, Kaplan G, Hamill AL, et al: Arachidonic acid metabolism by human monocytes, studies with platelet-depleted cultures. J Exp Med 1984;158:393–398.
72 Goldyne ME, Burrish GF, Poubelle P, et al: Arachidonic acid metabolism among

human mononuclear leukocytes. Lipoxygenase related pathways. J Biol Chem 1984; 259:8815–8820.
73 Fels AOS, Pawloski NA, Cramer EB, et al: Human alveolar macrophages produce leukotriene B_4. Proc Natl Acad Sci USA 1982;79:7866–7871.
74 Ochi KT, Yoshimoto S, Yamamoto K, et al: Arachidonate 5-lipoxygenase of guinea pig peritoneal polymorphonuclear leukocytes. Activation by adenosine 5-triphosphate. J Biol Chem 1983;258:5754–5758.
75 Rouzer CA, Samuelsson B: On the nature of the 5-lipoxygenase reaction in human leukocytes: enzyme purification and requirements for multiple stimulatory factors. Proc Natl Acad Sci USA 1985;82:6040–6044.
76 Veda N, Kaneko S, Yoshimoto T, et al: Purification of arachidonate 5-lipoxygenase from porcine leukocytes and its reactivity with hydroperoxyeicosatetraenoic acids. J Biol Chem 1986;261:7982–7988.
77 Hogaboom GK, Cook M, Newton JF, et al: Purification, characterization and structural properties of a single protein from rat basophilic leukemia (RBL-1) cells possessing 5-lipoxygenase and leukotriene A_4 synthetase activities. Mol Pharmacol 1986;30:510–519.
78 Rouzer CA, Matsumoto T, Samuelsson B: Single protein from human leukocytes possess 5-lipoxygenase and leukotriene A_4 synthase activities. Proc Natl Acad Sci USA 1986;83:857–861.
79 Shimizu T, Izumi T, Seyama Y, et al: Characterization of leukotriene A_4 synthase from murine mast cells: evidence for its identity to arachidonate 5-lipoxygenase. Proc Natl Acad Sci USA 1986;83:4175–4179.
80 Rouzer CA, Kargman S: Translocation of 5-lipoxygenase to the membrane in human leukocytes challenged with ionophore A23187. J Biol Chem 1988;263:10980–10988.
81 Feldberg W, Kellaway CH: Liberation of histamine and formation of lysocithin-like substance by cobra venom. J Physiol 1938;94:187–226.
82 Kellaway HC, Trethewie ER: The liberation of a slow reacting smooth-muscle stimulating substance in anaphylaxis. Q J Exp Physiol 1940;30:121–145.
83 Brocklehurst WE: The release of histamine and formation of slow reacting substance (SRS A) during anaphylactic shock. J Physiol 1960;151:416–435.
84 Brocklehurst WE: Slow-reacting substance and related compounds. Prog Allergy. Basel, Karger, 1962, vol 6, pp 539–558.
85 Bach MK, Brashler JR: In vivo and in vitro production of a slow reacting substance in the rat upon treatment with calcium ionophores. J Immunol 1974;113:2040–2044.
86 Jakschik BA, Kulczycki AA, Jr., MacDonald HH, et al: Release of slow reacting substance (SRS) from rat basophilic leukemia (RBL-1) cells. J Immunol 1977;119: 618–623.
87 Orange RP, Chang PL: The effect of thiols on immunologic release of slow reacting substance of anaphylaxis. II. Other in vitro and in vivo models. J Immunol 1976; 116:392–397.
88 Orange RP, Murphy RC, Austen KF: Inactivation of slow reacting substance of anaphylaxis (SRS-A) by arylsulfatases. J Immunol 1974;113:316–322.
89 Jakschik BA, Flakenhein S, Parker CW: Precursor role of arachidonic acid in release of slow reacting substance from rat basophilic leukemia cells. Proc Natl Acad Sci USA 1977;74:4577–4581.
90 Murphy RC, Hammarstrom S, Samuelsson B: Leukotriene C. A slow-reacting substance from murine mastocytoma cells. Proc Natl Acad Sci USA 1979;76:4275–4279.

91 Hammarstrom S, Murphy RC, Samuelsson B, et al: Structure of leukotriene C. Identification of the amino acid part. Biochem Biophys Res Commun 1979;91: 1266–1272.
92 Orning L, Hammarstrom S, Samuelsson B: Leukotriene A. A slow reacting substance from rat basophilic leukemia cells. Proc Natl Acad Sci USA 1980;77:2012–2017.
93 Morris HR, Taylor GW, Piper PJ, et al: Slow reacting substances (SRSs) from rat basophil leukemia (RBL-1) cells. Prostaglandins 1980;19:185–201.
94 Parker CW, Falkenhein SF, Huber MM: Sequential conversion of the glutathionyl side chain of slow reacting substance (SRS) to cysteinyl-glycine and cysteine in rat basophilic leukemia cells stimulated with A23187. Prostaglandins 1980;20:863–886.
95 Lewis RA, Drazen JM, Austen KF, et al: Identification of the C(6)-5-conjugate of leukotriene A with cysteine as a naturally occurring slow reacting substance of anaphylaxis (SRS-A). Importance of the 11-cis-geometry for biologic activity. Biochem Biophys Res Commun 1980;96:271–277.
96 Lee CW, Lewis RA, Corey EJ: Oxidative inactivation of LTC_4 by stimulated polymorphonuclear leukocytes. Proc Natl Acad Sci USA 1982;77:4166–4171.
97 Goetzl EJ: The conversion of leukotriene C_4 to isomers of leukotriene B_4 by human neutrophil peroxidase. Biochem Biophys Res Commun 1982;106:270–274.
98 Ford-Hutchinson AW, Bray MA, Doig MJH, et al: Leukotriene B, a potent chemokinetic and aggregating substance released from polymorphonuclear leukocytes. Nature 1980;286:264–265.
99 Hansson G, Lindgren J-A, Dahlen S-E, et al: Identification and biological activity of ω-oxidized metabolites of leukotriene B_4 from human leukocytes. FEBS Lett 1981; 130:107–112.
100 Powell WS: Properties of leukotriene B_4 20-hydroxylase from polymorphonuclear leukocytes. J Biol Chem 1984;259:3082–3087.
101 Shak S, Goldstein IM: Carbon monoxide inhibits ω-oxidation of leukotriene B_4 by human polymorphonuclear leukocytes: evidence that the catabolism of LTB_4 is mediated by a cytochrome P-450 enzyme. Biochem Biophys Res Commun 1984;123: 475–481.
102 Hamberg M, Samuelsson B: Prostaglandin endoperoxides: novel transformation of arachidonic acid in human platelets. Proc Natl Acad Sci USA 1974;71:3400–3404.
103 Nugteren DH: Arachidonic acid-12-lipoxygenase from bovine platelets. Methods Enzymol 1982;86:49–54.
104 Yokoyama C, Mizuno K, Mitachi H, et al: Partial purification and characterization of arachidonate 12-lipoxygenase from rat lung. Biochim Biophys Acta 1983;750: 237–243.
105 Narumiya S, Salmon JA, Cottee FH, et al: Arachidonic acid 15-lipoxygenase from rabbit peritoneal polymorphonuclear leukocytes. Partial purification and properties. J Biol Chem 1981;256:9583–9592.
106 Wiesner R, Hausdorf G, Anton M, et al: Lipoxygenase from rabbit reticulocytes. Iron content, amino acid composition and C-terminal heterogeneity. Biomed Biochim Acta 1983;42:431–436.
107 Maas RL, Turk J, Oats JA, et al: Formation of a novel dihydroxy acid from arachidonic acid by lipoxygenase-catalyzed double oxygenation in rat mononuclear cells and human leukocytes. J Biol Chem 1982;257:7056–7067.
108 Maas RL, Brash AR, Oates JA: A second pathway of leukotriene biosynthesis in porcine leukocytes. Proc Natl Acad Sci USA 1981;78:5523–5527.

109 Jubiz W, Radmark O, Lindgren JA, et al: Novel leukotriene product formed by initial oxygenation of arachidonic acid at C-15. Biochem Biophys Res Commun 1981;99:976–986.
110 Marcus A, Weksler B, Jaffe FA, et al: Synthesis of prostacylin from platelet-derived endoperoxides by cultured human endothelial cells. J Clin Invest 1980;66:979–986.
111 Wong PYK, Malik KV, Desiderio DM, et al: Hepatic metabolism of prostacyclin (PGI_2) in the rabbit. Formation of a potent novel inhibitor of platelet aggregation. Biochem Biophys Res Commun 1980;93:486–494.
112 Liston TE, Roberts LJ: Transformation of prostaglandin D_2 to 9α 11β-(15S)-trihydroxy-prosta-(5Z, 13E)-dien-1-oic acid (9α, 11β-prostaglandin F2). A unique biologically active prostaglandin produced enzymatically in vivo in humans. Proc Natl Acad Sci USA 1985;82:6030–6034.
113 Fitzpatrick F, Liggett W, McGee J, et al: Metabolism of leukotriene A_4 by human erythrocytes. A novel cellular source of leukotriene B_4. J Biol Chem 1984;259: 11403–11407.
114 McGee JE, Fitzpatrick FA: Erythrocytes-neutrophil interactions. Formation of leukotriene B_4 by transcellular biosynthesis. Proc Natl Acad Sci USA 1986;83:1349–1353.
115 Feinmark SJ, Cannon PJ: Endothelial cell leukotriene C_4 synthesis results from intercellular transfer of leukotriene A_4 synthesized by polymorphonuclear leukocytes. J Biol Chem 1986;261:16466–16472.
116 Maclouf JA, Murphy RC: Transcellular metabolism of neutrophil-derived leukotriene A_4 by human platelets. A potential cellular source of leukotriene C_4. J Biol Chem 1988;263:174–181.
117 Borgeat P, Fruteau de Laclos B, Picard S, et al: Studies on the mechanism of formation of the 5S, 12S-dihydroxy-618-6,8,10,14 (E,Z,E,Z)-icosatetraenoic acid in leukocytes. Prostaglandins 1982;23:713–724.
118 Marcus AJ, Broekman MJ, Safier LB, et al: Formation of leukotrienes and other hydroxy acids during platelet-neutrophil interactions in vitro. Biochem Biophys Res Commun 1982;109:130–137.
119 Marcus AJ, Safier LB, Ullman HL, et al: Platelet-neutrophil interactions. (12S)-Hydroxyeicosatetraen 1,20 dioic acid: a new eicosanoid synthesized by unstimulated neutrophils from (12S)-20-dihydroxyeicosatetraenoic acid. J Biol Chem 1988; 263:2223–2229.
120 Marcus AJ, Safier LB, Ullman HL, et al: 12S,20-0-Dihydroxyicosatetraenoic acid. A new isosanoid synthesized by neutrophils from 12S-hydroxyicosatetraenoic acid produced by thrombin- or collagen-stimulated platelets. Proc Natl Acad Sci USA 1984;81:903–907.
121 Henson PM: Release of vasoactive amines from rabbit platelets induced by sensitized mononuclear leukocytes and antigen. J Exp Med 1970;131:287–291.
122 Siraganian RP, Osler AG: Destruction of rabbit platelets in the allergic response of sensitized leukocytes. I. Demonstration of a fluid phase intermediate. J Immunol 1971;106:1244–1250.
123 Siraganian RP, Osler AG: Destruction of rabbit platelets in the allergic response of sensitized leukocytes. II. Evidence for basophil involvement. J Immunol 1971;106: 1252–1259.
124 Benveniste J, Henson PM, Cochrane CG: Leukocyte-dependent histamine release from rabbit platelets: the role of IgE, basophils, and a platelet-activating factor. J Exp Med 1972;136:1356–1360.

125 O'Flaherty JT, Wykle RL, Miller CH, et al: 1-0-Alkyl-sn-glyceryl-3-phosphorylcholines. A novel class of neutrophil stimulants. Am J Pathol 1981;103:70–78.
126 O'Flaherty JT, Miller CH, Lewis JC, et al: Neutrophil responses to platelet activating factor. Inflammation 1981;5:193–199.
127 Shaw JO, Pinckard RN, Ferrigni KS, et al: Activation of human neutrophils with 1-0-hexadecyl/octadecyl-2-acetyl-sn-glyceryl-3-phosphorylcholine (platelet activating factor). J Immunol 1981;127:1250–1256.
128 Stimler NP, Bloor CM, Hugli TE, et al: Anaphylactic actions of platelet-activating factor. Am J Pathol 1981;105:64–69.
129 Fisher RA, Shukla SD, Debuysere MS, et al: The effect of acetylglyceryl ether phosphorylcholine or glycogenolysis and phosphatidylinositol 4,5-biphosphate metabolism in rat hepatocytes. J Biol Chem 1984;255:5514–5520.
130 Shukla SD, Buxton DB, Olson MS, et al: Acetylglyceryl ether phosphorylcholine. A potent activator of hepatic phosphoinositide metabolism and glycogenolysis. J Biol Chem 1983;258:10212–10218.
131 Blank ML, Snyder F, Byers LW, et al: Antihypertensive activity of an alkyl ether analog of phosphatidylcholine. Biochem Biophys Res Commun 1979;90:1194–1199.
132 Braquet P, Touqui L, Shen TY, et al: Perspectives in platelet-activating factor research. Pharmacol Reviews 1987;39:97–145.
133 Benveniste J: Platelet-activating factor. A new mediator of anaphylaxis and immune complex deposition from rabbit and human basophils. Nature 1974;249:581–583.
134 Benveniste J, LeCouedic JP, Polonsky J, et al: Structural analysis of purified platelet-activating factor by lipases. Nature 1977;269:170–173.
135 Demopoulos CA, Pinckard RN, Hanahan DJ: Platelet-activating factor: evidence for 1-0-alkyl-2-acetyl-sn-glyceryl-3-phosphorylcholine as the active component (a new class of lipid chemical mediators). J Biol Chem 1979;254:9355–9361.
136 Benveniste J, Tence M, Varenne P, et al: Semi-synthèse et structure proposée du facteur activant les plaquettes (P.A.F.) PAF-acether, un alkyl ether analogue de la lysophosphatidylcholine. C R Acad Sci [D] 1979;289:1937–1942.
137 Hanahan DJ, Demopoulos CA, Liehr J, et al: Identification of platelet activating factor isolated from rabbit basophils as acetyl glyceryl ether phosphorylcholine. J Biol Chem 1980;255:5514–5520.
138 Tence M, Coeffier E, Keraly CL, et al: Effect of structural analogs of PAF-acether on platelet aggregation and desensitization; in Benveniste; J, Arnoux B (eds): Platelet-Activating Factor and Structurally Related Ether Lipids. INSERM Symp No 23. Amsterdam, Elsevier, 1983, pp 41–48.
139 Satouchi K, Pinckard RN, McManus LM, et al: Modification of the polar head group of acetyl glyceryl ether phosphorylcholine and subsequent effects upon platelet activation. J Biol Chem 1981;256:4425–4431.
140 Heymans F, Michel E, Borrel MC, et al: New total synthesis and high resolution HNMR spectrum of platelet activating factor, its enantiomer and racemic mixtures. Biochim Biophys Acta 1981;666:230–238.
141 Wykle RL, Miller CH, Lewis JC, et al: Stereospecific activity of 1-0-alkyl-2-0-acetyl-sn-glycero-3-phosphocholine and comparison of analogs in the degranulation of platelets and neutrophils. Biochem Biophys Res Commun 1981;100:1651–1656.
142 Mueller HW, O'Flaherty JT, Greene DG, et al: 1-0-alkyl-linked glycerophospholipids of human neutrophils. Distribution of arachidonate and other acyl residues in the ether-linked and diacyl species. J Lipid Res 1984;25:383–389.

143 Mueller HW, O'Flaherty JT, Wykle RL: Ether lipid content and fatty acid distribution in rabbit polymorphonuclear neutrophil phospholipids. Lipids 1982;17:72–79.
144 Sugiura T, Nakajima M, Sekiguchi N, et al: Different fatty chain compositions of alkenylacyl, alkylacyl and diacyl phospholipids in rabbit alveolar macrophages: High amounts of arachidonic acid in ether phospholipids. Lipids 1983;18:125–129.
145 Yoshioka S, Nakashima S, Okano Y, et al: Phospholipid (diacyl, alkylacyl, alkenylacyl) and fatty acyl chain composition in murine mastocytoma cells. J Lipid Res 1985;26:1134–1139.
146 Chilton FH, Ellis JM, Olson SC, et al: 1-0-alkyl-2-arachidonoyl-sn-glycero-3-phosphocholine. A common source of platelet-activating factor and arachidonate in the human polymorphonuclear leukocytes. J Biol Chem 1984;259:12014–12020.
147 Swendsen CL, Ellis JM, Chilton FH, et al: 1-0-alkyl-2-acyl-sn-glycero-3-phosphocholine: a novel source of arachidonic acid in neutrophils stimulated by calcium ionophore A23187. Biochem Biophys Res Commun 1983;113:72–78.
148 Albert DH, Snyder F: Release of arachidonic acid from 1-alkyl-2-acyl-sn-glycero-3-phosphocholine, a precursor of platelet-activating factor in rat alveolar macrophages. Biochim Biophys Acta 1984;796:92–99.
149 Touqui L, Jacquemin C, Dumarey C, et al: 1-0-alkyl-2-acyl-sn-glycero-3-phosphocholine is the precursor of platelet-activating factor in stimulated rabbit platelets. Evidence for an alkylacetyl-glycerophosphorylcholine cycle. Biochim Biophys Acta 1985;883:111–117.
150 Leslie CC, Detty DM: Arachidonic acid turnover in response to lipopolysaccharide and opsonized zymosan in human monocyte-derived macrophages. Biochem J 1986; 236:251–256.
151 Wykle RL, Malone B, Snyder F: Enzymatic synthesis of 1-0-alkyl-2-acetyl-sn-glycero-3-phosphocholine, a hypotensive and platelet-aggregating lipid. J Biol Chem 1980;255:10256–10261.
152 Albert DH, Snyder F: Biosynthesis of 1-alkyl-2-acetyl-sn-glycero-3-phosphocholine by rat alveolar macrophages; phospholipase A_2 and acetyl transferase activities during phagocytosis and ionophore stimulation. J Biol Chem 1982;258:97–102.
153 Ninio E, Mencia-Huerta JM, Benveniste J: Biosynthesis of platelet-activating factor (PAF-acether). V. Enhancement of acetyltransferase activity in murine peritoneal cells by calcium ionophore A23187. Biochim Biophys Acta 1983;751:298–303.
154 Lee T-C, Lenihan DS, Malone B, et al: Increased biosynthesis of platelet-activating factor in activated human eosinophils. J Biol Chem 1984;259:5526–5532.
155 Lee T-C, Malone B, Wasserman SI, et al: Activities of enzymes that metabolize platelet-activating factor (1-alkyl-2-acetyl-sn-glycero-3-phosphocholine) in neutrophils and eosinophils from humans and the effect of calcium ionophore. Biochem Biophys Res Commun 1982;105:1303–1308.
156 Renooij W, Snyder F: Biosynthesis of 1-alkyl-2-acetyl-sn-glycero-3-phosphocholine (platelet activating factor and a hypotensive lipid) by choline phosphotransferase in various rat tissues. Biochim Biophys Acta 1981;663:545–552.
157 Woodard DS, Lee T, Snyder F: The final step in the de novo biosynthesis of platelet-activating factor. Properties of a unique CDP-choline: 1-alkyl-2-acetyl-sn-glycerol-choline phosphotransferase in microsomes from the renal inner medulla of rats. J Biol Chem 1987;262:2520–2527.
158 Bussolino F, Gremo F, Tetta C, et al: Production of platelet-activating factor by chick retina. J Biol Chem 1986;261:16502–16508.

159 Satouchi K, Oda M, Saito K, et al: Metabolism of 1-0-alkyl-2-acetyl-sn-glycerol by washed rabbit platelets: formation of platelet activating factor. Arch Biochem Biophys 1984;2234:318–323.
160 Farr RS, Wardlow ML, Cox CP, et al: Human serum acid labile factor is an acetylhydrolase that inactivates platelet-activating factor. Fed Proc 1983;42:3120–3126.
161 Wardlow ML, Cox CP, Meng KE, et al: Substrate specificity and partial characterization of the PAF-acylhydrolase in human serum that rapidly inactivates platelet activating factor. J Immunol 1986;136:3441–3447.
162 Stafforini PM, McIntyre TM, Carter ME, et al: Human plasma platelet-activating factor acetylhydrolase: association with lipoprotein particles and role in degradation of platelet-activating factor. J Biol Chem 1987;262:4215–4221.
163 Blank ML, Lee TC, Fitzgerald V, et al: A specific acetylhydrolase for 1-alkyl-2-acetyl-sn-glycero-3-phosphocholine (a hypotensive and platelet-activating lipid). J Biol Chem 1981;256:175.
164 Chilton FH, O'Flaherty JT, Ellis MJ, et al: Metabolic fate of platelet-activating factor in neutrophils. J Biol Chem 1983;258:6357.
165 Pieroni G, Hanahan DJ: Metabolic behavior of acetyl glyceryl ether phosphorylcholine on interaction with rabbit platelets. Arch Biochem Biophys 1983;224:485.
166 Touqui L, Jacquemin C, Vargaftig BB: Conversion of ^{3}HPAF acether by rabbit platelets is independent from aggregation: evidence for a novel metabolite. Biochem Biophys Res Commun 1983;110:890.
167 Chilton FH, O'Flaherty JT, Ellis JM, et al: Selective acylation of lyso platelet activating factor by arachidonate in human neutrophils. J Biol Chem 1983;258: 7263.
168 Malone B, Lee T, Snyder F: Inactivation of platelet activating factor by rabbit platelets. Lyso-platelet activating factor as a key intermediate with phosphatidylcholine as the source of arachidonic acid in its conversion to a tetraenoic acylated product. J Biol Chem 1985;260:1531.
169 Robinson M, Snyder F: Metabolism of platelet-activating factor by rat alveolar macrophages: lyso-PAF as an obligatory intermediate in the formation of alkylarachidonoyl glycerophosphocholine species. Biochim Biophys Acta 1985;837:52.
170 Haroldsen PE, Voelkel NF, Henson PM, et al: Metabolism of platelet-activating factor in the isolated perfused rat lung. J Clin Invest 1987;79:1860.
171 Camussi G, Tetta C, Bussolino F, et al: Mediators of immune complex-induced aggregation of polymorphonuclear neutrophils. II. Platelet-activating factor as the effector substance of immune-induced aggregation. Int Arch Allergy Appl Immunol 1981;64:25–31.
172 Lynch JM, Lotner GZ, Betz SJ, et al: The release of a platelet-activating factor by stimulated rabbit neutrophils. J Immunol 1979;123:1219–1225.
173 Clark PO, Hanahan DJ, Pinckard RN: Physical and chemical properties of platelet-activating factor obtained from human neutrophils and monocytes and rabbit neutrophils and basophils. Biochim Biophys Acta 1979;628: 69–74.
174 Lotner GZ, Lynch JM, Betz SJ, et al: Human neutrophil-derived platelet activating factor. J Immunol 1980;124:676–682.
175 Betz SJ, Henson PM: Production and release of platelet-activating factor (PAF); dissociation from degranulation and superoxide production in the human neutrophil. J Immunol 1980;125:1756–1762.

176 Camussi G, Aglietta M, Coda R, et al: Release of platelet activating factor (PAF) and histamine. II. The cellular origin of human PAF. Monocytes, polymorphonuclear neutrophils and basophils. Immunology 1981;42:191–197.
177 Sanchez-Crespo M, Alonso F, Figido J: Platelet-activating factor in anaphylaxis and phagocytosis. I. Release from human peripheral polymorphonuclears and monocytes during the stimulation by ionophore A23187 and phagocytosis but not from degranulating basophils. Immunology 1980;40:645–649.
178 Arnoux B, Durand J, Rigaud M, et al: Release of platelet-activating factors (PAF-acether) and arachidonic acid metabolites from alveolar macrophages. Agents Actions 1981;11:555–562.
179 Arnoux B, Duval D, Benveniste J: Release of platelet-activating factors (PAF acether) from alveolar macrophages by calcium ionophore A23187 and phagocytosis. Eur J Clin Invest 1980;10:437–442.
180 Mencia-Huerta J-M, Roubin R, Morgat J-L, et al: Biosynthesis of platelet-activating factor (PAF-acether). Formation of PAF-acether from synthetic substrates by stimulated murine macrophages. J Immunol 1982;129:804–809.
181 Camussi G, Mencia-Huerta JM, Benveniste J: Release of platelet activating factor and histamine. I. Effect of immune complexes, complement and neutrophils on human and rabbit mastocytes and basophils. Immunology 1977;33:523–528.
182 Bussolino F, Benveniste J: Pharmacological modulation of platelet-activating factor (PAF) release from rabbit leukocytes. I. Role of c-AMP. Immunology 1980;40:367–372.
183 Betz SJ, Lotner GZ, Henson PM: Generation and release of platelet-activating factor (PAF) from enriched preparations of rabbit basophils; failure of human basophils to release PAF. J Immunol 1980;125:2749–2754.
184 Camussi G, Tetta C, Coda R, et al: Release of platelet-activating factor in human pathology. I. Evidence for the occurrence of basophil degranulation and release of platelet-activating factor in systemic lupus erythematosus. Lab Invest 1981;44:241–246.
185 Lewis RA, Goetzl EJ, Wasserman SI, et al: The release of four mediators of immediate hypersensitivity from human leukemic basophils. J Immunol 1975;114:87–92.
186 Mencia-Huerta JM, Lewis RA, Razin E, et al: Antigen-initiated release of platelet-activating factor from mouse bone marrow-derived mast cells sensitized with monoclonal IgE. J Immunol 1983;131:2958–2963.
187 Schleimer RP, MacGlashan DW, Peters, SP, et al: Characterization of inflammatory mediator release from purified human lung mast cells. Am Rev Respir Dis 1986;133:614–618.
188 Ninio E, Joly F, Hieblot C, et al: Biosynthesis of PAF-acether. IX. Role of phosphorylation-dependent activation of acetyltransferase in antigen-stimulated mouse mast cells. J Immunol 1987;139:154–160.
189 Camussi G, Aglietta M, Malavasi F, et al: The release of platelet-activating factor from human endothelial cells in culture. J Immunol 1983;131:2397–2401.
190 Prescott SM, Zimmermann GA, McIntyre TM: Human endothelial cells in culture produce platelet-activating factor (1-alkyl-2-acetyl-sn-glycero-3-phosphocholine) when stimulated with thrombin. Proc Natl Acad Sci USA 1984;81:3534–3539.
191 McIntyre TM, Zimmermann GA, Satoh K, et al: Cultured endothelial cell synthesize

both platelet-activating factor and prostacylin in response to histamine, bradykinin and adenosine triphosphate. J Clin Invest 1985;76:271–275.
192 Namm DH, High JA: Thrombin-induced formation of a polar lipid material within platelets that possesses platelet-activating activity. Thromb Res 1980;20:285–289.
193 Chingnard M, LeCouedic JP, Tence M, et al: The role of platelet-activating factor in platelet aggregation. Nature 1979;279:799–804.
194 Chingnard M, Vargaftig BB, Benveniste J, et al: L'agrégation plaquettaire et le platelet-activating factor. J Pharmacol 1980;11:371–376.
195 Vargaftig BB, Chingnard M, LeCouedic JP, et al: One, two, three or more pathways for platelet aggregation. Acta Med Scand 1980;642(suppl):23–29.
196 Chingnard M, LeCouedic JP, Vargaftig BB, et al: Platelet-activating factor (PAF-acether) secretion from platelets. Effect of aggregating agents. Br J Haematol 1980;46:455–460.
197 Chap H, Mauco G, Simon MF, et al: Biosynthetic labelling of platelet activating factor from radioactive acetate by stimulated platelets. Nature 1981;289:312–316.
198 Alam I, Smith JB, Silver MJ: Human and rabbit platelets form platelet-activating factor in response to calcium ionophore. Thromb Res 1983;30:71–77.
199 Oda M, Satouchi K, Yasunaga K, et al: Production of platelet activating factor by washed rabbit platelets. J Lipid Res 1985;26:1194–1298.
200 Kravis TC, Henson PM: IgE-induced release of platelet activating factor from rabbit lung. J Immunol 1975;115:1677–1682.
201 Pinckard RN, Farr RS, Hanahan DJ: Physiochemical and functional identity of rabbit platelet-activating factor (PAF) released in vivo during IgE anaphylaxis with PAF released in vitro from IgE sensitized basophils. J Immunol 1979;123:1847–1852.
202 Valone FH, Goetzl EJ: Immunologic release in the rat peritoneal cavity of lipid chemotactic and chemokinetic factors for polymorphonuclear leukocytes. J Immunol 1978;120:102–107.
203 Valone FH, Whitmer DI, Pickett WC, et al: The immunological generation of a platelet-activating factor and a platelet-lytic factor in the rat. Immunology 1979;37:841–846.
204 Camussi G, Tetta C, Deregibus MC, et al: Platelet-activating factor (PAF) in experimentally-induced rabbit acute serum sickness. Role of basophil-derived PAF in immune complex deposition. J Immunol 1982;128:87–90.
205 Lynch JM, Henson PM: The intracellular retention of newly synthesized platelet-activating factor. J Immunol 1986;137:2653–2658.
206 Cluzel M, Undem BJ, Chilton FH: Release of platelet activating factor and the metabolism of leukotriene B_4 by the human neutrophil when studied in a cell superfusion model. J Immunol, in press.
207 Robinson M, Blank ML, Snyder F: Acylation of lysophospholipids by rabbit alveolar macrophages: specificities of CoA-dependent and CoA-independent reactions. J Biol Chem 1985;260:7889–7894.
208 Kramer RM, Patton GM, Pritzker CR, et al: Metabolism of platelet activating factor in human platelets: transacylase-mediated synthesis of 1-0-alkyl-2-arachidonoyl-sn-glycero-3-phosphocholine. J Biol Chem 1984;259:13316–13321.
209 Chilton FH, Connell TR: 1-Ether-linked phosphoglycerides: major endogenous sources of arachidonate in the human neutrophil. J Biol Chem 1988;263:5260–5265.

210 Ramesha CS, Pickett WC: Platelet-activating factor and leukotriene biosynthesis is inhibited in polymorphonuclear leukocytes depleted of arachidonic acid. J Biol Chem 1986;261:7592–7599.
211 McIntyre TM, Reinhold SL, Prescott SM, et al: Protein kinase C activity appears to be required for the synthesis of platelet-activating factor and leukotriene B_4 by human neutrophils. J Biol Chem 1987;262:15370–15375.
212 Chilton FH: Potential phospholipid source(s) of arachidonate used for the synthesis of leukotrienes by the human neutrophil. Biochem J 1989;258:327–333.
213 Shaw JO, Printz MP, Hirabayashi K, et al: Role of prostaglandin synthesis in rabbit platelet activation induced by basophil-derived platelet-activating factor. J Immunol 1978;121:1939–1945.
214 Shaw JO, Klusick SJ, Hanahan DJ: Activation of rabbit platelet phospholipase and thromboxane synthesis by 1-0-hexadecyl/octadecyl-2-sn-glyceryl-3-phosphocholine (platelet-activating factor). Biochim Biophys Acta 1981;663:222–229.
215 McManus LM, Shaw JO, Pinchard RN: Thromboxane B_2 release during IgE anaphylaxis in the rabbit. J Immunol 1980;125:1950–1956.
216 Chilton FH, O'Flaherty JT, Walsh CE, et al: Platelet-activating factor: stimulation of the lipoxygenase pathway in polymorphonuclear leukocytes by 1-0-alkyl-2-0-acetyl-sn-glycero-3-phosphocholine. J Biol Chem 1982;257:5402–5408.
217 Voelkel NF, Worthen S, Reeves JT, et al: Nonimmunological production of leukotrienes induced by platelet-activating factor. Science 1982;218:286–289.
218 Lin AH, Morton DR, Gorman RR: Acetyl glyceryl ether phosphorylcholine stimulates leukotriene B_4 synthesis in human polymorphonuclear leukocytes. J Clin Invest 1982;70:1058–1063.
219 Schlondorff D, Satriano JA, Hagege J, et al: Effect of platelet-activating factor and serum-treated zymosan on prostaglandin E_2 synthesis, arachidonic acid release, and contraction of cultured rat mesangial cells. J Clin Invest 1984;73:1227–1232.
220 Kawaguchi H, Yasuda H: Effect of platelet-activating factor on arachidonic acid metabolism in renal epithelial cells. Biochim Biophys Acta 1986;875:525–531.
221 Yousufzai SYK, Abdel-Latif AA: Effects of platelet-activating factor on the release of arachidonic acid and prostaglandins by rabbit iris smooth muscle. Biochem J 1985; 228:697–699.
222 Hsueh W, Gonzalez-Crussi F, Arroyave JL: Release of leukotriene C_4 by isolated, perfused rat small intestine in response to platelet-activating factor. J Clin Invest 1986;78:108–114.
223 Billah MM, Bryant RW, Siegel MI: Lipoxygenase products of arachidonic acid modulate biosynthesis of platelet-activating factor (1-0-alkyl-2-acetyl-sn-glycero-3-phosphocholine) by human neutrophils via phospholipase A_2. J Biol Chem 1985; 260:6899–6904.
224 Tessner TG, O'Flaherty JT, Wykle RL: Stimulation of platelet-activating factor synthesis by a nonmetabolizable bioactive analog of platelet-activating factor and influence of arachidonic acid metabolites. J Biol Chem 1989;264:4794–4799.
225 Doebber TW, Wu MS: Platelet-activating factor (PAF) stimulates the PAF-synthesizing enzyme acetyl-CoA: 1-alkyl-sn-glycero-3-phosphocholine-acetyl transferase and PAF synthesis in neutrophils. Proc Natl Acad Sci USA 1987;84:7557–7563.

Floyd H. Chilton, PhD, Department of Medicine, Division of Clinical Immunology, School of Medicine, Baltimore, MD 21239 (USA)

Waksman BH (ed): 1939–1989: Fifty Years Progress in Allergy.
Chem Immunol. Basel, Karger, 1990, vol 49, pp 206–235

Antitumoral and Other Biomedical Activities of Synthetic Ether Lysophospholipids

Paul Gerhard Munder, Otto Westphal

Max-Planck-Institut für Immunobiologie, Freiburg i.Br., FRG

Introduction

In the late 1950s and early 1960s a research group around H. Fischer and P.G. Munder at the Max-Planck-Institut fuer Immunobiologie, Freiburg, was working on pathomechanisms of silicosis in connection with their general studies on the activation of macrophages [66]. They found that during phagocytosis of silicogenic quartz particles, phospholipase A_2 is activated, leading to a significant over-production of lysolecithin and its accumulation in the macrophage's cell membrane [68]. Because of its influence on membrane fluidity and biochemical membrane parameters [58, 67], the authors suggested that lysolecithin may be a common denominator of macrophage activation. And, indeed, they found that all known immune adjuvants would induce similar phospholipase A_2 activation in macrophages [69, 73]. Further, small amounts of exogenous lysolecithin strongly enhanced the phagocytic activity of peritoneal macrophages in vitro and in vivo [31]. These findings prompted a general survey on the possible role of lysolecithin in immune phenomena [56, 70–72].

In vivo lysolecithin – being a highly active 'detergent' [101] – has a short half-life time. It is biologically 'inactivated' either by retransformation into lecithin under the catalytic activity of acyltransferase, or by metabolization to glycero-phosphocholine through the action of phospholipase B [46, 65, 97]. This dynamic state of lysolecithin in vivo is shown in figure 1 [80]

In the early 1960s, we decided therefore to synthesize lysolecithin analogs with longer half-life times in vivo, and to test them in comparison to natural lysolecithin for biological activities, especially in suitable immune systems [102]. To our surprise, until that time (about 1965), no purely organic syntheses for lysolecithins or of any analogs had been described.

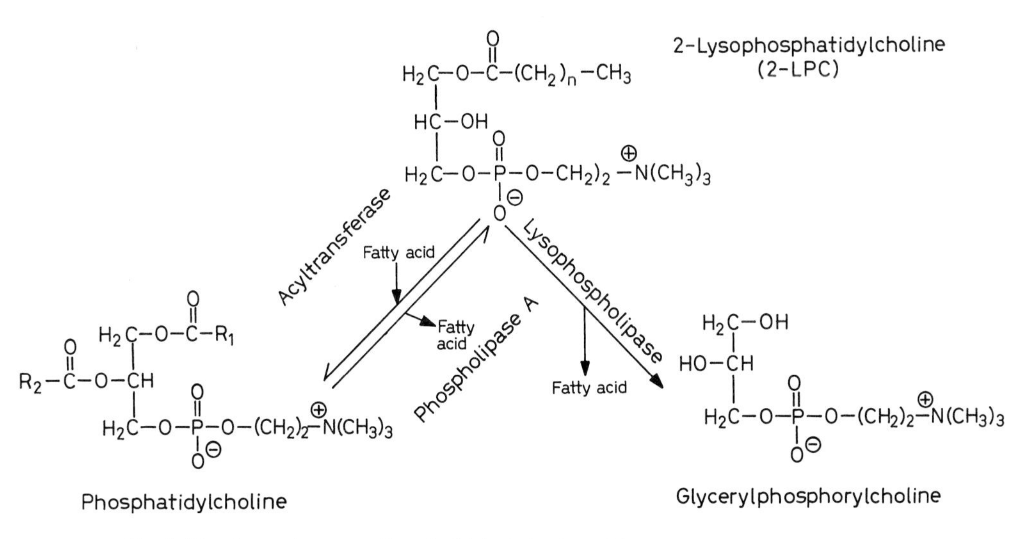

Fig. 1. The dynamic state of lysolecithin (after Lands).

Consequently, chemical diploma and doctoral candidates in O. Westphal's department undertook the elaboration of relevant synthetic procedures [9, 39, 100]. In the molecule of lysolecithin (fig. 2 (I)), especially the positions 1 and 2 in the glycerol backbone were taken into consideration. For example, we replaced the ester-function (position 1) and/or the hydroxyl-function (position 2) by other substituents in order to obtain analogs that would be no substrates for one or other, or both relevant enzyme systems of the Lands cycle (fig. 1). We thus replaced the ester in (I) by long-chain ether, ((II), (III), (V)); the >CHOH group was transformed into short-chain ether, such as $>CH\text{-}OCH_3$ (fig. 2 (III)). We also synthesized 1-O-acyl-(IV) and 1-O-alkyl-(V)-2-deoxy-lysolecithin analogs [53] (fig. 3). Since the carbon atom in position 2 of the glycerol backbone is asymmetric, synthetic compounds are being obtained as racemats (D-, L-), whereas 2-deoxy- are optically inactive. Asymmetric syntheses, beginning with structurally pure 3-carbon compounds like D- or L-serine, D- or L-glyceric acids, were later also performed [34–37].

Some of the ether analogs of lysolecithin so synthesized turned out to be potent immune modulators [56, 71, 73, 75]. In the course of these studies, Munder and co-workers, using in vitro and in vivo tumor models, found that some alkyl-lysophospholipids (ALP) displayed strong antitumor activities (70, 71, 74, 90). It became clear that we were dealing with a new class of immune modulating substances exerting antitumoral activities. Syntheses of

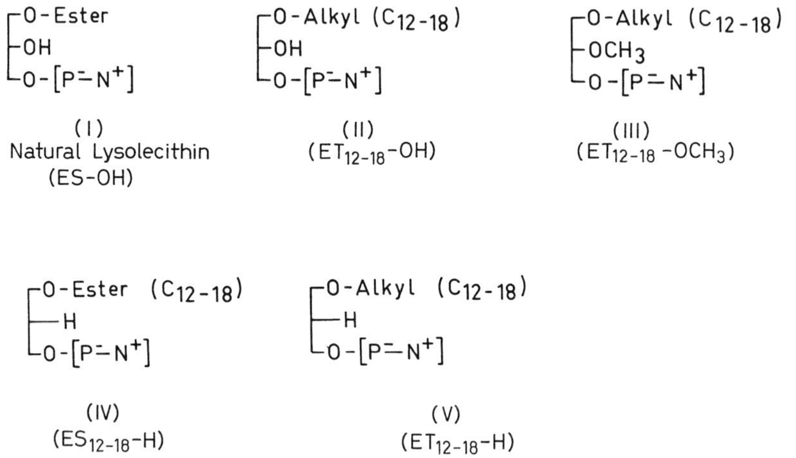

(I) Natural Lysolecithin (ES-OH): O-Ester / OH / O-[P⁻-N⁺]

(II) (ET_{12-18}-OH): O-Alkyl (C_{12-18}) / OH / O-[P⁻-N⁺]

(III) (ET_{12-18}-OCH_3): O-Alkyl (C_{12-18}) / OCH_3 / O-[P⁻-N⁺]

(IV) (ES_{12-18}-H): O-Ester (C_{12-18}) / H / O-[P⁻-N⁺]

(V) (ET_{12-18}-H): O-Alkyl (C_{12-18}) / H / O-[P⁻-N⁺]

Fig. 2. Schematic representation of some early synthetic lysolecithin analogs. –[P⁻ –N⁺] stands for –phosphocholine $-P(=O)(O^-)$ $O\text{-}CH_2\text{-}CH_2\text{-}N^+(CH_3)_3$.

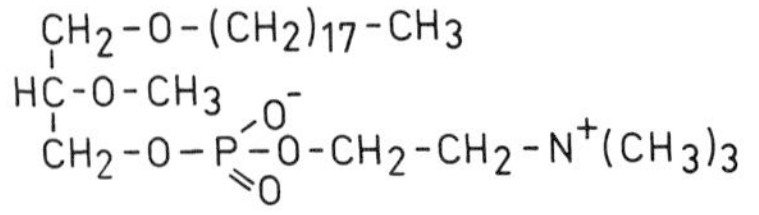

$$CH_2\text{-}O\text{-}(CH_2)_{17}\text{-}CH_3$$
$$HC\text{-}O\text{-}CH_3$$
$$CH_2\text{-}O\text{-}P(=O)(O^-)\text{-}O\text{-}CH_2\text{-}CH_2\text{-}N^+(CH_3)_3$$

Fig. 3. 1-Octadecyl-2-O-methyl-glycero-3-phosphocholine (ET_{18}-OCH_3).

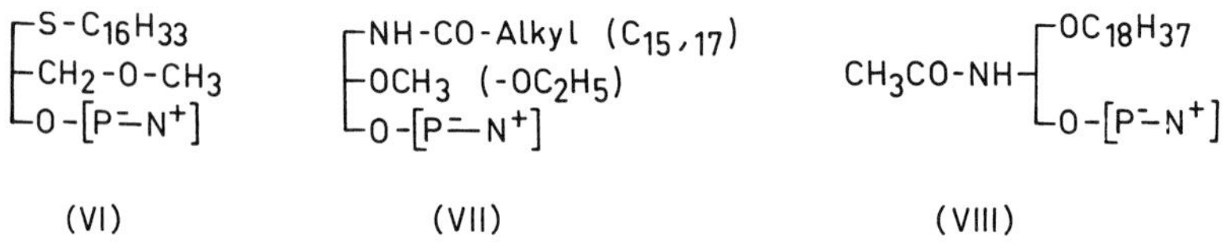

(VI): S-$C_{16}H_{33}$ / CH_2-O-CH_3 / O-[P⁻-N⁺]

(VII): NH-CO-Alkyl ($C_{15,17}$) / OCH_3 (-OC_2H_5) / O-[P⁻-N⁺]

(VIII): CH_3CO-NH– / $OC_{18}H_{37}$ / O-[P⁻-N⁺]

Fig. 4. Examples of lysolecithin analogs with hetero atoms (S, N).

model compounds worldwide were directed towards the aim of performing studies on structure/activity relations [8, 57, 64].

In many comparative tests one derivative, 1-O-octadecyl-2-O-methyl-gly-cero-3-phosphorylcholine, ET_{18}-OCH_3 (fig. 2, (III); fig. 3), turned out to be the most active antitumoral compound of our series.

Thus in the following period we concentrated our studies mainly on ET_{18}-OCH_3 and, indeed, it became a kind of world-wide standard in experiments and trials on antitumoral activities of such ether lipids. In 1987 a book was published containing a chapter about novel phospholipid analogs as antitumor agents [63] in which the authors stated: . . . "Thus ET_{18}-OCH_3 has become the effective standard against which other analogs of this type are evaluated for antineoplastic activity."

Since ET_{18}-OCH_3 has been introduced into antitumor research, many more or less related lysolecithin analogs have been synthesized. Certainly only some of them and their manufacture were published. At present, several hundreds of compounds of this series are known which is probably a minimum of what has really been achieved. Variations leading to promising types of antitumoral (and other) activities contain, for example, sulfur instead of the ether oxygen [28, 64] (fig. 4 (VI)) or amido nitrogen in the 1 position [34, 37, 62] (VII) or in the 2 position (VIII).

Of ET_{12}-H an isosteric compound, in which the ether oxygen -O-, in (V,C_{12}) is replaced by -CH_2 hexadecyl-phosphocholine (HPC), was proposed as an antitumor agent [54] – a structure, however, already rather remote from ether lysolecithin analogs. So far it is not known whether HPC has any biomedical advantages over ET_{12}-H.

Chemical and Biochemical Aspects

Synthesis of ET_{18}-OCH_3

The first synthesis of ET_{18}-OCH_3 was described by G. Kny in 1968 [53] on the basis of the experience of Arnold and Weltzien with 1-O-alkyl- and 2-O-methyl derivatives of glycerol [9]. In principle, 2-O-methyl-glycerol (fig. 5 (IX)) was transformed into 1-O-octadecyl-2-O-methyl-glycerol (X) into which the phosphorycholine radical was finally introduced by the reaction with phosphodichloric acid-2-bromoethylester and trimethylamine according to the method of Hirt and Berchtold [47] which for a long time was, and still is, widely applicable in phospholipid syntheses [48].

Principally the same procedures were followed for the synthesis of ET_{18}-OCH_3 by Berchtold [11] and Tsushima et al. [93]. See also [48]. With Berchtold's procedure a pure and chromatographically uniform end product is being obtained. However, in the meantime, more efficient procedures were developed after the introduction of the cyclic 2-chloro-2-oxo-1.3.2-oxaphospholane (fig. 6 (XII)) which is obtained by the reaction of the respective phosphorochloridite (XI) with oxygen [38].

2-Chloro-2-oxo-1.3.2-oxaphospholane (XII) reacts with primary alcoholic functions to give the respective organic phosphoric ring compounds (fig. 7), as first shown by Chabrier et

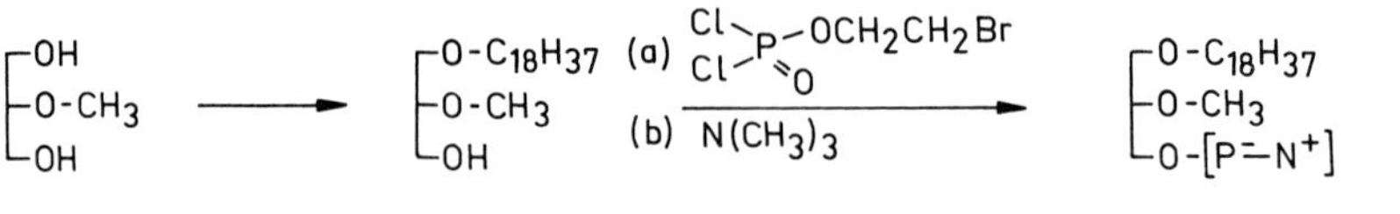

(IX) (X) (III, C_{18}) Et_{18}-OCH_3

Fig. 5. Synthesis of ET_{18}-OCH_3[Kny] [53].

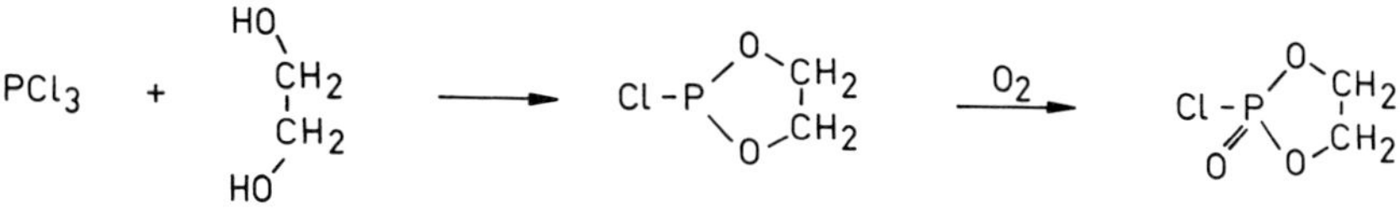

(XI) (XII)

Fig. 6. Synthesis of the 2-chloro-2-oxo-1.3.2-oxaphospholane reagent.

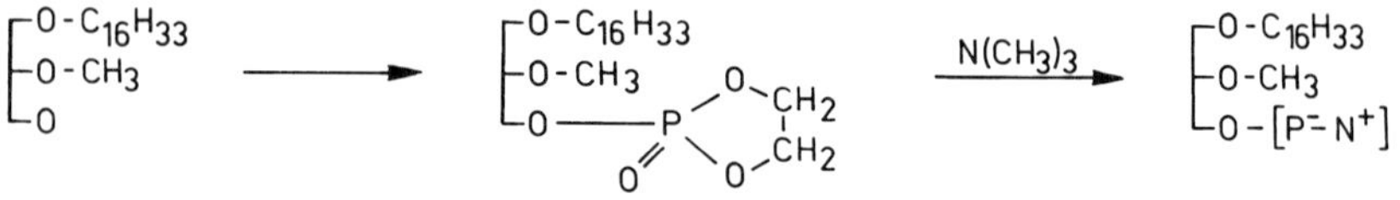

(XIII) (XII) (XIV) (III, C_{16}) ET_{16}-OCH_3

Fig. 7. Synthesis of ET_{16}-OCH_3.

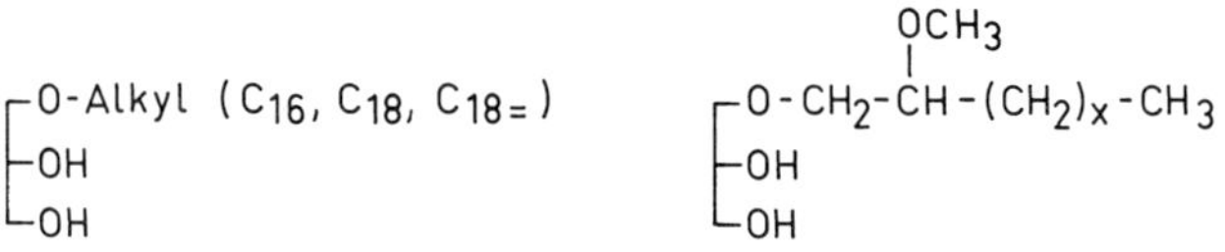

Long chain Glyceryl Ethers — 2-Methoxy-substituted Glyceryl Ethers

Fig. 8. Glyceryl ethers derived from fish liver oil.

al. [33, 77, 79]. The reagent has since been widely used for phospholipid syntheses. As an example, the stero-specific synthesis of the ET_{16}-analog (III, C_{16}) of ET_{18}-OCH_3 (fig. 7) was performed by Bhatia and Hajdu [23] with an overall yield of 92% of the pure end product starting from the glycerol-diether (XIII).

The same procedure can be applied for the synthesis of ET_{18}-OCH_3. The authors of the above-mentioned publication [23] claim that employing chlorodioxaphospholane, according to their experience, should pave the way to many new alkoxyphospholipids. Also Marx et al. [62], comparing all known methods, claim that 'the cyclic phosphorylating agent (XII) which is much less reactive' – than $POCl_3$ + glycols [40] – 'gave by far the best results'. The more aggressive $POCl_3$ method is applicable for the introduction of $-[P^- -N^+]$ into relatively stable primary alcoholic functions [48, 103]. Thus, the use of the prefabricated pure phospholane (XII) is more widely applicable and seems to lead to pure end products in comparatively high yield [23, 28, 29, 62].

Another crucial step comprises the introduction of the long-chain alkyl group. Originally alkyl iodides and bromides were used. They can be replaced by the corresponding (and cheaper) long-chain alkyl-methansulfonates [10, 23].

Thus, since its introduction as a biologically interesting substance, the synthesis of ET_{18}-OCH_3 and related compounds has continuously been improved.

Natural Glycerol Ethers

Long-chain alkyl and alkenyl ethers of glycerol are known to occur in nature. The first identified were the plasmalogens as being derivatives of Alk-1-enyl-glycero-phosphorylcholines [45]. 1-Alkyl-2.3-diacylglycerols were found in fish liver oil, and 1-alkyl-glycerols were prepared from the non-saponifiable fraction. Three characteristic compounds were isolated, analyzed and later synthesized: 1-O-octadecyl-(batylalcohol), 1-O-hexadecyl-(chimylalcohol) and 1-O-octadecenyl-glycerol (selachylalcohol) [44]. 1-O-(2-methoxy)alkyl-glycerols were found in larger quantities in Greenland shark liver oil [26, 27] (fig. 8).

In extensive studies since 1958 [25, 27, 59], several clinical research groups have raised biomedical interest in these substances. According to Hallgren [44], these compounds exhibit bacteriostatic properties, hemopoietic effects [78], and neuromuscular activities; they have been reported to protect against radiation injury [25, 30]. In high doses they inhibit neoplastic growth [1, 25–27, 32] and accelerate wound healing.

In relation to biological activities of ET_{18}-OCH_3 and other ether-lysolecithin analogs, the reported effects of natural ether glycerols on tumor growth and metastasis and on protective effects against irradiation [25] are of interest. For the observed beneficial effects, the glycerylethers are being administered orally with food in doses of the order of 500–1,000 mg/day over long periods of time [44]. Whether the glycerylethers have a direct action or whether, in vivo, they are first biochemically transformed into the active principle, might be worthwhile investigating. According to F. Snyder [81, 83, 84] the following sequence of enzymatic reactions would be relevant: 1. phosphorylation via an ADP-dependent phosphotransferase (kinase). The 1-alkyl-2-lyso-sn-glycero-3-phosphate obtained can be 2. acylated in the 2-position to form the alkyl analog of phosphatidic acid (1-alkyl-2-acyl-sn-glycero-3-P). The latter is then 3. dephosphorylated to provide alkyl-acyl-glycerols which are 4. substrates for choline phosphortransferase. Alkyl-acyl-phosphocholines could then 5. be converted to alkyl-lysophospholipids by a phospholipase A_2.

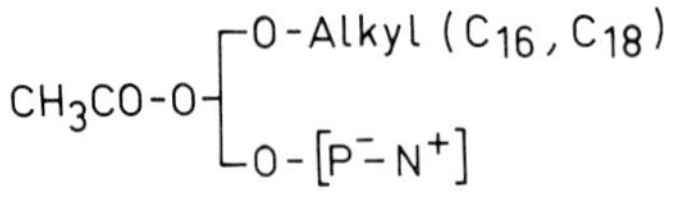

$$CH_3CO\text{-}O\text{-}\left[\begin{array}{l}O\text{-Alkyl }(C_{16}, C_{18})\\ \\ O\text{-}[P^{-}\text{-}N^{+}]\end{array}\right.$$

(XV)

Fig. 9. The structure of platelet activating factor (PAF).

Of great importance was the finding in 1979 that the platelet-activating factor (PAF), which had been extensively studied by Benveniste et al. [12], is an alkyl-phospholipid and could be identified as 1-O-alkyl-2-O-acetyl-sn-glycero-3-phosphorylcholine (fig. 9 (XV)) [12, 81–83].

This prompted a wave of additional interest in ether phospholipids, their synthesis, biosynthesis, and activities in biological systems [12, 76, 81–83]. The number of synthetic PAF-like and anti-PAF compounds soon exceeded even related compounds in the antitumor field. Most of these new substances were also routinely tested in suitable in vitro and in vivo systems. Research on the antitumor activities of ether lipids was certainly greatly stimulated by the new 'PAF era'. We may remind the reader that ET_{18}-OCH_3 had been conceived long before the structure of PAF was established. PAF, in tolerable doses, has no antitumor activity in vitro or in vivo, and ET_{18}-OCH_3 does not exert significant PAF or anti-PAF activity in vivo. Thus only by chance is ET_{18}-OCH_3 a 2-O-methyl 'analog' of PAF. For reviews on PAF and related ether lipids see [24, 82].

The Fate of ET_{18}-OCH_3 in vivo

In 1964, Kennedy et al. [92] discovered an enzyme – now termed O-alkyl-cleavage enzyme or O-alkyl-glycerol-monooxygenase – which catalyses the oxidation of glycerolethers to the respective long-chain aldehydes which, in turn, are then transformed into the corresponding fatty acids. Snyder, who has elaborated on the enzyme reactions involved in the synthesis of ethergly-cero-phosphocholines and their metabolism, showed that in certain tumor cell membranes the O-alkyl-cleavage enzyme is either lacking or of very low activity [85, 88]. This finding prompted investigations into the biochemistry of ether lipids, especially in tumor cells. Snyder et al. [86–88] had found a relative accumulation of ether lipid in tumor cell membranes. This seemed to provide an explanation for the selective cytotoxicity of ether phospholipids, like ET_{18}-OCH_3, for tumor cells. However, more recently Snyder has shown 'that normal (non-cancer) cells that are resistant to the cytotoxicity of the methoxy-analog (ET_{18}-OCH_3) may also have low levels of the alkyl-cleavage enzyme, like cancer cells that are highly sensitive to the killing of the methoxy-analog' [84].

Recently, Kötting et al. [54, 94], based on indirect evidence, proposed a phospholipase C-dependent cleavage of ET_{18}-OCH_3 as being required for its tumor-cytotoxic activity [43]. Thus, ET_{18}-OCH_3 and similar compounds would be a 'prodrug', the active principle being 1-octadecyl-2-O-methyl-glycerol (X). This compound has been found in trace amounts as an ET_{18}-OCH_3 metabolite in various tumor cells [95, 96, 99]. However, neither we (submitted for publication) nor Snyder et al. [84, 85] were able to demonstrate any antitumor activity of octadecyl-methyl-glycerol in vitro or in vivo comparable to ET_{18}-OCH_3.

Compound X (fig. 5) can be taken as an O-methyl-derivative of batylalcohol (fig. 8). For the described activities of batylalcohol [44] in man, considerably higher doses have to be given (orally) compared to ET_{18}-OCH_3. In their relevant report, Snyder et al. [95] finally state: ... 'We believe these data rule out any role of a phospholipase C activity as a factor in the cytotoxic mechanisms of the antitumor ether-linked phospholipids'. In the same publication [95] the authors stress their earlier concept about the selective tumor-cytotoxic action of ET_{18}-OCH_3 stating: ... 'the inhibition of the uptake of ^{3}[H]-choline observed in the undifferentiated cells treated with alkylmethoxy-GPC (ET_{18}-OCH_3) and the lack of inhibition in the differentiated cells further supports that ... the antitumor action of alkylmethoxy-GPC is, at least in part, due to an impaired transport of molecules across the membrane of sensitive cells'.

At present we are inclined to believe that it is the whole etherphospholipid molecule which exerts the selective antitumoral effects. The further elucidation of the metabolism of antitumorally active lysolecithin analogs in higher animals and the human, which were conceived not to be substrates of the main enzyme systems of the Lands cycle (fig. 1), therefore, is still open to extended biochemical studies.

Results in vitro

Antitumoral Activity in vitro

The antitumoral effects of etherlysophospholipids were discovered when the influence of these compounds on cell mediated immunity was studied. The interaction of macrophages and/or lymphocytes with tumor cells was investigated in the presence of various etherlysophospholipids [3, 4, 7, 15, 16, 17]. Although a remarkable enhancement of the antitumoral activity of macrophages could be observed after incubation with e.g. ET_{18}-OCH_3, the tumor cells in the control groups (but not normal cells) were also destroyed by

ET_{18}-OCH_3 in a dose dependent manner in the absence of immunocompetent cells. Thus these compounds also have a direct cytotoxic action on tumor cells [5, 6, 14, 18–20, 49, 50, 52, 55, 60, 89, 91].

Direct Cytotoxicity

Table 1 summarizes the results of the biological activity of 2-lysophosphatidyl-choline (lysolecithin), of synthetic rac.1-octadecyl-2-methoxy-glycero-3-phosphoryl-choline (ET_{18}-OCH_3) and ET_{18}-OH on various established tumor cell lines. Normal cells, like 3T3 or ST-V (table 1), are not destroyed.

Structure/Activity Studies

For the cytotoxic activity of a given lysophospholipid (figs. 2, 3, 4) on malignant cells certain prerequisites were found to be of importance: substitution of O-acyl in the 1-position of lysolecithin by the non-ester (i.e. ether-) linked alkyl groups with chain lengths preferably between C_{12} and C_{18}; substitution of the OH group in the 2-position by short-chain alkyl groups (for example -OCH_3) or by -H (2-deoxy-). In principle, the two positions can also be interchanged. A certain proportion between the lipophilic (long-chain alkyl) and the lyophilic (phosphocholine) characteristic of the amphipathic molecule, is essential for biologic activity in relation to tolerability of etherlysophospholipids. As an example, ET_{12}-H (table 2) is about 10 times less toxic, but also about 10 times less tumoricidal in various systems.

Tumoricidal analogs of natural lysolecithin lead to inhibition of the phospholipid metabolism primarily in the plasma membrane of malignant cells. As etherlysophospholipids are very slowly, if at all, degraded by tumor cells after they have been taken up from carrier proteins like serum albumin, they accumulate in the phospholipid bilayer. This accumulation leads to a breakdown of the permeability barrier of the cell as the adsorbed etherlipids, having only one long aliphatic side chain on the glycerol backbone, are highly surface-active. While disrupting the ordered bilayer structure of the phospholipids in the cellular membrane, it can be assumed that clusters (micelles) of adsorbed etherlipids will give rise to holes in the tumor cell membrane. Morphologically, this process can, indeed, be followed by scanning-electron-microscopy. A first alteration is the loss of microvilli beginning about 6 h after the addition of active compounds. After 24 h all villi have disappeared and in some areas small defects are about to develop. After 48 to 72 h large parts of the cellular membrane have been shed accompanied by cell death. These morphological changes are concentration-dependent. The process

Table 1. Direct cytotoxicity of various lysophospholipids: cells were incubated in DMEM + 10% fetal calf serum in the presence of 15 μg of lysophospholipids for 48 h; each value is a mean of 8 cultures (SD 9%)

Tumor cells		Lysolecithin	ET_{18}-OH	ET_{18}-OCH_3
EL4		110.4[1]	54.9	0.2
L 1210		114.6	10.6	0.1
P815		107.0	20.2	1.2
Abelson	8.1	103.8	9.1	0.2
WEH1	7.1	97.5	92.4	31.6
–	22.1	88.9	24.1	0.2
–	231.1	108.2	89.1	1.4
–	279.1	103.8	3.4	0.5
EBV-transformed human cells				
Raji		90.6	0.6	0.1
Kaplan		95.7	62.2	1.6
B 95/8		109.3	0.1	0
Virus-transformed murine fibroblasts				
3T3[2]		104.0	109.7	79.8
SV 3T3[3]		107.2	101.5	5.1
ST-U[2]		107.0	90.8	90.3
FV-STU[4]		109.2	91.1	19.5

[1]Percentage of thymidine incorporation into tumor cells in the absence of lysophospholipids. [2]Embryonic fibroblasts. [3]SV-40-transformed fibroblasts. [4]Fried leukemia virus-transformed fibroblasts.

Table 2. Direct cytotoxicity of various lysophospholipids

	Lysophospholipid μg/ml	% ^{3}H-thymidine incorporated in Metha-A cells in the presence of lysophospholipids after		
		24h	48 h	72 h
Control		100	100	100
Lysophosphatidylcholine	10	102 ± 13	101 ± 12	90 ± 5
	5	98 ± 11	97 ± 12	93 ± 10
ET_{12}-H	20	92 ± 8	65 ± 7	42 ± 7
	10	80 ± 24	106 ± 23	96 ± 12
	5	110 ± 16	127 ± 27	103 ± 14
ES_{18}-H	10	104 ± 5	98 ± 3	198 ± 12
ET_{18}-H	5	34 ± 14	9 ± 3	3 ± 1
ET_{18}-OCH_3	5	32 ± 13	16 ± 7	2 ± 0.9
	3	43 ± 8	39 ± 13	25 ± 9
	2	67 ± 14	54 ± 16	41 ± 14

described has been observed with 10 µg ET_{18}-OCH_3/10^5/ml freshly explanted human leukemic cells. The culture medium was supplemented with 10% ABO-serum and the usual antibiotics.

During the first 24 h the morphological changes can be correlated to various biochemical alterations. Thus, thymidine incorporation and oxygen consumption start to decrease after 5 to 6 h; lactatedehydrogenase, increasingly, leaks out of the cytoplasm with concommitant trypan-blue uptake. The described phenomena vary quantitatively somewhat from tumor to tumor but the final result of cellular destruction will be the same. This breakdown of the permeability barrier can only be observed when the adsorbed lysophospholipids have the structure of an etherlysophospholipid due to their stability towards the usual metabolizing enzymes – in contrast to the parent lysolecithin, 2-LPC, which is rapidly metabolized after its adsorption onto cellular membranes or after intracellular uptake. Thus, etherlysophospholipids remain unchanged in tumor cells. The disarrangement of the phospholipid bilayer caused by the insertion of etherlipid is probably the main cause of the destruction of tumor cells. However, as the normal phospholipids of cellular membranes are continuously turned over to renew the phospholipid moieties (fig. 1) [42, 46, 65], the presence of nonmetabolizable etherlysophospholipids obviously leads to an interference with this vital dynamic cellular process. From the scheme shown in fig. 1 it can be seen that synthetic etherlysophospholipids, like ET_{18}-OCH_3, are no substrates for at least two enzymes of this circular renewing pathway (Lands pathway) [65]. In addition, the lysolecithin analogs inhibit the action of these membrane enzymes, thus interfering with the normal phospholipid cycle. It could be shown that lysophospholipase is inhibited by etherlipid compounds [63].

Moreover, if the OH group in the 2-position is blocked, the (re)acylation of physiological lyso-compounds by an acyltransferase is strongly inhibited while their formation by phospholipase A_2 continues. The inhibition of lysophospholipase and acyltransferase versus the normal unimpaired activity of phospholipase A_2 leads to a marked transient increase of potentially cytotoxic endogenous (ester)-lysophospholipids which – locally – enhance, to some extent, the cytotoxic effect of the etherphospholipid. In addition, the normal activity of the phospholipase A_2 leads to a breakdown and decrease of cellular lecithin which cannot be compensated for by reacylation of the lysolecithin formed. All these alterations together finally contribute to the death of the tumor cell.

There is no direct evidence as to whether phospholipase C or D play any significant role in the metabolism of etherlysophospholipids in tumor cells as

Table 3. Meth-A growth inhibition by the combined application of ET_{18}-OCH_3 and cis-platinum

	Day 7		Day 21		Day 28		Weight[1]		Tumor free/ total
	vol[2]	%	vol[2]	%	vol[2]	%	g	%	
Control	13.503	100	47.262	100	142.816	100	22.3	100	0/10
12.5 μg ilmofosin[3]	10.980	81	11.967	25	47.508	33	7.6	34	0/10
12.5 μg ET_{18}-OCH_3	7.668	57	9.939	21	36.205	25	6.2	28	1/10
5μg cis-platinum	11.47	85	16.679	35	65.729	46	10.8	48	1/10
10 μg cis-platinum	10.915	81	12.618	27	50.871	36	7.3	33	1/10
6 μg cis-platinum + 12.5 μg ilmofosin	7.873	58	4.529	10	16.631	12	2.5	11	6/10
10 μg cis-platinum + 12.5 μg ilmofosin	8.087	60	6.250	13	26.192	18	3.8	17	5/10
5 μg cis-platinum + 12.5 μg ET_{18}-OCH_3	5.992	44	9.650	20	34.330	24	5.7	26	5/10
10 μg cis-platinum + 12.5 μg ET_{18}-OCH_3	9.223	68	6.103	13	21.432	15	3.6	16	7/10

[1]Weight of tumors excised after death at day 28
[2]Tumor volume/group of 10 mice.
[3]See figure 4 (VI)

proposed by Unger et al [94]. As already mentioned, no products were hitherto known to be formed by the activity of phospholipase C and/or D, exerting any significant tumor cytotoxicity ('prodrug' phenomenon; see [54]).

The available data indicate that the target for etherlysophospholipids is the tumor cellular membrane. This is unique because all known antineoplastic agents act via the nuclear or cytoplasmic apparatus of the tumor cell. Consequently, etherlysophospholipids are not mutagenic. As etherlysophospholipids increase the permeability of the cellular membrane of tumors one can envisage an antitumor therapy combining etherlysophospholipids (like ET_{18}-OCH_3) with known antitumor agents like Cis-platinum, cyclophosphamide, and others. Preliminary animal experiments show such synergistic effects (table 3).

Cell-Mediated Antitumoral Effects

As mentioned above, the antitumoral effect of etherlysophospholipid was discovered when these compounds were studied in systems of cell-mediated immunity. At that time, tumor cell destruction was thought to be due to a direct interaction of immune competent cells which also mediate comparable immunological phenomena like rejection of transplanted organs. It took some time until we were able to differentiate between the direct cytotoxicity of etherlysophospholipids and the etherlipid-mediated tumor cytotoxicity by immunocompetent cells. It was clear from the beginning that neither B- nor T-cells, but rather macrophages and possibly natural killer cells (NK cells) [51], are stimulated by etherlysophospholipids in their tumorcytotoxic activity. Preincubation of potentially tumorcytotoxic cells with etherlysophospholipids followed by washing the cells lead only to a minor antitumoral response when incubated with (most) tumor target cells. On the other hand, direct cytotoxicity of the compounds could be demonstrated, but it was not as strong as observed in combination with macrophages and/or NK cells. Finally, an experimental set-up was developed which allows measurement of the cytotoxic effect of etherlipids and of activated macrophages in one and the same culture separately. For this purpose macrophages and/or NK cells are first preincubated with 5–10 μg/ml of the tumor-cytotoxic lysophospholipid, washed after 24 h and further co-cultivated with tumor target cells in a concentration < 1 μg/ml – a concentration which has no direct cytotoxic effect on tumor cells, at least not in the presence of 10% fetal calf serum. If, however, the so-preactivated macrophages and tumor target cells are incubated in the presence of 0.2–1 μg/ml of etherlipid for 1–3 days, the tumor cells will be completely destroyed. This result can only be interpreted by the assumption that high, but non-lytic concentrations of etherlipids trigger a process in macrophages which is strongly enhanced when the so-activated defense cells are further exposed to very low concentrations of etherlysophospholipids. Figure 10 demonstrates these phenomena.

The biochemical mechanisms of the transformation of normal, non-cytotoxic macrophages to highly effective cytotoxic cells is as yet not understood. Generally, among all etherlysophospholipids tested, only those compounds induced cytotoxic macrophages that have also a direct cytotoxic action on tumor cells. However, a few exceptions were noticed. Some etherlipids had a moderate to strong tumor cytotoxicity which was even inhibited after the addition of macrophages, spleen or NK cells. On the other hand, there are a few compounds which display only weak or moderate tumor cytotoxicity, but very strongly activate macrophages. Future studies may show whether these exceptional effects can be explained by differences in the above mentioned phospholipid metabolism of the target cell membrane.

Another aspect for the understanding of this activation process seems to evolve on the production of IL-1 and TNF by macrophages after incubation

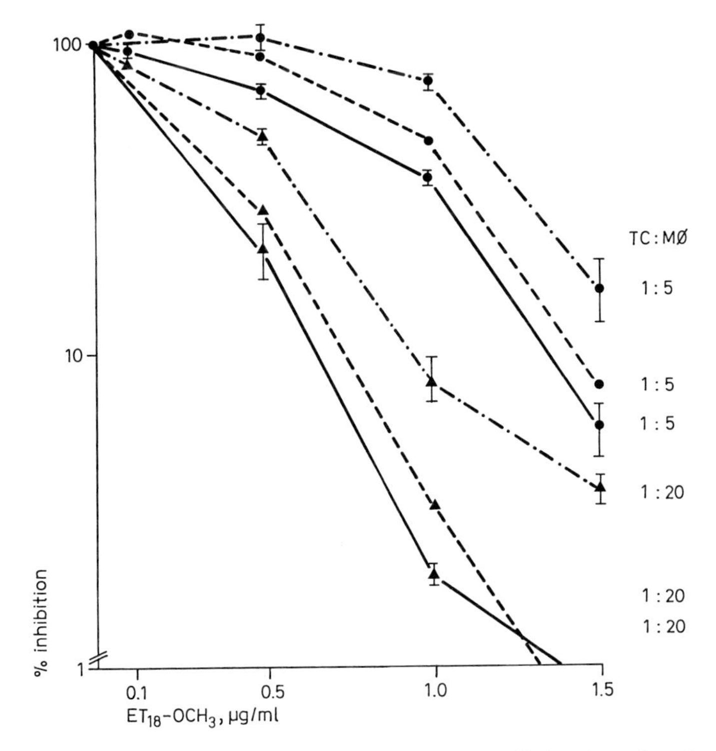

Fig. 10. Destruction of Abelson 8.1 lymphoma cells in vitro after the incubation (3 days) with ET_{18}-OCH_3-activated macrophages. Surviving cells: ·——· Thymidin incorporation: ·-----· Alkaline phosphatase: —·—·—

with effective etherlysophospholipids. Furthermore, there seems to be an effect of some of these compounds on the protein kinase C system. High-concentrations (> 10 µg/ml) actually inhibit this system [96]; but recent studies by our group indicate that 0.11 µg already have an activating effect. It is at present not known which step in this interplay of signals, leading to an overall macrophage activation, is either inhibited or stimulated by etherlipids, such as ET_{18}-OCH_3.

Therapeutic Index in vitro

As mentioned above, all tested malignant cells are destroyed within 48 h by incubation with 10 µg/ml ET_{18}-OCH_3. In our hands there was only one exception, a freshly transplanted human melanoma. In contrast to these findings, a significant difference between the cytotoxic activity on malignant

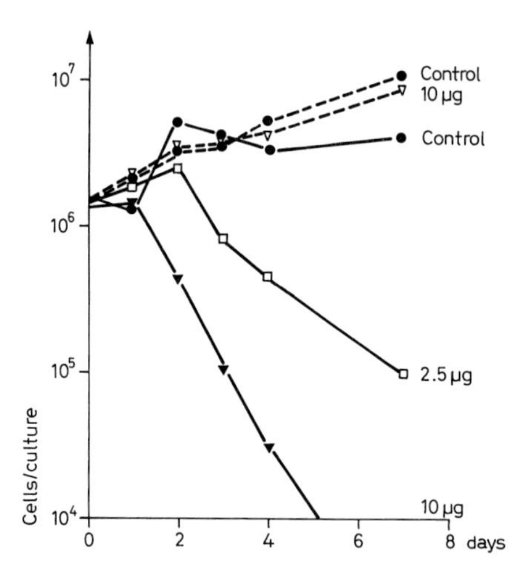

Fig. 11. The effect of ET_{18}-OCH_3 on Meth-A sarcoma cells and normal L929 fibroblasts. The number of vital cells was determined daily by the trypan blue exclusion test. Controls ·———·, ·-----·; L929 fibroblasts ▽———▽; Meth-A sarcoma □———□ (2.5 μg); Meth-A sarcoma ▼———▼ (10 μg).

cells and normal cells was found. Non-tumorigenic L929 fibroblasts grow normally and are not destroyed by etherlysophospholipids in concentrations that are already highly tumoricidal (fig. 11). The same is true for freshly transplanted mouse liver cells. See also table 1. The only change which was observed (morphologically) is that fibroblasts round up and grow in suspension when compounds like ET_{18}-OCH_3 are added (10 μg/ml) to the culture medium. A similar phenomenon can be observed when mature bone marrow macrophages are co-cultivated with non-metabolizable etherlysophospholipids. It should be noted that immature progenitor bone marrow cells are inhibited in vitro in the presence of etherlipids like ET_{18}-OCH_3 (20 μg/ml). However, no bone marrow cytotoxicity has ever been found in vivo, although the presence of the compound could clearly be demonstrated in the blood forming tissue (to be published).

Experiments and Results in vivo

Cancer Therapy in Animal Models

In parallel to studies in vitro, the antitumor effect of etherlysophospholipids has been studied in different tumor models in animals. In general,

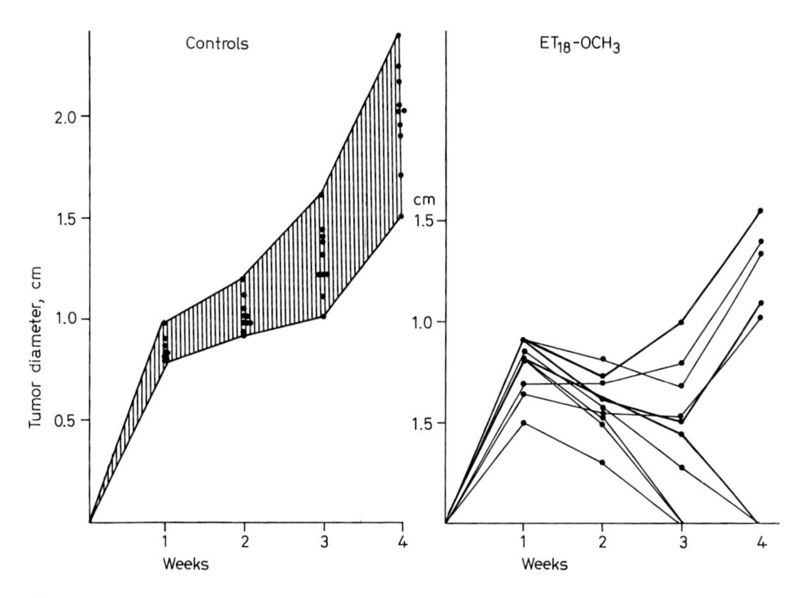

Fig. 12. Influence of ET_{18}-OCH_3 on the growth of Meth-A sarcoma cells in vivo.

syngeneic transplantable tumors in various strains of mice were used. The effect of etherlysophospholipids on the growth of tumors in vivo was studied in the following tumor models: Meth-A sarcoma in (C57-black$_6$-Balb/c)F_1 mice, 3-Lewis-lung-carcinoma in C57-black$_6$ mice, myeloma X5563 in Balb/c mice, 20 autochtonous methylcholantren induced tumors in Balb/c mice and L1210 in DBA mice. Studies on the i.p. growth of Ehrlich's ascites tumor cells in NMRI mice and of EL4 lymphoma cells in C57-black$_6$ mice were also performed. The growth of all these tumors in vivo is more or less inhibited by the oral application of etherlysophospholipids like ET_{18}-OCH_3. The lowest effective dose was found to be 10 μg/mice/day. Doses higher than 200 μg/animal did not usually improve the result. It has to be pointed out, however, that no apparent significant toxicity has ever been observed in mice even if the doses were near the LD50. The LD50 in mice, rats, guinea pigs and subhuman primates was found to be between 40 and 55 mg/kg, given i.v. once. If the compound is given orally the LD50 in mice and rats is about fourfold higher.

Figure 12 shows some typical results of many experiments performed in the Meth-A tumor system. 10–50 μg of ET_{18}-OCH_3 were given 5 times/week beginning 3 days after s.c. implantation of 10^5 tumor cells. The treatment was continued for 3 weeks. The figure demonstrates: 1. All tumor cells inoculated grow to a measurable tumor within the first week. 2. The treatment with etherlysophospholipids retards the growth of the tumor in the second week. 3. Between 50–80% of the established tumors will disappear during the third week. 4. Some of the tumors will sneak through during the fourth week. Increasing the dose or adding other classical antitumor drugs like Cyclophosphamide does not prevent this 'breakthrough' of the tumor.

It should be pointed out, however, that the results can be greatly improved when the etherlysophospholipids are given together with low doses of classical chemotherapeutic compounds right from the beginning (table 3). As shown in this table the simultaneous application of low doses of etherlysophospholipids (ET_{18}-OCH_3 and Ilmofosin, s.VI (in figure 4) with low doses of Cis-platinum are most effective in enhancing the antitumoral effect in vivo. This is demonstrated by the decreased tumor volume, tumor weight and by the increase in the number of surviving animals. Whether this synergistic effect can be explained by the two different sites of drug attack on the tumor cell or whether the addition of membrane active etherlysophospholipids increases the uptake of the chemotherapeutic substance like Cis-platinum remains to be established. Similar findings have been obtained with Cyclophosphamide and with Vinca-alkaloids. (P.G. Munder et al., unpublished). Studies with Doxorubicin are under way.

The metastatic spread of tumor cells from the primary tumor is the most prominent problem in treating tumor patients. Therefore a series of studies was performed over the years using the transplantable 3-Lewis-lung carcinoma [15]. Besides ET_{18}-OCH_3, a number of slowly metabolizable etherlysophospholipids have been used in this tumor model. The lysophospholipids were given orally as we had found that radioactive ET_{18}-OCH_3 is optimally absorbed from the gastrointestinal tract [...].

Figure 13 represents one typical experiment. 10^5 tumor cells (more than 90% trypan blue negative) were implanted into the right hind foot pad of C57 $black_6$ mice. After the primary tumor had reached a diameter between 6–8 mm the tumor-bearing footpad was amputated, the animals randomized and treated 5 days per week p.o. with ET_{18}-OCH_3 for twenty days. In the control groups the animals start to die around day 20. At day 50 all untreated animals have died. The surviving mice are observed until day 120 in all groups. All animals are then sacrificed, the lung removed and injected with black ink. Late and very small metastases can thus be detected. As shown in figure 13, about 50–60% of the treated mice will survive. In this system the lowest effective dose is between 20–50 μg/mouse/day. Ilmofosin VI and ET_{18}-OCH_3 have about the same antitumoral activity. If the etherlysophos-

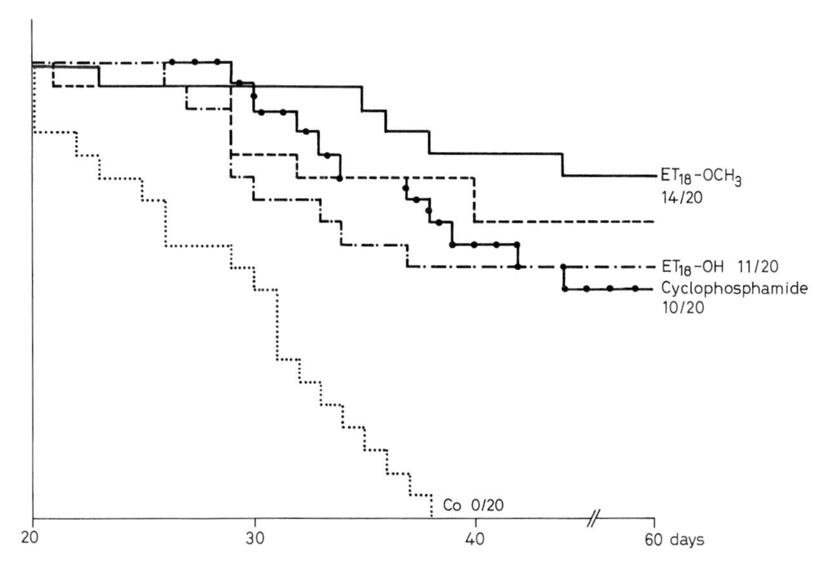

Fig. 13. The influence of Alkylysophospholipids and Cyclophosphamide on the metastasis of 3 Lewis-Lung carcinoma.

pholipids are given together with other antineoplastic agents like Cyclophosphamide, the survival rate can be increased up to 80–90% (see above). Usually etherlysophospholipids will not influence the growth of the primary tumor to a significant degree. However, if combined with other antitumoral compounds, an effect also on the primary tumor can be observed. Under this regimen the development of metastases will also be inhibited due to the retarded growth of the primary tumor and, most likely, a decreased spread of tumor cells into the circulation. Therefore the interpretation of the data is more complicated compared to the results obtained when the treatment began after the removal of the primary tumor.

To explain the antimetastatic effect several possibilities can be proposed: (a) The seeded tumor cells from the primary tumor are destroyed by the direct cytotoxicity of ET_{18}-OCH_3; (b) Circulating tumor cells are destroyed in the blood; (c) Micrometastases in the lung are recognized and attacked by activated macrophages [7, 15]; (d) Circulating tumor cells are prevented from invading the lung tissue. This latter phenomenon has been described in an in vitro system [89].

Tumors that grow in the peritoneal cavity like Ehrlich-ascites-tumor L 1210, and EL 4 are much more difficult to treat [90]. Only a limited number of inoculated tumor cells can be eliminated successfully. This, however, is not surprising as already 5 Ehrlich-ascites tumor cells will kill a NMRI or A/J mouse [14]. The same is true for the ascites forms of the other tumors. If the same tumors are injected s.c. a good therapeutic response can be observed; however, spontaneous regression of the established tumors may occur.

It is a well-known fact that transplantable tumors have their drawback as some of the originally syngeneic tumor cells drift away into slight allogenicity. Therefore the effect of etherlysophospholipids on chemically induced tumors was studied in mice and rats [22]. Mice were inoculated with 1 mg methylcholanthrene intradermally at the back. Treatment started either immediately or at various times (10–40 days) after inoculation. The results can be summarized as follows: (a) All untreated mice, usually 20–30/group, developed autochtonous tumors. (b) Immediate start of treatment after implantation of the cancerogen was suboptimal. (c) Mice with an established tumor at day 40–90 can still be treated (2.5 mg/kg/twice weekly). In about 50% of the animals the tumors will disappear. (d) If treatment (2.5 mg/kg) is stopped at day 90 tumors will reappear. (e) If treatment is continued until day 120–160 no regrowth of the chemically induced tumor will occur.

A possible dose-response relationship has so far not been studied in this system. There is, however, some indication for a dose response in autochthonous tumors in rats [22].

Purging of Leucemic Bone Marrow

Experimental studies were recently intensified to investigate the possibility of purging leucemic bone marrow by incubating these cells with ET_{18}-OCH_3 in order to eliminate the malignant cells [21, 98]. Figure 14 provides an example of several experiments. Preincubation of bone marrow mixed with L 1210 leucemic cells with various concentrations of ET_{18}-OCH_3 prolongs the life of mice transplanted with this bone marrow. Unpurged bone marrow will kill the mice within two to three weeks, whereas preincubated bone marrow does not transfer the leucemia to normal animals [98]. If the recipient animals are first irradiated with a lethal dose (1000 rad) reconstitution with purged bone marrow not only keeps the animals alive, but also prevents the development of a leucemia in the transplanted mice. In a pilot study a similar procedure was used before transplanting back autologous bone marrow of patients with chronic myelocytic leucemia from the remission period. No conclusions can yet be drawn about the value of this preincubation in the human clinical situation.

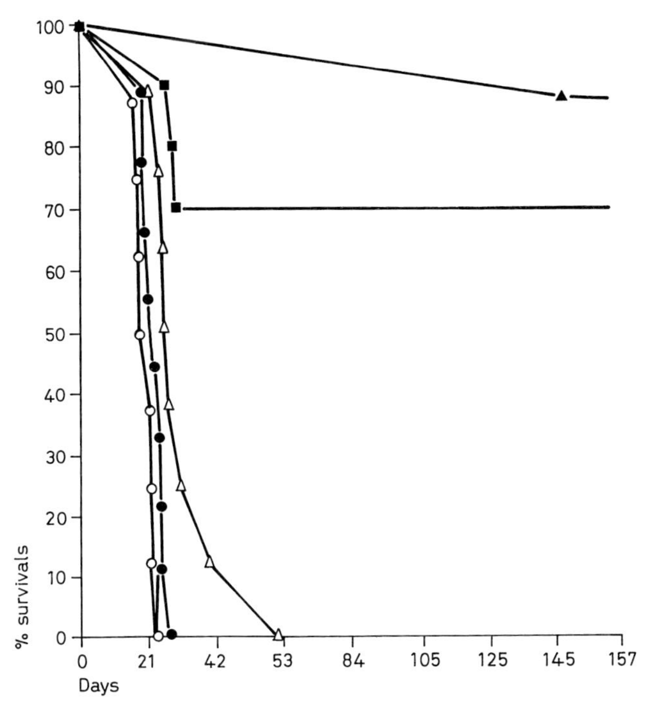

Fig. 14. Survival rate of Balb/c-mice in % after the transplantation of leucemic bone marrow incubated in vitro with various concentrations of ET_{18}-OCH_3 Control ○——○; 2.5 µg/ml □——□; 5 µg/ml ●——●; 10 µg/ml △——△; 20 µg/ml ■——■; 100 µg/ml ▲——▲.

ET_{18}-OCH_3 in Human Cancer Therapy

Pilot studies were carried out with ET_{18}-OCH_3 in patients who could no longer be treated with an established surgical, radiotherapeutic or chemotherapeutic regimen. In a limited number of patients with metastatic disease caused by various primary tumors (carcinoma of the breast or prostate, renal carcinoma etc.) some regression was seen, and in some cases there was no progression for several months [19]. Under these conditions, complete remissions were not seen. In a controlled multicentric clinical trial the effect of ET_{18}-OCH_3 on non-small-cell lung cancer (NSLC) has been studied for more than a year (study director: Prof. P. Drings, Thoraxklinik Heidelberg) and is being continued intensely. The data so far indicate that ET_{18}-OCH_3 is at least as effective in prolonging life as the most effective combination therapy with classical chemotherapeutics. There were no serious side effects, allowing the patients to remain at home during the treatment (P. Drings, personal communication).

The Role of Macrophages in vivo

It is of course almost impossible to study the antitumoral effect of etherlysophospholipid-activated macrophages in vivo. Thus the role of activated macrophages was elaborated indirectly. Mice were inoculated with 3-Lewis-lung carcinoma cells as in [15]. The primary tumor was removed as described. Two days after amputation mice were injected with ten-days' old bone marrow-derived macrophages incubated for 24 h with 10 μg/ml ET_{18}-OCH_3. Figure 15 demonstrates the results. There is a definite increase in the number of surviving animals when injected with activated macrophages. There is, however, also a slight effect of non-pretreated bone marrow macrophages indicating their role in the antitumoral defense of the organism. It should be stressed that only macrophages were transferred. It seems therefore worthwhile investigating the role of peripheral monocytes in the antitumoral defense after activation. This approach is currently under way (R. Andreesen, personal communication).

Other Biomedical Effects of Etherlysophospholipids

The Effect on the Induction and Course of Experimental Allergic Encephalomyelitis (EAE)

EAE is considered to be an autoimmune disease which reflects to a certain extent the clinical picture of multiple sclerosis in the human. Although one cannot fully compare the two diseases, it seems reasonable to assume that in both cases autoaggresive T-lymphoblasts play a key role.

This could be confirmed in two different experimental set-ups [41]: (a) active immunization of Lewis rats with purified myelin basic protein (MBP) in Freund's complete adjuvans; (b) passive transfer of specifically sensitized T-cells in vivo and restimulated in vitro 2–3 times with MBP before transfer into normal healthy Lewis rats, after [41].

As shown in figure 16, MBP-stimulated T-lymphoblasts as determined by FACS analysis are completely destroyed in the presence of ET_{18}-OCH_3 whereas metabolizable lysophospholipids have no effect.

Previous studies have shown that resting lymphocytes are not affected by ET_{18}-OCH_3. Lectin-stimulated lymphocytes are, however, also damaged to a certain degree [2].

If the animals in these two systems are treated orally with ET_{18}-OCH_3, the onset, duration and severity of the disease could be changed in a positive manner. At present we can state that daily application of ET_{18}-OCH_3 (1–5

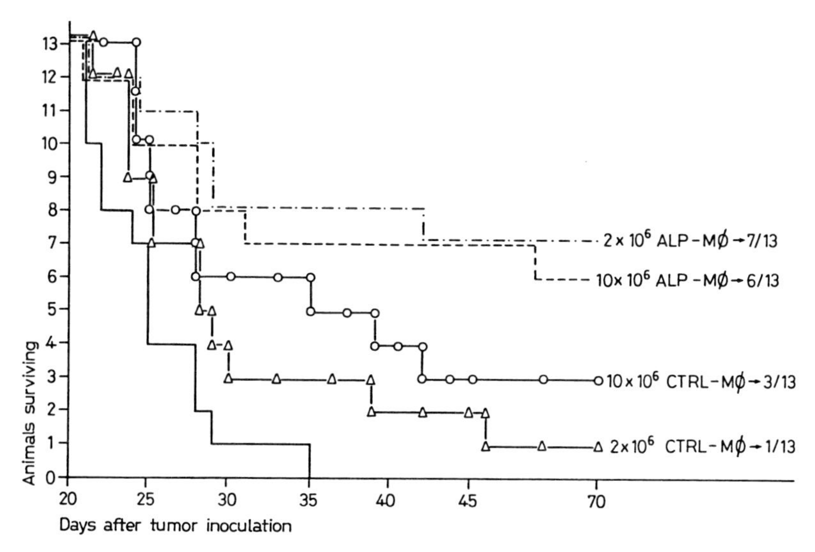

Fig. 15. Adoptive immunotherapy of 3-Lewis-Lung metastases with ET_{18}-OCH_3 activated bone marrow-derived macrophages. For the description of the tumor model see IV.1. Therapy with $2{\times}10^6$ macrophages per mouse was started 3 days after the removal of the primary tumor and continued twice weekly for 6 weeks. ALP-MO = ET_{18}-OCH_3-activated macrophages CTRL-MO = Control macrophages, Survivors/total.

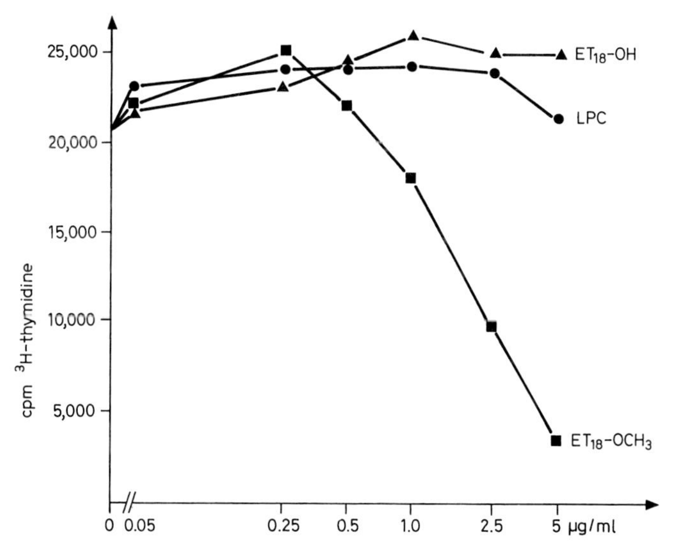

Fig. 16. Destruction of myelin basic protein specific T-cell blasts by ET_{18}-OCH_3 in vitro. T-cell blasts were induced in vivo and restimulated with MBP in vitro as described in [41].

mg/rat) completely either inhibits the development of EAE or decreases the clinical symptoms (paralysis of the tail, paraplegia, incontinence), thus preventing the death of the animal. Surprisingly, there is no significant effect on the EAE-induced weight loss. An optimal regimen for the treatment of EAE, induced either way, has still to be worked out.

Multiple Sclerosis

In a pilot study, a limited number of patients with a definite diagnosis of multiple sclerosis (MS) were given ET_{18}-OCH_3 (150–300 mg twice/week). The patients were at different stages of the disease. Patients in early stages of M.S. generally showed a slight improvement of clinical symptoms. The number of patients is still too few and the time of treatment too short to judge a possible therapeutic value for this indication. Nevertheless, it should be mentioned that three patients taking the drug now for more than four years showed no signs of serious side effects and are in a stable disease condition. Patients with the slowly progressing form of MS do not seem to benefit as much as patients with the relapsing course of the disease.

Clinical phase II studies are planned in the near future, as soon as the data from the phase I/II study with tumor patients are fully available. If the data in vivo so far obtained can be confirmed, ET_{18}-OCH_3 would fulfill some of the requirements necessary for a drug which might be used for a long period of time to fight this disease: (a) It seems to be effective; (b) It has no severe side effects such as bone marrow depression, nor any carcinogenicity; (c) It is not mutagenic; (d) It does not affect any vital organ like liver or kidney. Experimental studies on other autoimmune diseases are at present under way.

Conclusions

(1) Etherlipids, like ET_{18}-OCH_3, are potent immune modulators. They activate macrophages which acquire cytotoxicity against tumor cells, but not against normal cells.

(2) Such etherlipids exert a direct and selective destructive action on tumor cell membranes by interfering with their phospholipid metabolism. Thus the tumor cell, already disordered by the direct action of the ether lipid, will further be subject to the cytotoxic attack of etherlipid-activated macrophages.

(3) In contrast with other established antitumoral drugs, etherlipids do not act over the nuclear apparatus of the cell, but over the tumor cell membrane. Etherlipids are not mutagenic; they can be given over long periods of time – also with interruptions – without signs of the development of resistance by selection.

(4) Because of the cell membrane attack of etherlipids, while the known antitumoral drugs act over the cell nucleus, a combination of the two types of antitumoral principles in clinical therapy is indicated and is being underlined by respective animal experiments.

(5) Etherlipids, like ET_{18}-OCH_3 are fully active over the oral route, which is even to be preferred due to a certain depot effect in passing through the digestive tract. They may be administered together with milk, milk products, or in the form of suitable capsules. Many cancer patients can therefore be treated ambulantly over long periods of time.

(6) Etherlipids will pass the blood brain barrier, probably due to their lyophilic/lipophilic (amphiphilic) character. This may explain successes in the treatment of certain brain cancers.

(7) Etherlipids show anti-metastatic and also prophylactic activities in suitable animal tumor experiments.

(8) Small doses of etherlipids induce the outdifferentiation – 'demalignization' – of undifferentiated human myeloic leukemia cells, as shown by Honma et al. [52].

(9) Etherlipids, like ET_{18}-OCH_3 exert a selective destructive action on many virus-transformed cells.

(10) Etherlipids were shown to act selectively destructive on certain autoimmune lymphocytes. Treatment of multiple sclerosis patients over several years, for example, gave promising results.

We can state, therefore, that some synthetic etherlipids like ET_{18}-OCH_3, as developed in the 1960s – more than twenty years ago – turned out to be a new type of immune modulator with selective antitumoral activity and additional potencies of biomedical interest.

Acknowledgement

The authors want to thank the Jung-Stiftung für Wissenschaft und Forschung, Hamburg, for the generous grant supporting research described in this review.

References

1 Ando K, Kodama K, Kato A, et al: Antitumor activity of glyceryl ethers. Cancer Res 1972;32:125–129.

2 Andreesen R, Modolell M, Weltzien HU, et al: Alkyl-lysophospholipid induced

suppression of human lymphocyte response to mitogens and selective killing of lymphoblasts. Immunobiology 1979;156:498.
3 Andreesen R, Modolell M, Speth V, et al: Human macrophage activation by alkyl-lysophospholipids. Immunobiology 1979;156:255.
4 Andreesen R, Osterholz J, Luckenbach GA, et al: Tumor cytotoxicity of human macrophages after incubation with synthetic analogues of 2-lysophosphatidylcholine. J Natl Cancer Inst 1984;72:53.
5 Andreesen R, Modolell M, Weltzien HU, et al: Selective destruction of human leukemic cells by alkyl-lysophospholipids. Cancer Res 1978;38:3894.
6 Andreesen R, Modolell M, Munder PG: Selective sensitivity of chronic myelogenous leukemia cell populations to alkyl lysophospholipids. Blood 1979;54:519.
7 Andreesen R, Munder PG: Ether-lysophospholipids and cellular immunity: A potential role for antitumor activity; in Braquet P, Le Plessis R (eds): New Trends in Lipid Mediators Research. Karger, Basel 1988, pp 16–29.
8 Arnold B, Reuther R, Weltzien HU: Distribution and metabolism of synthetic alkyl analogs of lysophosphatidylcholine in mice. Biochim Biophys Acta 1978;530:47.
9 Arnold D, Weltzien HU, Westphal O: Synthesen von Cholinphosphatiden. III. Über die Synthese von Lysolecithinen und ihren Ätheranaloga. Liebigs Ann Chem 1967; 709:234–239.
10 Baumann WJ, Mangold HK: Reactions of aliphatic methanesulfonates I. Syntheses of long-chain glyceryl-(1) ethers. J Org Chem 1964;29:3055–3057.
11 Berchtold P: Preparation of 1-O-octadecyl-2-O-methyl-sn-glycero-3-phosphocholine. Chem Phys Lipids 1982;30:389–392.
12 Benveniste J, Vargaftig BB: Platelet-activating factor: an ether lipid with biological activity; in Mangold HK, Paltauf F (eds): Ether Lipids, Biochemical and Biomedical Aspects. Academic Press, New York 1983, pp 356–376.
13 Berdel WE, Korth R, Reichert A, et al: Lack of correlation between cytotoxicity of agonists and antagonists of platelet-activating factor (Pafacether) in neoplastic cells and modulation of ^{3}H-Pafacether binding to platelets from humans in vitro. Anticancer Res 1987;7:1181–1188.
14 Berdel WE, Bausert WRE, Fink U, et al: Antitumor action of alkyllysophospholipids. Anticancer Res 1981;1:345.
15 Berdel WE, Munder PG: Metastatic growth of 3-Lewis lung carcinoma in mice treated with alkyl-lysophospholipids and lysophospholipid-induced peritoneal macrophages. Anticancer Res 1981;1:397.
16 Berdel WE, Bausert WR, Weltzien HU, et al: The influence of alkyl-lysophospholipids and lysophospholipid-activated macrophages on the development of metastases of 3-Lewis lung carcinoma. Eur J Cancer 1980;16:1199.
17 Berdel WE, Fink U, Egger B, et al: Growth inhibition of malignant hypernephroma cells by autologous lysophospholipid incubated macrophages obtained by a new method. Anticancer Res 1981;1:135.
18 Berdel WE, Fink U, Egger B, et al: Inhibition by alkyl-lysophospholipids of tritiated thymidine uptake in cells of human malignant urologic tumors. J Natl Cancer Inst 1981;66:813.
19 Berdel WE, Fink U, Maubach PA, et al: Response of acute myelomonocytic leukemia to alkyl-lysophospholipids. A case report. Blut 1982;44:177.
20 Berdel WE, Fromm M, Fink U, et al: Cytotoxicity of thioetherlysophospholipids in leukemias and tumors of human origin. Cancer Res 1983;43:5538.

21 Berdel WE, Okamoto S, Vogler R, et al: The effect of purging bone marrow cells and cryopreservation on the growth of marrow progenitor cells from leucemic patients. Blood 1988;72 (suppl 2) p 379a.
22 Berger M, Munder PG, Schmähl D, et al: Influence of the alkyl-lysophospholipid ET_{18}-OCH_3 on methylnitrosourea-induced rat mammary carcinomas. Oncology 1984;41:109.
23 Bhatia SK, Hajdu J: A new approach to the synthesis of ether phospholipids. Preparation of 1.2-dialkyl-glycero-phosphocholines from L-glyceric acid. Tetrahedron Letters 28(3):271–274.
24 Bittman R, Witzke NM, Lee TC, et al: Synthesis and biochemical studies of analogs of platelet-activating factor bearing a methyl group at C2 of the glycerol backbone. J Lipid Res 1987;28:733–738.
25 Boeryd A: Effects of alkoxyglycerols and especially selachylalcohol on the bone marrow in connection with irradiation treatment and in leukemia treatment. Nature 1958;182:1484–1485.
26 Boeryd B, Hallgren B, Ställberg G: Studies on the effect of methoxy-substituted glycerol ethers on tumor growth and metastasis formation. Brit J Exp Pathol 1971; 52(2):221–230.
27 Boeryd B, Hallgren B: Action on various experimental tumor host systems of methoxy-substituted glycerol ethers incorporated into the feed. Acta path Microbiol Scand 1980;A88:11–18.
28 Bosies E, Herrmann DBJ, Bicker U, et al: Synthesis of thioether phosphocholine analogues. Lipids 1987;22(11):947–954.
29 Bosies E: The introduction of the phosphorylcholine radical in the synthesis of VI(13) is now performed in yields of 85–90% crude and 75% of pure end product. Personal Commun, June 1989.
30 Brohult A: Alkoxyglycerols and their use in radiation treatment. Acta Radiologica, 1963; suppl 223.
31 Burdzy K, Munder PG, Fischer H, et al: Steigerung der Phagozytose von Peritoneal makrophagen durch Lysolecithin. Zeitschr Naturforsch 1964;19b:1118–11.
32 Cabot MC, Welsh CJ, Snyder F: Increasing the levels of ether-linked lipids in M-L cells by glyceryl ether supplementation depresses growth and choline utilization. Biochim Biophys Acta 1982;713:16–22.
33 Chabrier P, Thong NT, Le Maitre D: Action de quelques agents nucleophiles sur les dioxaphospholanes et les dioxaphosphorinanes. Cr Acad Sci Paris (Ser C) 1969; 268:1802–1804.
34 Chandrakumar NS, Hajdu J: Synthesis of enzyme-inhibitory phospholipid analogs. II. Preparation of chiral 1-acyl-2-acylamido-2-deoxy-glycerophosphorylcholines from serine. Tetrahedron Letters 1981;22(31):2949–2952.
35 Chandrakumar NS, Hajdu J: Synthesis of enzyme-inhibitory phospholipid analogs. J Org Chem 1982;47:2144–2147.
36 Chandrakumar NS, Hajdu J: A new method for the stereo-specific synthesis of ether phospholipids. Preparation of the amide analog of platelet-activating factor and related derivatives. Tetrahedron Letters 1982;23:1043–1046.
37 Chandrakumar NS, Hajdu J: Stereospecific synthesis of etherphospholipids. Preparation of 1-alkyl-2 (acylamino)-2-deoxy-glycero-phosphorylcholines. J Org Chem 1983; 48:1197–1202.

38 Edmundson RS: Oxidation of phosphorochloridites. Chem Ind (London) 1962; 1828–1829.
39 Eibl HJ, Arnold D, Weilzien HU, et al: Synthesen von Cholinphosphatiden. I. Zur Synthese von α- und β-Lecithinen und ihren Ätheranaloga. Liebigs Ann Chem 1967;709:226–230. See also Eibl, HJ: Doctor thesis, Univ Freiburg Germany, 1964.
40 Eibl HJ: Phospholipid synthesis: Oxazaphospholanes and dioxaphospholanes as intermediates. Proc Natl Acad Sci USA 1978;75:4074.
41 Engelhardt B, Diamantstein T, Wekerle H: Immunotherapy of experimental autoimmune encephalomyelitis (EAE): Differential effect of anti-IL-2 receptor antibody therapy on actively induced and T-line mediated EAE of the Lewis rat. J Autoimmun 1989;2:61–73.
42 Ferber E: Phospholipid dynamics in plasma membranes, in Biological Membranes. Vol II. Chapman D, Wallach DFH (eds). Academic Press, New York 1973;221.
43 Fleer EAM, Unger C, Kim D-J, et al: Metabolism of ether phospholipids and analogs in neoplastic cells. Lipids 1987;22(11):856–861.
44 Hallgren B: Therapeutic effects of ether lipids; in Mangold HK, Paltauf F (eds): Ether Lipids, Biochemical and Biomedical Aspects. Academic Press Inc, New York 1983; 15:261–275.
45 Harrocks LA, Sharma M: Plasmalogens and O-alkyl-glycero-phospholipids; in Hawthorne JN, Anschl GB (eds): Phospholipids. Elsevier, Amsterdam 1982, pp 51–93.
46 Hill EE, Lands WEH: Phospholipid metabolism; in Wakil SJ (ed): Lipid Metabolism. Academic Press Inc, New York 1970, pp 185–267.
47 Hirt R, Berchtold R: Zur Synthese der Phosphatide. Eine neue Synthese der Lecithine. 2. Mitt Pharmac Acta helv 1958;33:31–38.
48 Hirth G, Barner R: Synthese von Glycerylätherphosphatiden. I. Mitt Helv Chim Acta 1982;65:1059–1084.
49 Hoffman DR, Stanley JD, Berchtold R, et al: Cytotoxicity of ether-linked phytanyl phospholipid analogs and related derivatives in human H1-60 leukemia cells and polymorphonuclear neutrophils. Res Comm Chem Pathol Pharmacol 1984;44:293–306.
50 Hoffman DR, Hoffman LH, Snyder F: Cytotoxicity and metabolism of alkyl phospholipid analogues in neoplastic cells. Cancer Res 1986;46:5803–5809.
51 Hoffman T, Hirata F, Bougnoux P, et al: Phospholipid methylation and phospholipase A_2 activation in cytotoxicity by human natural killer cells. Proc Natl Acad Sci USA 1981;78:3839.
52 Honma Y, Kasukabe T, Okabe-Kado J, et al: Antileukemic effect of alkyl-phospholipids. I. Inhibition of proliferation and induction of differentiation of cultured myeloid leukemia cells by alkyl ethyleneglycolphospholipids. Cancer Chemother Pharmacol 1983;11:73.
53 Kny G: Über Lysolecithin-Analoga – Synthese und biologische Eigenschaften. Chemical Diploma Thesis, Univ Freiburg Germany 1969, esp. pp 55–64. In this thesis many relevant alkyl-lysophospholipids are listed.
54 Kötting J, Fleer EAM, Unger C, et al: Synthetische Alkyllysophospholipide als Antitumormittel – Struktur-Verwandte des 'Platelet-Activating Factors'. Fat Sci Technol 1988;90(3):345–351.
55 Kudo I, Nojima S, Wook Chang H, et al: Antitumor activity of synthetic alkylphospholipids with or without PAF activity. Lipids 1987;22(11):862–867.

56 Langer W, Munder PG, Weltzien HU, et al: Der Einfluβ von Lysophosphatidanaloga auf die Immunantwort. 4. Congr Ges Immunologie, Bern, Oct 1972. Zf Immunitäts-forsch 1973;145:20.
57 Lee T-C, Snyder, FL: Synthetic alkyl-phospholipid analogs: A new class of antitumor agents. In Phospholipid and cellular regulation. Vol II, Kuo LF (ed). CRC Press, Boca Raton 1984.
58 Lehninger AL: Supramolecular organization of enzyme and membrane systems. Naturwiss 1966;53:57–63.
59 Linman JW, Long MJ, Korst DR, et al: Studies on the stimulation of hemopoiesis by batylalcohol. J Lab Clin Med 1959;54:335–343.
60 Maistry L, Robinson KM, Evers P, et al: Morphologic effects of an antitumor agent on human esophageal carcinoma cells in vitro. Scanning Electron Microsc 1980;3:109.
61 Mangold HK, Paltauf F (eds): Etherlipids, Biochemical and Biomedical Aspects. Academic Press, New York 1983, pp 439.
62 Marx MH, Piantadosi C, Noseda A, et al: Synthesis and evaluation of neoplastic cell growth inhibition of 1-n-alkylamide analogues of glycero-3-phosphocholine. J Med Chem 1988;31:858–863.
63 Modest EJ, Daniel LW, Wyckle RL, et al: in Bristol-Meyers' Cancer Symp, Harrap R, Connors T (eds). Vol 8, 387. Academic Press Inc, New York 1987.
64 Morris-Natschke S, Inoles JR, Daniel LW, et al: Synthesis of sulfur analogues of alkyl lysophospholipid and neoplastic cell growth inhibitory properties. J Med Chem 1986;29:2114–2117.
65 Mulder E, van Deenen LLM: Metabolism of red cell lipids III. Pathways for phospholipid renewal. Biochem Biophys Acta 1965;106:348–356.
66 Munder PG, Fischer H: Über die polarographische Bestimmung des Sauerstoffver-brauchs von Leukocyten und Makrophagen und deren Beeinflussung durch siliko-gene Stäube. Beitr Silikoseforsch Grundlagen de Silikose-forsch 1963;5:21.
67 Munder PG, Ferber E, Fischer H: Lysophosphatide und Zellmembran. Unter-suchungen über die Abhängigkeit der cytolytischen Wirkung des Lysolecithins von Membranenzymen. Zeitschr Naturforsch 1965;20b:1048–1058.
68 Munder PG, Modolell M, Ferber E, et al: Phospholipide in quarzgeschädigten Makrophagen. Biochem Zf 1966;344:310–313.
69 Munder PG, Modolell M, Fischer H: The influence of various adjuvants on the metabolism of phospholipids in macrophages. Int Arch Allergy 1969;36:117–1.
70 Munder PG, Fischer H, Weltzien HU, et al: Lysolecithin analogs: a new class of immunopotentiators with antitumor activity. Proc Am Assoc Cancer Res 1976;17: 174.
71 Munder PG, Weltzien HU, Modolell M: Lysolecithin analogs: a new class of immunopotentiators; in Miescher PA (ed): Immunopharmacology, 7th Internat Symp, Bad Schachen (Germany), pp 411–424. Schwabe and Co., Basel 1977.
72 Munder PG, Modolell M, Ferber E, et al: The relationship between macrophages and adjuvant activity; in Van Furth R (ed): Mononuclear Phagocytes. Blackwell Scien-tific, Oxford 1970, p 445.
73 Munder PG, Modolell M: Adjuvant induced formation of lysophosphatides and their role in the immune response. Int Arch Allergy 1973;45:133.
74 Munder PG, Modolell M, Bausert W, et al: Alkyl-lysophospholipids in cancer therapy; in Hersh EM, et al (eds): Augmenting Agents in Cancer Therapy. Raven Press, New York 1981, p 441.

75 Munder PG, Modolell M, Andreesen R, et al: Lysophosphatidylcholine (lysolecithin) and its synthetic analogues. Immune modulating and other biologic effects. Springer Semin Immunopathol 1979;2:187.
76 Murphy RC, Clay KL: Chemical synthesis and mass spectrometry of PAF; in Snyder F (ed): Platelet-Activating Factor and Related Lipid Mediators. Plenum Press, New York 1987, pp 9–31.
77 Ngyen HP, Thuong NT, Chabrier P: Action de la tri-methylamine et de la triphenylphosphine sur les dixaphospholanes-1.3.2 et dioxaphosphorinanes-1.3.2. Cr Acad Sci Paris 1970;271:1465–1467.
78 Osmond DG, Roylance PJ, Webb AJ, et al: The action of batyl alcohol and selachyl alcohol on the bone marrow of the guinea pig. Acta haematol 1963;29:180–186.
79 Revel M, Navech J: Action de la dimethylamine sur quelques oxo-2-dioxaphospholanes-1.3.2. Cr Acad Sci Paris (C) 1969;266:121–124.
80 Robertson AF, Lands WEM: Metabolism of phospholipids in normal and spherocytic human erythrocytes. J Lipid Res 1964;5:88.
81 Snyder F: Enzyme pathways of platelet-activating factor, related alkylglycerolipids, and their precursors; in Snyder F (ed): Platelet-Activating Factor and Related Lipid Mediators. Plenum Press, New York 1987, pp 89–116, especially Fig. 16, p 108.
82 Snyder F: Introduction. Historical aspects of alkyl lipids and their biologically active forms (ether lipids, platelet activating factor, and antihypertensive renal lipids); in Snyder F (ed): Platelet-Activating Factor and Related Lipid Mediators. Plenum Press, New York 1987, pp 1–5.
83 Snyder F: Biochemistry of platelet-activating factor. A unique class of biologically active phospholipids. Proc Soc Exp Biol Med 1989;190:125–135.
84 Snyder F: Private communication (April 10, 1989).
85 Snyder F, Malone B, Piantadosi C: Enzymatic studies of glycol and glycerol lipids containing O-alkyl bonds in liver and tumor tissues. Arch Biochem Biophys 1974;161:402–407.
86 Snyder F, Wood R: The occurrence and metabolism of alkyl and alk-1-enyl ethers of glycerol in transplantable rat and mouse tumors. Cancer Res 1968;28:972.
87 Snyder F, Wood R: Alkyl and alk-1-enyl ethers of glycerol in lipids from normal and neoplastic human tissues. Cancer Res 1969;29:251.
88 Soodsma JF, Piantadosi C, Snyder F: The biocleavage of alkyl glyceryl ethers in morris hepatomas and other transplantable neoplasms. Cancer Res 1970;30:309.
89 Storme G, Berdel WE, von Blitterswijk WJ, et al: Anti-invasive effect of racemic-1-octadecyl-2-methoxy-glycero-3-phosphocholine and other lysophospholipid analogs in MO_4 mouse fibrosarcoma cells in vitro. Cancer Res 1985;41:351.
90 Tarnowski GS, Mountain SM, Stock SC, et al: Effect of lysolecithin and analogs on mouse ascites tumors. Cancer Res 1978;38:339–344.
91 Tidwell T, Guzman G, Vogler WR: The effects of alkyl-lysophospholipids on leukemic cell lines. I. Differential action on two human leukemic cell lines, HL60 and K562. Blood 1981;57:794.
92 Tietz A, Lindberg M, Kennedy EP: A new pteridin requiring enzyme system for the oxidation of glyceryl ethers. J Biol Chem 1964;239:4081–4090.
93 Tsushima S, Yoshioka Y, Tanida S, et al: Syntheses and antimicrobial activities of alkyl-lysophospholipids. Chem Pharm Bull 1982;30(9):3260.
94 Unger C, Eibl HJ, Kim DJ, et al: Sensitivity of leukemia cell lines to cytotoxic alkyl-

lysophospholipids in relation to O-alkyl cleavage enzyme activities. J Natl Cancer Inst 1987;78:219–222.
95 Vallari DS, Smith UL, Snyder F: HL-60 cells become resistant towards antitumor ether-linked phospholipids following differentiation into a granulocytic form. Biochem Biophys Res Comm 1988;156(1):1–8.
96 Van Blitterswijk HJ, Van der Bend RL, Kramer JM, et al: A metabolite of an antineoplastic ether phospholipid may inhibit transmembrane signalizing via protein kinase C. Lipids 1987;22(11):842–846.
97 Van Deenen LLM, Holman R: Progress in the chemistry of fats and other lipids. VII. Part I, 1–127. Pergamon Press Ltd, London 1965.
98 Vogler WR, Winton EF, Boggs R, et al: The effect of alkyl-lysophospholipid (ET_{18}-OCH_3) on granulocyte-macrophage progenitor cells from normal and chronic myelocytic leukemic marrows. Exp Hematol 1982;10(suppl 11):15.
99 Weber N, Benning H: Metabolism of ether glycolipids with potentially antineoplastic activity by Ehrlich ascites tumor cells. Biochim Biophys Acta 1988;959:91–94.
100 Weltzien HU, Westphal O: Synthesen von Cholinphosphatiden. IV. O-Methylierte und O-acetylierte Lysolecithine. Liebigs Ann Chem 1967;709:240–243.
101 Weltzien HU: Cytolytic and membrane-perturbing properties of lysophosphatidylcholine. Biochim Biophys Acta 1979;559:259.
102 Westphal O, Fischer H, Munder PG: Adjuvanticity of lysolecithin and synthetic analogues. 8th Internat Congress Biochem, Interlaken/Switzerland, 3–9 Sept 1970. Abstr pp 319–320.
103 Witzke NM, Bittman R: Convenient synthesis of racemic mixed-chain ether glycerophosphocholines from fatty alkyl-alcyl ethers: useful analogs for biophysical studies. J Lipid Res 1986;27:344–351.

Prof. Paul Gerhard Munder, Max-Planck-Institut für Immunobiologie,
D–7800 Freiburg i.Br. (FRG)

Waksman BH (ed): 1939–1989: Fifty Years Progress in Allergy.
Chem Immunol. Basel, Karger, 1990, vol 49, pp 236–244

From Protective Immunity to Allergy: The Cellular Partners of IgE

André Capron, Jean-Paul Dessaint

Centre d'Immunologie et de Biologie Parasitaire, Unité Mixte
INSERM U167-CNRS 624, Institut Pasteur, Lille, France

Extensive studies initiated in parasitic diseases have unequivocally established that IgE antibodies can interact directly with mononuclear phagocytes, eosinophils, and platelets through specific surface receptors now identified as Fc_ERII [1]. Genes encoding for B cell, monocyte/macrophage and, more recently, eosinophil IgE receptor have been cloned and, although studies on molecular structure indicate a close homology between Fc_ERII on inflammatory cells and on B cells, there are also indications emerging of some degree of heterogeneity among the second class of receptors for IgE. The important consequence of this interaction is cytotoxicity against the target parasite recognized by cell-bound IgE, i.e. the schistosome larvae (schistosomula) or the filarial larvae (microfilariae). These observations thus introduced IgE antibody and its unusual cellular partners, beside IgG antibody and lymphoid K cells, in the general concept of antibody-dependent cell-mediated cytotoxicity reactions (ADCC). The essential role of IgE antibody in these ADCC systems was indicated by selective absorption of IgE, resulting in the abrogation of parasite attrition, whereas schistosome targets could be killed by macrophages, eosinophils or platelets in the presence of an anti-schistosome monoclonal antibody of the IgE class in the rat model [2]. The essential function of the new class of receptor for IgE that was discovered during these studies has been demonstrated by using a monoclonal antibody to Fc_ERII (BB10), which blocked IgE antibody-dependent killing by any of the three effector cells, eosinophils, macrophages, and platelets [3]. Confirmation of the in vivo relevance of such observations has been brought by passive transfer of in vivo IgE antibody-sensitized eosinophils, macrophages, and platelets, or of anti-schistosome monoclonal IgE antibody, which each conferred on naive rats resistance against a challenge schistosome infection

[2]. Thus, an alternative function of IgE antibody can be demonstrated in protective immune defense against helminth parasites, against which no other effector mechanism is known to be efficient.

In parallel, considerable knowledge has been accumulated on the classical protagonists of immediate-type hypersensitivity reactions, with the precise identification of the region of the ε chain of IgE that binds to the high-affinity receptor (Fc_ERI), and the cloning of this four-chain receptor selectively expressed by mast cells and basophils [4]. However, inflammatory reactions are complex and result from the participation of many bioactive mediators and many different cell types. Although most mediators responsible for bronchoconstriction, bronchial hyperreactivity and infiltration with inflammatory cells have initially been considered as derived from mast cells, considerable doubt now exists about their cellular sources [5]. Thus, the precise role of inflammatory cell types, whether resident or recruited, has to be considered all the more so since release of most of the mediators of anaphylaxis by macrophages, eosinophils, and even platelets has been demonstrated. The expression of Fc_ERII by subsets of these cells leads one to reconsider their participation, not only as recruited secondary effectors, but also as primary actors in the allergic drama.

The Second Receptor for IgE (Fc_ERII) on Monocytes/Macrophages, Eosinophils and Platelets

Various experimental procedures, including rosetting with IgE-coated erythrocytes, labelled IgE binding, use of polyclonal and monoclonal anti-Fc_E receptor antibodies, have allowed the unequivocal demonstration of specific receptors for IgE on macrophages, eosinophils, and platelets in man and various animal species. This new receptor for IgE differs from the classical mast cell or basophil receptor by a lower affinity (Ka 10^7 vs. 10^9 M^{-1}) and a distinct antigenicity. The absence of cross-reactivity between mast cell receptors on the one hand and inflammatory cell receptors on the other, whereas polyclonal and monoclonal (BB10) antibodies to the latter receptor bind similarly to macrophages, eosinophils, and platelets and specifically inhibit their interaction with IgE molecules, justify the individualization we have proposed of a second class of receptors for IgE (Fc_ERII) that is expressed by inflammatory cells [1].

Genes encoding for B cell monocyte/macrophage [4, 6] and more recently eosinophil IgE receptor [7] have been cloned and, although studies on molecular structure indicate a close homology between $Fc_{E}RII$ on inflammatory cells and on B cells, there are also emerging indications of some degree of heterogeneity among the second class of receptors for IgE. Recent studies performed on eosinophils indicate that their IgE receptors contain a sequence commonly involved in the primary structure of adhesion proteins [8]. Likewise, interaction between $Fc_{E}RII$ and a member of the integrin superfamily (gpllb IIIa) has been evidenced for platelets [9].

Cytophilic Binding of IgE to Monocytes/Macrophages, Eosinophils and Platelets in Allergic Patients

Although the $Fc_{E}RII$ is of lower affinity than the mast cell or basophil IgE receptor, the increased extrinsic affinity of $Fc_{E}RII$ for IgE dimers or complex (10^8 M^{-1}) [10] gives this class of receptors a particular significance in all situations where IgE complexes are produced. This is the case not only of parasitic infections [2] but also of various allergic diseases including asthma and rhinitis [11].

The number of $Fc_{E}RII$-bearing cells increases in all pathological or experimental situations associated with elevated IgE levels [1, 12], and evidence for the intervention of $Fc_{E}RII$ in allergy is also provided by the observation that macrophages, eosinophils, and platelets from allergic patients bear IgE on their surface.

In the case of eosinophils for instance, cytophilic IgE was detected by flow cytofluorometry with anti-IgE on approximately 75% of the cells from patients with hypereosinophilia. A significant increase in the percentage of IgE-bearing cells was found in those patients with elevated serum-IgE levels ($>$240 kU/l). In patients with the PIE syndrome, eosinophils collected by bronchoalveolar lavage showed higher percentages of surface-bound IgE than blood cells. Similarly, IgE could be detected on a fraction of nasal eosinophils [13].

Similar observations were made with bronchoalveolar macrophages from asthmatic patients and with platelets from allergic asthmatics [9, 13].

A striking result of these studies, directly demonstrating in vivo binding of IgE to inflammatory cells, is the inconstant and low detection of cytophilically-bound IgG on these cells, thus conferring an operational superiority on surface-bound IgE in triggering the cell. Besides this direct evidence, ex vivo studies have also demonstrated the possibility of IgE-dependent activation of inflammatory cells in allergy.

Activation of Monocytes/Macrophages, Eosinophils and Platelets by IgE in Allergic Patients

Following the binding of IgE of the appropriate molecular form, i.e. at least dimers, signs of cell activation can be detected early (10–30 min) in the various populations concerned.

Monocytes/macrophages respond to IgE-dependent triggering by releasing lysosomal enzymes, neutral proteases, interleukin-1, reactive oxygen metabolites, and also potent mediators of anaphylaxis such as sulphidopeptide leukotrienes, leukotriene B4, prostaglandins and platelet-activating factor (PAF) [1, 12–14]. Interestingly, these observations initially made with purified monoclonal IgE were confirmed after passive sensitization of normal alveolar macrophages with serum from allergic patients by adding either antihuman IgE or the specific allergen. The release was dependent on IgE antibody in the patient's sera, as shown by the disappearance of the anti-IgE-induced changes after heating at 56 °C or after the depletion of IgE of the sensitizing serum. Moreover, alveolar macrophages collected from patients allergic to house mite or to grass pollen release lysosomal enzymes selectively on addition of the specific allergen, of anti-IgE, or the F(ab′)2 fragments therefrom. In the same conditions, PAF was found to be released by macrophages from allergic patients exposed in vitro to the allergen [14].

Further evidence of the participation of alveolar macrophages in allergic reactions is the decrease in macrophage concentrations of lysosomal enzymes observed after a local provocation test with the specific allergen, while the enzyme levels in bronchoalveolar lavage fluids were shown to be significantly higher than in the control lung [15].

By measuring the extracellular release of eosinophil peroxidase (EPO), it was shown that IgE-anti-IgE induces the release of up to 26% of the cellular content of the enzyme. Other basic proteins from eosinophil granules are also released, mainly the major basic protein (MBP), but the eosinophil cationic protein (ECP) is hardly liberated by IgE-dependent reactions. PAF is also generated by human eosinophils in response to activation by IgE [13].

In fact, this apparent selectivity in mediator release may be related to eosinophil heterogeneity: the particular subset of human eosinophils, sometimes found in blood in hypereosinophilic patients, but prominent in tissues, characterized by low density on metrizamide gradients, high expression of IgE receptors, and binding of the BB10 monoclonal antibody also has the selective capacity of releasing EPO and MBP by anaphylactic antibody challenge. In contrast, eosinophils with normal density are activated by IgG antibodies to release mainly cationic proteins (MBP, ECP) [13]. The selectivity of the mediators released after IgE- or IgG-dependent stimulation is consistent with in vivo observations showing that release of MBP is linked to hypersensitivity states including allergic asthma, whereas ECP levels are reportedly lower in asthmatic patients than in hypereosinophilic individuals without bronchospasms [16]. This selectivity in the release of granule proteins might lend support to the participation of eosinophils through allergic reactions when pathological studies show signs of selective release of these proteins, as recently found in eosinophil gastroenteritis [17].

Oxygen metabolite generation was found to be stimulated by successive incubation of normal platelets with IgE and anti-IgE or with serum from allergic patients and anti-IgE. In the latter situation, previous absorption of IgE from allergic serum prevents platelet triggering by anti-IgE. Furthermore, platelets collected from allergic donors were triggered to release oxygen metabolites and histamine by the addition of the corresponding allergen or anti-IgE antibody, but not by unrelated antigens or anti-IgG. However, IgG-dependent activation triggers the release of serotonin and platelet aggregation, which is not observed after cross-linking of platelet-bound IgE [9, 13]. Ex vivo studies have also indicated the possibility of IgE-dependent activation of platelets. An example is provided by platelets from hymenoptera venom-sensitized individuals. In these patients, platelets can be triggered directly by the purified allergen (yellow jacket or honeybee venom) to produce cytocidal factors and oxygen metabolites, a characteristic feature of activation by IgE. The allergen, anti-IgE or its F(ab') fragment could also trigger the release of histamine from platelets in such patients [Divry et al., unpubl. observation]. Interestingly, platelet reactivity was found to be significantly decreased (76.8%) after rush desensitization. In the case of two polysensitized patients, hymenoptera venom desensitization led to a decrease in platelet response not only to vespula venom but also to grass pollen [18].

This series of observations clearly indicates that platelets have the capacity to be activated, during local or systemic allergic reactions, by the interaction of membrane-bound IgE with the allergen. One can therefore consider that this circulating Fc_ERII-bearing cell subpopulation can be involved in the network of inflammatory cells which participate in allergic reactions. Although the involvement of platelets in systemic anaphylaxis is easy to postulate, their interaction with the allergen during localized tissue anaphylactic reactions can be taking place in the vessels, with allergens leaking into the blood, or at the endothelium-tissue interface, through gaps forming between endothelial cells, possibly as a result of anaphylactic discharge of vasoactive amines by basophils or mast cells. Indeed, platelet activation, together with platelet accumulation within the pulmonary vessels, has been reported to be associated with anaphylaxis or bronchoconstriction in patients and in animal models of allergy [19, 20].

Opposed to mast cell or basophil for which the direct effect of antiallergic drugs such as disodium cromoglycate (DSCG) or nedocromil is poorly demonstrated, it is worth mentioning that these drugs exert a direct inhibitory activity on IgE-dependent activation of cell populations expressing $Fc_{E}RII$. Alveolar macrophages, eosinophils, and platelets were, indeed, tested in allergic diseases, and a significant inhibition of their response to IgE-dependent stimulation was observed with a dose and kinetics of DSCG and nedocromil similar to their in vivo activity, providing new insights into the mechanisms of these commonly prescribed drugs in allergic diseases [21].

Cellular Networks in Allergic Reactions

The demonstration that the expression of the second class of receptors for IgE is increased in allergic patients and that $Fc_{E}RII$ can possibly be upregulated by IgE indicates that immediate-type hypersensitivity can no longer be viewed as a mere breakthrough in IgE production, but instead that multiple changes in many reactive cells are in fact involved in allergy, resulting in an upgraded response potential.

The production of many proinflammatory mediators by mononuclear phagocytes, eosinophils and platelets in response to IgE immune complexes points to the primary effector function of cells until now considered to be involved only in nonspecific inflammatory responses, and places them, along with $Fc_{E}RI$-bearing mast cells or basophils, as the direct cellular partners of IgE. In allergy, where there is an increased production of IgE and the formation of IgE complexes, $Fc_{E}RII$ behaves like an essential signalling structure. Local formation of IgE immune complexes capable of binding with higher affinity to $Fc_{E}RII$, and activating proinflammatory cells may as well be assumed, resulting from local exposure to allergens when (free) IgE antibodies circulate. Thus, although monomeric IgE rapidily dissociates from the cell surface, preformed or newly formed IgE immune complexes will persist and confer a direct role in the pathophysiology of allergic disorders on $Fc_{E}RII$-positive cells.

The selective activation of inflammatory cell subpopulations through their Fc_ERII does not however exclude any cooperation with cells bearing the Fc_ERI. In this respect, eosinophil-mediated cytotoxicity in the rat requires accessory mast cells or their mediators, among which the eosinophil chemotactic factor of anaphylaxis (ECF-A) plays an essential role. ECF-A tetrapeptides have indeed been shown to enhance the expression of Fc_ERII on rat and human eosinophils [22].

This network of multiple cells triggered by IgE through one or the other Fc_E receptors is even more complex: Besides the interaction of IgE with the lymphocytes that bear low-affinity IgE receptors that contributes to the regulation of the production of this antibody isotype, the effector cells themselves are controlled by an array of lymphokines which upgrade or reduce their response. Various interleukins indeed control the differentiation of so-called thymus-dependent mast cells and basophils, when interleukin-5 (IL-5) enhances the effector capability of eosinophils [23]. Platelet responses are also modulated by antagonistic lymphokines, since interferon-γ and TNFα both potentiate, and a platelet activation suppressive lymphokine (PASL) inhibits their IgE-induced responses [24, 25]. These various lymphokines are produced by separate T cell subsets, and it is worth mentioning that, early during rush immunotherapy by hymenoptera venom, the abrogation of IgE-dependent platelet responses is preceded by marked changes in the balance of various CD4 inducer T cells, with increase in the suppressor inducer subset [26] and detectable serum levels of the suppressive PASL [27].

The induction of allergic reactions, which now appears to involve multiple effector cells triggered by preformed (circulating) or locally formed IgE-allergen complexes through two types of IgE Fc receptors, is thus influenced by a network of cytokine cross-talk. This points to a direct intervention of various T cell subsets, not only at the level of the regulation of IgE production, but also of the modulation of the various cellular targets of IgE by acting on their expression of Fc_ERII and/or their capability of mediator release.

It should be stressed that such a complex network of IgE-dependent effector cells and immunoregulatory cells that amplify or decrease their effector capacity is evidenced not only in allergic disorders, but also in protective immune responses against helminth parasites [28]. It is perhaps not surprising that IgE, an immunoglobulin class that appeared late in phylogeny likely to provide mammals with the possibility of mounting protective reactions towards helminth parasites, and that can trigger (through one or the other of its Fc_E receptors) the discharge of many potent bioactive mediators, has emerged during evolution with a complex array of cellular partners which control the induction and regulation of its production as well as the regulation of the expression of the function of its target cells. The parallelism in the interventions of IgE either in protective or in autoaggressive reactions is even reinforced by the recent demonstration that, much as in allergic disorders, blocking antibodies that counteract the protective effect of anaphylactic antibody are produced in those individuals who fail to develop immunity against schistosomes [28]. This mirror image strengthens the idea that allergists still have more to learn in the study of parasitic models.

References

1 Capron A, Dessaint JP, Capron M, et al: From parasites to allergy: The second receptor for IgE (Fc_ERII). Immunol Today 1986;7:15–18.
2 Capron A, Dessaint JP: Effector and regulatory mechanisms in immunity to schistosomes. A heuristic view. Ann Rev Immunol 1985;3:455–476.
3 Capron M, Jouault T, Prin L, et al: Functional study of a monoclonal antibody to IgE Fc receptor (Fc_ERII) of eosinophils, platelets and macrophages. J Exp Med 1986; 164:72–89.
4 Metzger H: Molecular aspects of receptors and binding factors for IgE. Adv Immunol 1988;43:277–312.
5 Holgate ST, Hardy C, Robinson C, et al: The mast cell as a primary effector cell in the pathogenesis of asthma. J Allergy Clin Immunol 1986;77:274–282.
6 Delespesse G, Sarfati M, Hofsetter H, et al: Structure, function and clinical relevance of the low affinity receptor for IgE. Immunol Invest 1988;17:363–387.
7 Yokota A, Kikutani H, Tanaka T, et al: Two species of human Fc epsilon receptor II (Fc epsilon RII/CD23). Tissue-specific regulation of gene expression. Cell 1988;55: 611–618.
8 Grangette C, Gruart A, Ouaissi MA, et al: IgE receptor on human eosinophils (Fc_ERII). Comparison with B cell CD23 and association with adhesion molecule. J Immunol 1989; in press.
9 Capron A, Ameisen JC, Joseph M, et al: New function for platelets and their pathological implications. Int Arch Allergy Appl Immunol 1985;77:107–114.
10 Finbloom DS, Metzger H: Binding of immunoglobulin E to the receptor on rat peritoneal macrophages. J Immunol 1982;129:2004–2008.
11 Stevens WJ, Bridts CH: IgG-containing and IgE-containing circulating immune complexes in patients with asthma and rhinitis. J Allergy Clin Immunol 1984;73: 276–282.
12 Spiegelberg HL: Structure and function of Fc receptors for IgE on lymphocytes, monocytes and macrophages. Adv Immunol 1984;35:61–68.
13 Capron A, Dessaint JP, et al: IgE receptors on inflammatory cells. 12th Forum in Immunology. Ann Inst Pasteur/Immunol 1986;137C:353–377.

14 Joseph M, Tonnel AB, Torpier G, et al: Involvement of IgE in the secretory processes of alveolar macrophages from asthmatic patients. J Clin Invest 1983;71:221–230.
15 Tonnel AB, Joseph M, Gosset P, et al: Stimulation of alveolar macrophages after local provocation test in asthmatic patients. Lancet 1983;8339:1406–1408.
16 Gleich H, Adolphson CR: The eosinophil leukocyte. Adv Immunol 1986;39:177–253.
17 Torpier G, Colombel JF, Mathieu-Chandelier C, et al: Eosinophilic gastroenteritis. Ultrastructural evidence for a selective release of eosinophil major basic protein. Clin Exp Immunol 1988;74:404–408.
18 Tsicopoulos A, Tonnel AB, Wallaert B, et al: Decrease of IgE-dependent platelet activation in hymenoptera hypersensitivity after specific rush desensitization. Clin Exp Immunol 1988;71:433–438.
19 Lellouch L, Tubiana A, Lezfort S, et al: Ultrastructural evidence for extravascular platelet recruitment in the lung upon intravenous injection of PAF to guinea pig. Br J Exp Pathol 1985;66:345–355.
20 Page CP, Paul W, Morley J: Platelets and bronchospasm. Int Arch Allergy Appl Immunol 1984;74:347–350.
21 Tsicopoulos A, Lassalle P, Joseph M, et al: Effect of disodium cromoglycate on inflammatory cells bearing the Fc epsilon receptor type II (Fc_ERII). Int J Immunopharm 1988;10:227–236.
22 Capron M, Capron A, Joseph M, et al: IgE receptors on phagocytic cells and immune response to schistosoma infection. Monogr Allergy 1983;18:33–44.
23 Silberstein DS, David JR: The regulation of human eosinophil function by cytokines. Immunol Today 1987;8:380–385.
24 Pancré V, Joseph M, Mazingue C, et al: Induction of platelet cytotoxic functions by lymphokines. Role of Interferon-γ. J Immunol 1987;138:4480–4495.
25 Pancré V, Auriault C, Joseph M, et al: A suppressive lymphokine of platelet cytotoxic functions. J Immunol 1986;137:585–591.
26 Tilmant L, Dessaint JP, Tsicopoulos A, et al: Concomitant augmentation of $CD4^+CD45^+$ suppressor inducer subset and diminution of $CD4^+CDw29^+$ helper inducer subset during hyposensitization. Clin Exp Immunol 1989;767:13–18.
27 Tsicopoulos A, Tonnel AB, Wallaert B, et al: A circulating suppressive factor of platelet cytotoxic functions after rush immunotherapy in Hymenoptera venom hypersensitivity. J Immunol 1989;142:2683–2688.
28 Capron A, Dessaint JP, Capron M, et al: Immunity to schistosomes. Progress toward vaccine. Science 1987;238:1065–1072.

André Capron, MD, Centre d'Immunologie et de Biologie Parasitaire,
Unité Mixte INSERM U167-CNRS 624, Institut Pasteur,
1 rue du Professeur A. Calmette, F-59019 Lille Cedex (France)

Waksman BH (ed): 1939–1989: Fifty Years Progress in Allergy.
Chem Immunol. Basel, Karger, 1990, vol 49, pp 245–263

Complement Deficiency and the Pathogenesis of Autoimmune Immune Complex Disease

Peter J. Lachmann

Molecular Immunopathology Unit, Medical Research Council Centre, Cambridge, UK

It can confidently be asserted that autoimmunity does not arise from any single cause. Different autoimmune diseases are caused by different pathogenetic mechanisms and have different genetic associations. Table 1 shows a tentative classification of some major autoimmune diseases into several types. The type with which this paper is concerned is the last in the table: the autoimmune immune complex diseases of which systemic lupus erythematosus (SLE) is the prototype.

This disease has a number of particular characteristics. Its pathogenesis is in large part due to immune complexes affecting the vessels of the skin, serous membranes, joints, kidneys and, indeed, virtually any part of the body; although a few manifestations may be due to a direct effect of antibody, e.g. antiphospholipid antibodies causing recurrent abortion. The autoantibodies are directed predominantly against non-organ-specific cellular components and, most characteristically against the constituents of the DNA nucleo-protein particle (DNA, histone and non-histone proteins) and of the RNA translation particle (various RNA nucleo-proteins as well as some specific tRNA transferases). The third and striking characteristic of this disease is the strong association that it has with genetic complement deficiencies of the components of the early classical complement pathway. This is shown in table 2. Here it can be seen that homozygous deficiencies of C1 and of C4 are virtually sufficient causes for the development of SLE or a related autoimmune immune complex disease even though such complete homozygous complement-deficient subjects make up only a tiny proportion of the patient population. C2 deficiency, which among Caucasians is the commonest form of homozygous complement component deficiency, shows a lesser

Table 1. Autoimmune disease

Type	Examples	Proposed aetiological mechanism	Genetic association
Anti-receptor antibody-mediated disease	myasthenia gravis Graves' disease	'internal image'; idiotype network derangement	A1 B8 DR3 (female only)
'Sequestered antigen' disease	Goodpasture's disease phaco-anaphylaxis	abnormal release of antigen: solvent exposure (or infection) / surgery	DR2
Autoimmune endocrinopathies	IDDM Hashimoto's disease	aberrant class II MHC expression on hormone producing cells	A1 B8 DR and DR4
'T cell'-mediated disease	multiple sclerosis/ EAE rheumatoid arthritis	unknown; ? response to unknown infectious agent	DR2 DR4
Autoimmune immune complex disease	SLE glomerulonephritis	defect in immune complex handling	complement deficiency

Table 2. Complement deficiencies of early classical pathway components

Compo-nent	Chromo-somal locus	Number	Healthy	I/C disease	(SLE)	Infec-tions	(Neis-seria)	Other
Clq	Ip	17	2	15	(7)	5	(2)	skin lesions
Clr/Cls	12	9	2	7	(6)	–	–	–
C4	6p	16	2	14	(12)	6	–	–
C2	6p	77	15	43	(23)	30	(3)	–
C1-ina	11	>500	–	2–5%		? (not reported)		hereditary oedema (all)

Data collected from that given in Rother and Rother [26].

Table 3. Complement deficiency among Osaka blood donors [27]

	No	Homozygote frequency	Gene frequency	Heterozygote frequency
Tested	52,175			
Absence of complement activity	283			
C9 deficient	55	0.105%	3.25%	6.28% or c. 1/16
C7 deficient	5	0.0096	0.98%	1.94% or c. 1/52
C8 deficient	2	0.0038%	0.62%	1.23% or c. 1/81
C2 deficient	0			
Uncharacterised C5~8 deficiency	5			
Evidence of complement activation	198			
Reason for deficiency unknown	17			

C9 deficiency among random Japanese patient group: 48/49,577 or 0.097%
n.b.

Average CH50 of C9-deficient blood donors	12 U
Average CH50 of C9-deficient random patients	within normal range (130–50 U)
Average CH50 of 1 C9-deficient patient with chronic bronchitis and actute phase reaction	63 U

degree of association with SLE, probably about a third of the patients having this disease. The figure in table 2 seems higher but is almost certainly slanted by the fact that healthy C2-deficient patients are no longer reported whereas those with the disease still are. It seems likely that the reason for the lower disease association with C2 is that in C2-deficient subjects, complement fixation proceeds as far as the fixation of C4 and that bound C4 does show, albeit to a lesser degree, some of the important functions normally carried out by fixed C3. These will be discussed later.

When the surprising finding that patients with SLE include those who are genetically complement deficient was first made, there was some reluctance to admit that this association was causal and two further explanations were considered. The first was that it was an ascertainment artefact and that a large pool of undiagnosed early complement deficiencies would be found in a healthy population. This has been disproved in extensive studies carried out in many parts of the world and it is clear that there is no substantial number of these deficiencies unidentified in the healthy population. The most striking data come from Japan (table 3) where enormous surveys have been made and where indeed it has been found that C9 deficiency is extremely common. Early classical pathway component deficiencies were, however, virtually never encountered. Our own much less substantial data at Cambridge also came to the conclusion that only rarely does population screening show up unexpected deficiency of these components.

The second explanation was that the complement deficiency was acting as a marker gene for the true disease susceptibility gene. This explanation is unlikely since deficiencies of both C1q coded on chromosome 1, C1r and s coded on chromosome 12 and C4 and C2 coded in the major histocompatibility complex (MHC) on chromosome 6, are all associated with the same disease. Even for the MHC-linked complement loci, where the explanation has the most plausibility, it seems to be excluded since Caucasian C2 deficiency is nearly always found on the basis of one particular HLA haplotype, A10,B18,C4A2B4,DR2, whereas the complete C4 deficiency haplotypes are variable and quite different. Finally, C1 inhibitor deficiency which gives a secondary subtotal deficiency of C2 and C4 is also associated with an increased incidence of this form of autoimmune disease. It therefore does seem to be the case that deficiency of the early components of the classical pathway does itself predispose to the development of autoimmune immune complex disease.

As already stated, however, these homozygous, deficient patients are a trivial component of the sizeable population of patients with SLE (estimated to be about 1 in 2,000 for women in the United Kingdom and much higher in certain other populations, for example in the West Indies). The question therefore arises whether lesser degrees of defective complement function, either genetic (e.g. heterozygous complement deficiency) or acquired due to complement activation might have similar pathogenetic effects.

In the case of heterozygous C4 deficiency, which is common, there is indeed an increased risk of developing lupus. Data have been obtained both by Fielder et al. [1] in London and by Dunckley et al. [2] on a group of people of various ethnic origins in Australia and the Far East. In all cases an average relative risk of above 2 and going up to about 5 has been found for C4A deficiency but not for C4B deficiency. C4A is the isotype of C4 which binds particularly to immune complexes and this therefore again lends plausibility to the idea that the interaction of complement with immune complexes plays a role in preventing the development of immune complex disease. Rather more than half of all lupus patients in all these studies carried a C4AQ0 allele. If there are complement defects in the other half of the patients they are presumably acquired rather than genetic.

By What Mechanism May an Ineffective Complement System Give Rise to Disease?

Does complement deficiency interfere with the immune response? There is a quite extensive body of work suggesting that complement deficiencies in animals are associated with impairment of the antibody response, particularly to small concentrations of antigen in the absence of adjuvant. This is most severe in C3 deficiency but also occurs in C4 deficiency and C2 deficiency to a roughly comparable amount (all in guinea pigs) [3]. Nevertheless, these animals do not become agammaglobulinaemic and therefore they seem to be able to produce an adequate antibody response to their normal environmental antigenic stimulation.

It is also possible that complement deficiency influences the nature of the antibody response. It has recently been found [4] that whereas IgG1, IgG2 and IgG3 levels in patients with early component deficiencies are normal, IgG4 levels are quite markedly depressed (fig. 1). Since IgG4 is a non-complement-fixing antibody its deficiency in patients without complement seems paradoxical but the most likely reason is that IgG4 is the last IgG isotype made in the immune response. Although the mechanism by which complement deficiency affects an antibody response is not known, it seems likely the most important route is via the failure to localise antigen on follicular dendritic cells in the germinal centres of lymph nodes which seems to require both antibody and complement and this localisation is required for the generation of proper B cell memory [5]. In addition to this mechanism, there is the effect on B lymphocytes by reaction through their CR2. Here it has been claimed by Erdei et al. [6] that complexed C3dg is an up-regulatory stimulus whereas free C3dg is a down-regulatory one and this mechanism too may play some part. It does, however, seem that these effects on the induction of the immune response play a minor role compared with the effects on the effector side which are about to be discussed.

Role of Complement in the Handling of Immune Complexes

Effect of Complement on the Solubility of Immune Complexes

It has been known since the work of Heidelberger [7] that complement delays antigen-antibody precipitation and makes the precipitate less flocculent. The phenomenon of solubilisation was, however, first studied in detail by Miller and Nussenzweig [8] who demonstrated that an intact *alternative*

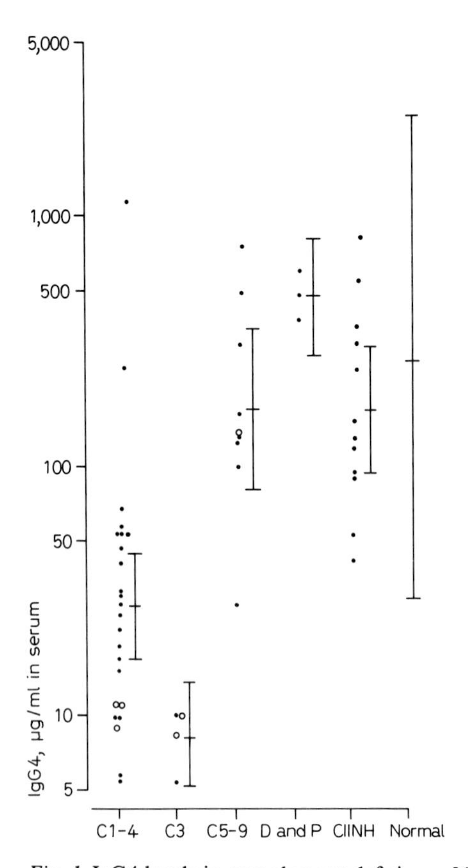

Fig. 1. IgG4 levels in complement deficiency [4].

complement pathway was required for the resolubilisation of insoluble immune complexes. This is a somewhat unphysiological situation and the reverse experiments done by Schifferli and Peters [9] demonstrated that in preventing the initial insolubilisation when antigen antibody and complement are mixed it is the *classical* pathway that is particularly important. This is perhaps not unexpected since soluble immune complexes do not activate the alternative complement pathway.

How this phenomenon works is not intuitively obvious but an explanation can be derived from the Goldberg theory of immune precipitation. Goldberg [10, 11] postulated that in an antigen-antibody interaction all potential antigen-antibody bonds are equally likely to form. From this he concluded that the complex most likely to form was that which could be

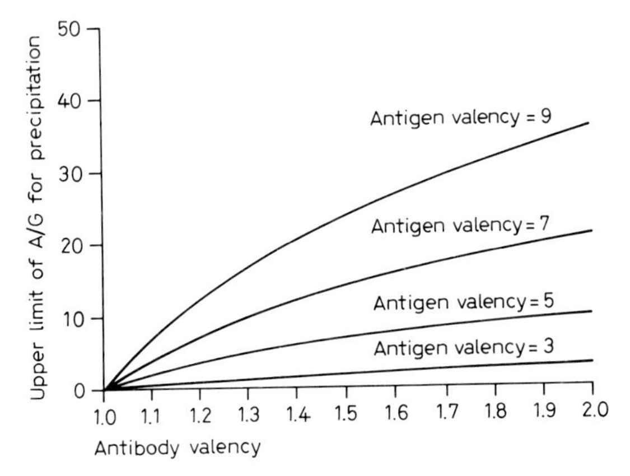

Fig. 2. Effect of antigen and antibody valency on formation of large aggregates [10, 11].

formed in the maximum number of different ways. This number can be calculated, albeit with great difficutly, if one knows the valency of the antigen and antibody and their concentration. The Goldberg calculations showed that in antigen excess it is large numbers of small complexes that are most likely to be formed and that as the equivalence zone is approached the probability changes quite sharply to the formation of a small number of very large complexes – which gives rise to precipitation. If the Goldberg equations are recalculated to show the likelihood of precipitation at different antigen-antibody rations (fig. 2), it can be seen that there is a sharp fall in this probability as both antigen and antibody valency fall. What seems likely to happen when complement is fixed on the antigen-antibody lattice is that C3b (and C4b) – binding, as they tend to do, to the Fd portion of an antibody molecule already incorporated into the lattice by one valency – will thereby render this antibody molecule univalent and that the effect of complement fixation is substantially to reduce antibody valency and probably to a lesser extent antigen valency (since antigen makes up only 20% of an equivalence precipitate). To this extent it should be possible to mimic some of the effects of complement by introducing univalent antibody fragments into the lattice.

Experiments of this description have been carried out on an ovalbumin-anti-ovalbumin system and using the monovalent Fab/c fragment [Barbosa et al., unpubl. observations]. It is of course clear that if Fab/c is used as the only antibody no precipitation at all will result but it can be seen (fig. 3) that as increasing amounts of Fab/c are added to intact IgG precipitation is

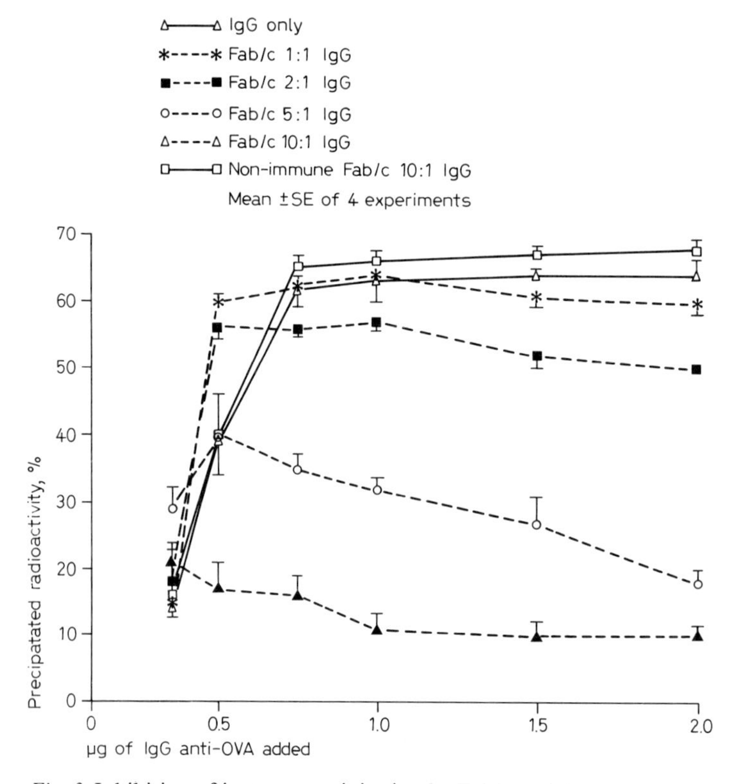

Fig. 3. Inhibition of immuneprecipitation by Fab/c anti-OVA.

progressively reduced. However, and paradoxically, at some concentrations there is an initial enhancement of precipitation in antigen excess. This is not predicted by the Goldberg equations but one may guess that the introduction of univalent antibody in antigen excess will actually increase the bulk of the precipitate before it reduces the precipitability, i.e. there are free antigen sites available which can be occupied by univalent antibody, thereby enhancing the amount of antibody present. When these Fab/c-containing complexes are formed in the presence of complement, solubilisation occurs; and in order to see differential effects the complement concentration must be reduced to approximately 5%. One then sees that solubilisation is again enhanced by Fab/c except in the antigen excess zone where more precipitation occurs. Such results are compatible with accepting the essential correctness of the Goldberg hypothesis but do not

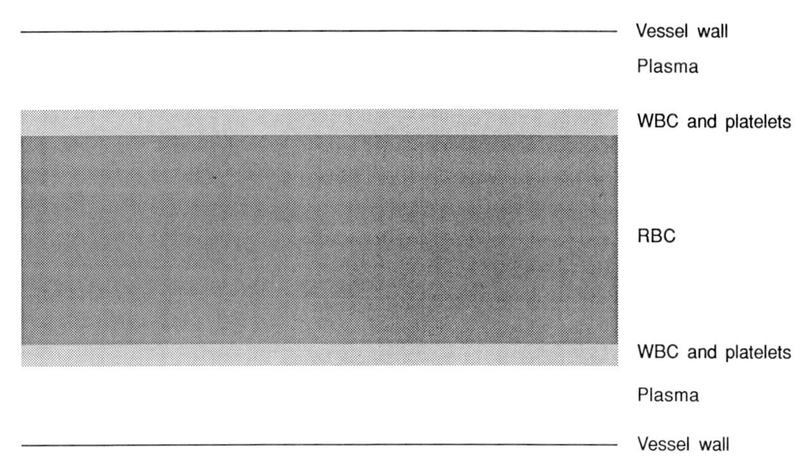

Fig. 4. Streamline flow in blood vessels.

themselves explain why smaller complexes should be less likely to cause immune complex disease.

Carriage of Immune Complexes on Erythrocyte CR1

Medof and Oger [12] first pointed out that immune complexes in whole primate blood are carried substantially attached to the formed elements of the blood rather than free in the plasma and that the great majority of those are attached to erythrocyte CR1. CR1 is the only complement receptor on primate erythrocytes and although it is present in lower numbers (varying from about 200 to 1,200 per erythrocyte on a genetic basis) it nevertheless represents the great majority of CR1 in the blood. Furthermore, it has been shown by Schifferli et al. [13] that CR1 on erythrocytes is clustered and therefore binds immune complement-coated immune complexes much more efficiently than is the case of leucocyte CR1 which although present in much larger amounts is uniformly distributed and therefore less able to bind immune complexes. The biological significance of carriage of immune complexes on CR1 rather than free in the plasma can be inferred from the fact that in streamline flow in blood vessels (fig. 4) it is normally the plasma that is in contact with the endothelium while the red cells travel in the centre of the stream and it is only at sites of turbulence or in the macrophage-lined sinusoids of the liver and spleen that red cells will normally come into contact with the lining of the vessels. It is known that most immune

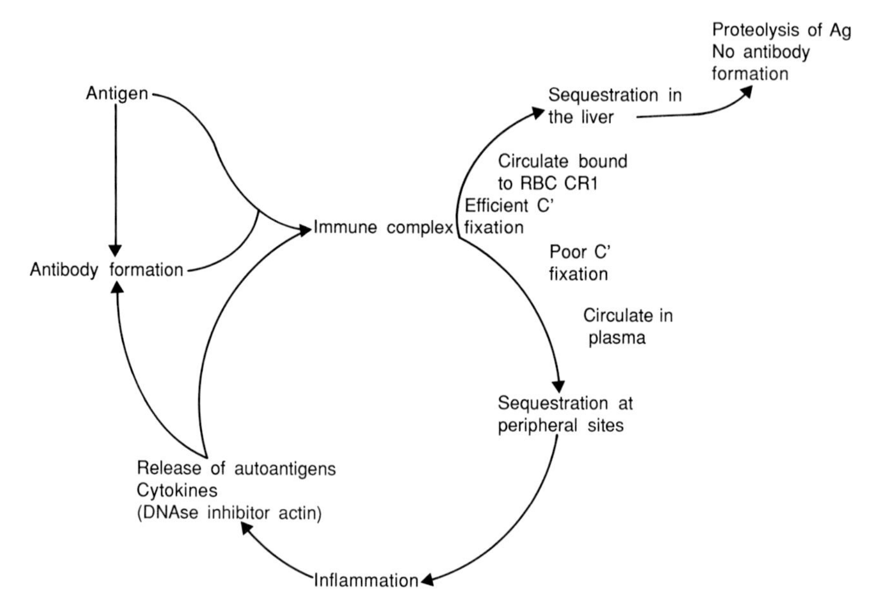

Fig. 5. Schema for the handling of immune complexes.

complexes are cleared in the liver and since the normal liver has no lymphoid tissue, clearance at this site is not accompanied by feedback antibody formation to either the antigens in the immune complexes or to further antigens released as a result of the inflammatory reaction.

On the other hand, immune complexes cleared across the endothelia can give rise to inflammation and further antibody formation. The schema envisaged is shown in figure 5 where two routes of immune complex handling are shown. In the first, immune complexes well coated with complement are carried to the liver and satisfactorily removed; in the second, they are inadequately complement coated and removed from the circulation at other sites where they can give rise to inflammation and feedback antibody formation and the generation of immune complexes establishing a vicious circle which in turn leads to the formation of immune complex disease.

In vivo evidence for this type of schema comes from various sources. The first is the work of Waxman et al. [14] who demonstrated using baboons that whereas in normal animals immune complexes are cleared predominantly in the liver and relatively slowly, in decomplemented animals they are cleared more rapidly and at extrahepatic sites. Walport et al. [personal commun.] have demonstrated the transfer in vitro of immune complexes

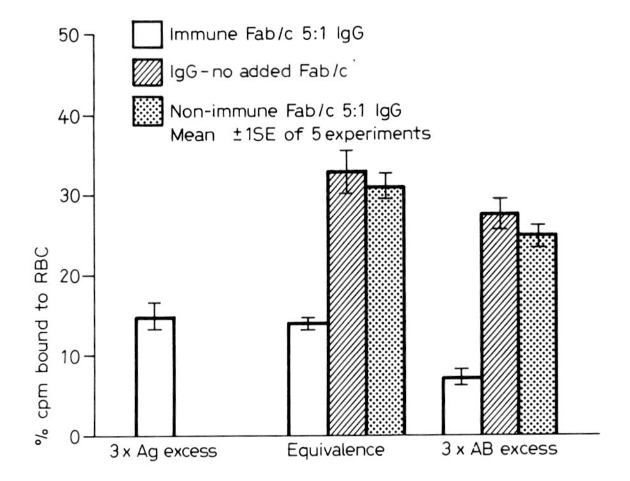

Fig. 6. Binding to CR1 of preformed immune complexes made with Fab/c at different antigen-antibody ratios and treated with 5% complement.

bound to red cells onto macrophages coating beads on a column and Emlen et al. [15] have demonstrated a similar phenomenon on monolayers of monocytes.

Does Increased Solubility of Complexes Promote Better Carriage on Red Cell CR1?

Experiments with Fab/c-containing complexes treated with complement to show to what extent they bind to CR1 (fig. 6) demonstrate that the case of preformed complexes treated with complement binding to CR1 is directly related to complex size and therefore not to solubility. These larger complexes formed in antigen excess with Fab/c bind to CR1 better than comparable complexes made with whole IgG. Although it is conceivable that this mirrors an effect that may be seen at critical complement fixation levels in vivo it seems unlikely that it is a phenomenon of major importance since when complexes are formed in the presence of complement this enhancing phenomenon is not seen and the Fab/c-containing complexes do not bind to CR1 under any conditions whereas those made with whole IgG do bind at equivalence or in antibody excess though not in antigen excess. One may therefore take the view that solubilisation itself is not of importance in getting better immune complex transport and it may be that it is an epiphenomenon of binding more C3 into the antigen-antibody lattice and it is this which makes the transport on CR1 more efficient [16].

Evidence that Immune Complex Transport on CR1 Is Important in Systemic Lupus Erythematosus

The original observation that this mechanism was operating in immune complex disease in vivo was that in SLE (and subsequently in many other immune complex diseases) levels of CR1 on erythrocytes are reduced. This finding was originally interpreted as demonstrating a genetic predisposition to the disease [17, 18] and there is indeed a genetically determined distribution of CR1 numbers on human erythrocytes. However, following the cloning of the CR1 gene by Klickstein et al. [19] this group discovered a restriction fragment length polymorphism associated with CR1 numbers so that it became possible to allotype subjects by their DNA to ascertain their CR1 number genotype. It was then possible to measure CR1 numbers for each phenotype and in this way it was shown that CR1 numbers are reduced in SLE in both the homozygous 'high-number' phenotype and the heterozygous 'intermediate-number' phenotype [20].

Interestingly enough, no patients with the homozygous low number phenotype with SLE were found either in the study by Moldenhauer or in two other studies [21; Yount and Lachmann, unpubl. observations]. This last study was selected to include patients with severe lupus since it had been suggested that failure to find these patients might occur because they had particularly severe disease and therefore were lost early from the study group. There is no overall excess of the low number allele among patients with lupus and it is perhaps therefore likely either that this finding is a statistical artefact or that for reasons that are unclear there is some degree of protection from SLE in genetically low CR1 numbers.

Nevertheless, it is clear that CR1 numbers decrease in immune complex disease in vivo. This decrease is not due to occupancy of the receptors interfering with the assay since the monoclonal antibodies used to assay CR1 are not affected by occupancy. It is also – at least in the great majority of patients – not due to autoantibodies to the receptor – and it does not occur if cells are mixed in vitro with lupus plasma. An ingenious experiment by Walport et al. [22] infusing compatible high CR1 number cells into patients with low CR1 numbers has shown that the CR1 numbers on the injected cells fall towards the numbers found in vivo over a period of up to 72 h and it has been suggested that the CR1 is cleaved from the cell along with the immune complexes that it carries by Kupffer cell enzymes in the liver.

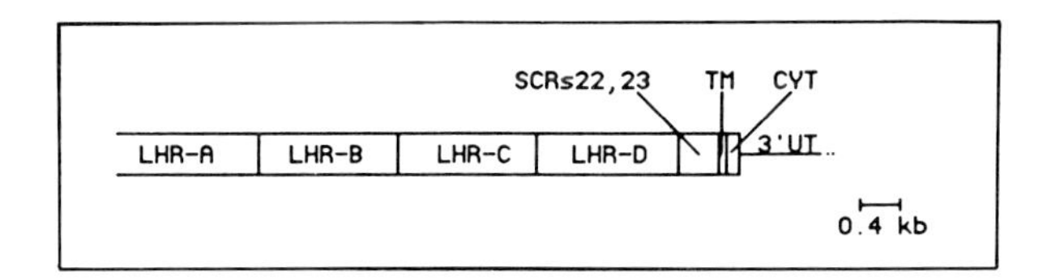

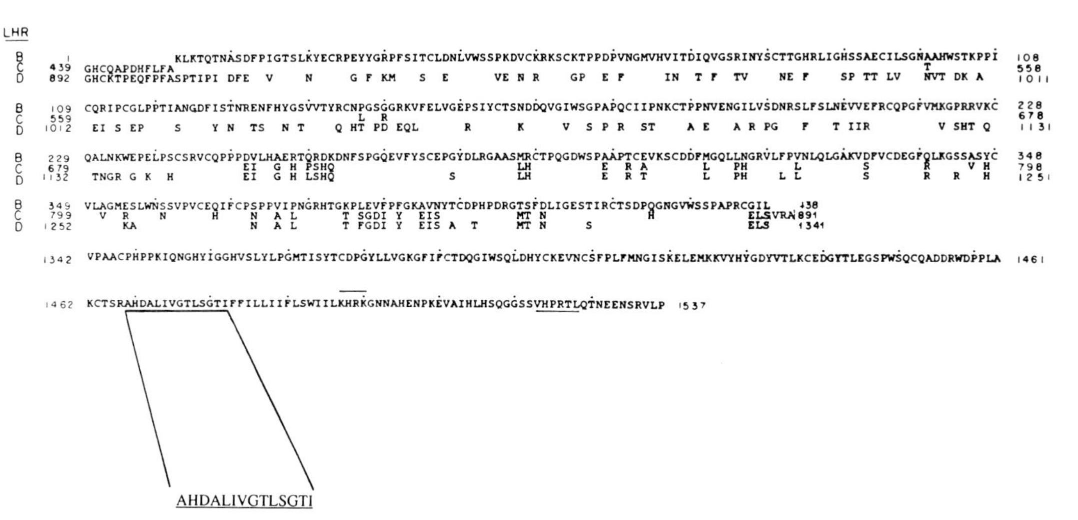

Fig. 7. Sequence of the CR1-STP. One cleavage site between the transmembrane portion and the first consensus repeat. From Klickstein et al. [19].

We have recently approached this problem in another way which is to try to obtain an antibody recognising the cleaved CR1 on red cells – the 'CR1 stump'. The nature of the stump can be predicted from the sequence where there is just one trypsin cleavage site between the transmembrane portion and the first short consensus repeat (fig. 7). A peptide corresponding to this sequence has been synthesised, coupled to PPD and used to raise antibody in rabbits. The antibodies were affinity purified on a column of the immunising peptide and were then shown to react specifically with trypsin-treated as opposed to normal erythrocytes. However, only a very modest proportion of the radiolabelled anti-peptide antibody would bind to trypsin-treated erythrocytes. The antibody was therefore repurified by affinity on trypsin-treated human erythrocyte ghosts. This second affinity purified antibody now shows a much higher degree of uptake onto trypsinised erythrocytes and can be used to measure the CR1 stump. When used on trypsinised cells, it can be shown that

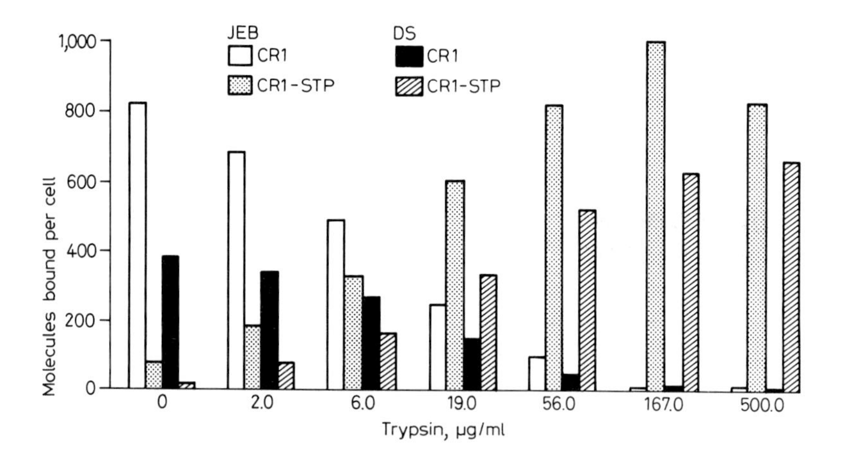

Fig. 8. CR1 and CR1-STP levels on red cells, treated with different amounts of trypsin, from 2 individuals with different genetic CR1 levels.

as the CR1 level declines the CR1 stump level increases [16]. Furthermore, higher levels were achieved in CR1 high subjects than in CR1 low subjects (fig. 8). Unfortunately, however, the affinity of the antibody is biphasic and mostly low and it is therefore difficult to achieve conditions which are suitable for measuring the CR1 stump on cells ex vivo with confidence. However, initial experiments have shown that patients with lupus do show a higher percentage of their total CR1 as stump compared to normals. Further antisera are now being raised to try to achieve better affinity antibodies.

Immune Complex Formation in Humans in vivo

A series of very interesting observations has been made by Walport's group [23] on a series of patients with ovarian carcinoma being treated with highly iodinated mouse monoclonal antibodies for therapeutic purposes. To prevent bone marrow toxicity by these antibodies the patients were injected intravenously with a human monoclonal antibody to mouse Ig to trap the mouse immunoglobulin leaking into the circulation. In this way complexes, albeit of a highly non-physiological kind but with very high radioactivity, were formed in vivo and this has allowed the formation of complexes and their

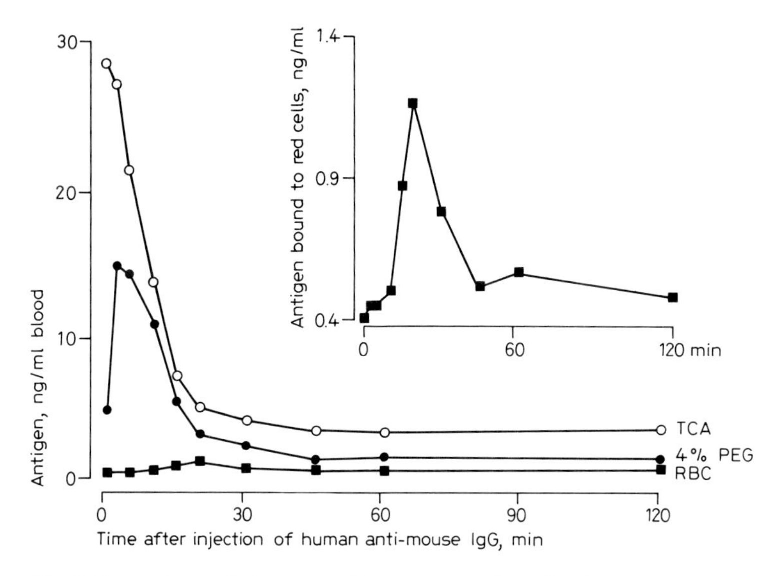

Fig. 9. Formation and clearance of immune complexes in vivo.

attachment to CR1 to be studied. It can be seen in figure 9 that in these patients a proportion of the complexes does fix to CR1 although the great majority does not.

In studies on different subjects [13] immune complexes were formed with tetanus toxoid in vivo and their clearance observed. This was done both in normal and in complement-deficient subjects. 'Trapping' of immune complexes (i.e. rapid removal of a proportion of complexes within the first minutes after injection) was shown to be absent in the complement-deficient subjects. It may be that this 'trappable' population of complexes are those that are particularly important in generating immune complex disease and that bind particularly to CR1.

Conclusions

A hypothesis is put forward suggesting that the essential defect leading to the formation of autoimmune immune complex disease is a failure to properly handle immune complexes so that instead of being harmlessly removed in the liver they give rise to inflammatory reactions at extrahepatic sites, thereby

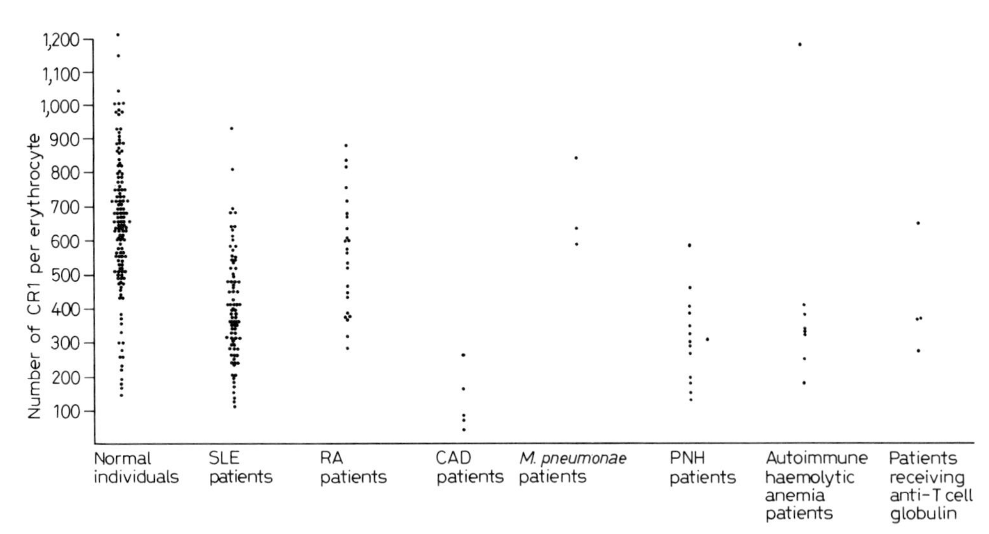

Fig. 10. Number of CR1 per erythrocyte in normal subjects and patients with autoantibodies and/or complement activation.

giving rise to feedback antibody formation and the generation of more immune complex. This applies only in an autoimmune situation since it is only then that further antibody formation to intracellular sequestered autoantigens ensues and it is for this reason presumably that patients with these handling defects do not develop immune complex disease from all exogenous antigens. There may well be other genetic factors affecting how readily such antibodies are made; but the fact that virtually all homozygous C1- and C4-deficient subjects develop lupus suggests that these other factors are not of overriding importance.

Investigation of the mechanism by which complement deficiency leads to autoimmune complex disease suggests that particular importance attaches to the capacity of a proportion of the immune complexes to be carried on red cell CR1. The observation that red cell transport of complexes does occur in lupus is demonstrated by the depression of CR1 numbers and this is indeed found also in a variety of other diseases accompanied by complement fixation on red blood cells whether by immune complexes or directly by red cell antibodies (fig. 10).

There follows from this hypothesis the corollary that autoantibody production in lupus is antigen-driven rather than primarily due to polyclonal activation or idiotypic disorganisation and indeed there is good evidence that this is so. This derives in part from the fact that the antibodies are formed against defined antigenic particles. For example, antibodies are made to DNA and histone as well as non-histone proteins in the DNA nucleoprotein particle and the 'LE cell antibody' itself is directed against conformational epitopes in DNA nucleohistone rather than to any individual on its own [24]. There is also evidence that the formation of anti-nuclear antibodies is not in itself an unusual event. As early as 1967 Biro [25] demonstrated that two-thirds of patients with pulmonary infarcts developed a brief spike of IgM anti-nuclear antibody. This reinforces the idea that it is not the formation of these autoantibodies but their persistence which is abnormal and reinforces the conclusion that it is on the effector side of the immune response that the essential defect must be sought.

This contrasts sharply with other types of autoimmunity, for example the autoimmune endocrinopathies, where it is believed that the expression of class II MHC antigens on hormone-producing cells causes them to become autoantigenic; or autoimmunity due to anti-receptor antibodies which is believed to be due to problems with the idiotypic network allowing internal image antibodies to be formed. In the case of the autoimmune immune complex disease it is, however, apparently the effector side that is at fault and this does suggest that it may be advisable to treat asymptomatic patients with deficiencies of the early components of the complement pathway before they develop disease and perhaps to try to avoid prolonged complement depression occurring in disease although it must be confessed that safe and effective methods of so doing are not yet available.

References

1 Fielder AHL, Walport MJ, Batchelor JR, et al: Family study of the major histocompatibility complex in patients with systemic lupus erythematosus: importance of null alleles of C4A and C4B in determining disease susceptibility. B Med J 1983; 286: 425–428.

2 Dunckley H, Gatenby PA, Hawkins B, et al: Deficiency of C4A is a genetic determinant of systemic lupus erythematosus in three ethnic groups. J Immunogenet 1987; 14:209–218.

3 Bitter-Suermann D, Berger R: Guinea pigs deficient in C2, C4, C3 or the C3a receptor; in Rother E, Rother U, (eds): Hereditary and Acquired Complement Deficiencies in Animals and Man. Basel, Karger, 1986, vol 39, pp 134–158.

4 Bird P, Lachmann PJ: The regulation of IgG subclass production in man: low serum IgG4 in inherited deficiencies of the classical pathway of C3 activation. Eur J Immunol 1988;18:1217–1222.

5 Klaus GGB, Humphrey JH: The generation of memory cells. I. The role of C3 in the generation of B memory cells. Immunology 1977;33:31–40.

6 Erdei A, Melchers F, Schulz T, et al: The action of human C3 in soluble or cross-linked form with resting and activated murine B lymphocytes. Eur J Immunol 1985; 15:184–188.
7 Heidelberger, M: Quantitative chemical studies on complement or alexin. J Exp Med 1941;73:681.
8 Miller GW, Nussenzweig V: A new complement function: solubilization of antigen-antibody aggregates. Proc Natl Acad Sci USA 1975;72:418–422.
9 Schifferli JA, Peters DK: Complement, the immune-complex lattice, and the pathophysiology of complement-deficiency syndromes. Lancet 1983;i:957.
10 Goldberg RJ: A theory of antibody-antigen reactions. I. Theory for reactions of multivalent antigen with bivalent and univalent antibody. J Am Chem Soc 1952; 74:5715–5725.
11 Goldberg RJ: A theory of antibody-antigen reactions. II. Theory for reactions of multivalent antigen with multivalent antibody. J Am Chem Soc 1953;75:3127–3131.
12 Medof ME, Oger JJ-F: Competition for immune complexes by red cells in human blood. J Lab Clin Immunol 1982;7:7–13.
13 Schifferli JA, Ng YC, Paccaud J-P, et al: The role of hypocomplementaemia and low erythrocyte complement receptor type 1 numbers in determining abnormal immune complex clearance in humans. Clin Exp Immunol 1989;75:329–335.
14 Waxman FJ, Hebert LA, Cornacoff JB, et al: Complement depletion accelerates the clearance of immune complexes from the circulation of primates. J Clin Invest 1984; 74:1329–1340.
15 Emlen W, Burdick G, Carl V, et al: Binding of model immune complexes to erythrocyte CR1 facilitates immune complex uptake by U937 cells. J Immunol 1989; in press.
16 Barbosa JE, Emlen W, Brown DL, et al: Effect of specific monovalent (Fab/c) antibodies on the formation and handling of immune complexes (IC) in vitro. Abstract from the Second European Meeting on Complement in Human Disease. Complement 1988;5:203–204.
17 Miyakawa Y, Yamada A, Kosaka K, et al: Defective immune adherence receptor (C3b) on erythrocytes from patients with systemic lupus erythematosus. Lancet 1981;ii:493–497.
18 Wilson JG, Wong WW, Schur PM, et al: Mode of inheritance of decreased C3b receptors on erythrocytes of patients with systemic lupus erythematosus. Engl J Med 1982;307:981–986.
19 Klickstein LB, Wong WW, Smith JA, et al: Human C3b/C4b receptor (CR1). Demonstration of long homologous repeating domains that are composed of the short consensus repeats characteristic of C3/C4 binding proteins. J Exp Med 1987;165: 1095–1112.
20 Moldenhauer F, David J, Fieder AHL, et al: Inherited deficiency of erythrocyte complement receptor type 1 (CR1) does not cause disease susceptibility to systemic lupus erythematosus. Arthritis Rheum 1987;30:961–966.
21 Cohen JHM, Caudwell V, Levi-Strauss M, et al: Genotypic analysis of CR1 on erythrocytes in an SLE population from French hospitals. Abstract from the Second European Meeting on Complement in Human Disease. Complement 1988;5:206–207.
22 Walport MJ, Ng YC, Lachmann PJ: Erythrocytes transfused into patients with SLE and haemolytic anaemia lose complement receptor type 1 from their cell surface. Clin Exp Immunol 1987;69:501–507.

23 Davies K, Hird V, Stewart S, et al: A study of in vivo immune complex formation and clearance in man. 1989;in preparation.
24 Lachmann PJ: An attempt to characterize the lupus erthematosus cell antigen. Immunology 1961;4:153–163.
25 Biro CE: Algunos conceptos immunologicos en relacion con la patogenia del lupus eritematoso diseminado. Arch Inst Cardiol Mex 1967;37:669–674.
26 Rother K, Rother U: In Kallos P, Lachmann PJ, et al (eds): Hereditary and Acquired Complement Deficiencies in Animals and Man. Prog Allergy. Basel, Karger, 1987, vol 39.
27 Inai S, Akagaki Y: The incidence of C9 deficiency in Japan. 'Proc Int Symp Frontiers of Complement Research, Kashigojima, 1984.

Peter J. Lachmann, MD, Molecular Immunopathology Unit,
Medical Research Council Centre, Hills Road, Cambridge, CB2 2QH (UK)

Waksman BH (ed): 1939–1989: Fifty Years Progress in Allergy.
Chem Immunol. Basel, Karger, 1990, vol 49, pp 264–277

Glucocorticoids

Carl G.A. Persson, Ulf Pipkorn

Departments of Clinical Pharmacology and Oto-Rhino-Laryngology,
University Hospital, Lund, Sweden

Background

After several years of chemical-biochemical research by Kendall [30] and others and, thanks to the support of the pharmaceutical industry, adrenocorticotropic hormone and cortisone became available towards the end of the 1940s. A period of extensive clinical experimentation with these much needed hormones followed. Already in 1950 Thorn et al. [67] could review the results of trials covering about 50 different diseases.

Soon after the demonstration of the beneficial effect in the treatment of rheumatoid arthritis [24], these hormones were tested and found to be of great benefit in a variety of other diseases such as eczema, asthma and rhinitis [10, 12]. Since then dermal diseases have emerged as the disease group in which the greatest amount of synthetic glucocorticoids is used. It was soon apparent that there are limitations in the use of these hormones as pharmacological agents. They suppressed endogenous cortisol production through influence on the hypothalamic-pituitary-adrenal (HPA) axis, and symptoms frequently recurred when these drugs were withdrawn.

Physical Therapies and Drugs from Plants and Animal Material May Have Provided Glucocorticoid Actions

The medical history of asthmatic remedies suggests that glucocorticoid effects may have been achieved a long time ago [44]. Liquorice is known to contain substances such as glycyrrhetinic acid which structurally resembles hydrocortisone. This acid may have anti-inflammatory properties, but it is not known to what extent these are glucocorticoid-like. Liquorice was included in treatments of asthma recommended by Jacobus Sylvius in 1545, Sir Floyer in 1698, and William Withering in 1786. In Withering's letter, liquorice is used interchangeably with testaceous powder, indicating that it was considered a suitable constituent rather than an active component.

Withering, in 1786, wisely recommended 'constant bodily exercise in the coldest season' as well as 'frequent use of the cold bath'. He might have read an account by the asthmatic Dr. Smollett in 1766 who, desperate because of his incapacitating disease, plunged into the cold open sea and improved in health. Laennec in 1819 also described 'constant exercise in the open air' and 'uninterrupted use of the cold bath' as important remedies for asthma. In fact, Laennec's best cure of his own asthma was hours of physical exercise. It cannot be excluded that these beneficial stresses were in part effective due to increased cortisol levels. Other experimental treatments may also fall in this category of endogenous cortisol stimulants. There were reports on fever-induced 'cure' of asthma [14], and in 1923 Schiff noted that reactions induced by repeated milk injections were quite effective in both allergic and nonallergic asthma. Incidentally, glucocorticoid drugs of today do not distinguish between 'intrinsic' and 'extrinsic' forms of asthma [45].

Desiccated adrenal glands ('adrenal substance') were provided by Burroughs and Welcome in the early 1900s. 'Adrenal substance' was first given to asthmatics by Solis-Cohen [57]. At that time adrenal extracts had been demonstrated to produce prompt vasoconstriction in experimental systems (presumably due to its content of adrenaline). Solis-Cohen gave 'adrenal substance' orally (up to 6 g daily) and any fraction of adrenaline that could have been absorbed would rapidly have been catabolized. It is therefore likely that the gradual improvements in asthma observed by Solis-Cohen reflected glucocorticoid actions of the repeatedly ingested 'adrenal substance'. From the results of this trial, Solis-Cohen's belief that asthma was a 'vasomotor ataxia of the relaxing variety' was strengthened. He emphasized that the ingredient responsible for the effect remained to be identified: 'What the active agent is and how much or how little of that active agent is absorbed, I must leave to laboratory students to determine. Clinically I have watched closely and critically enough to satisfy myself that neither the susceptibility of patients to suggestion nor the activity of the observer's imagination are sufficient in themselves to account for the whole of the results.'

Pituitrin was used in asthma treatment during the first decades of this century [14]. If successful it would have further supported a pathogenetic role for vasodilation. Had anterior pituitary injections been employed, this would have been even more interesting.

The first report on the anti-asthma action of ACTH is not well covered in the current literature. In 1949, Bordley et al. [10], from the Departments of Otorhinolaryngology and Medicine at Johns Hopkins Hospital reported on glucocorticoid actions in inflammatory airway diseases. Their observations are original and their study is a milestone in the medical history of asthma and rhinitis [47]. Bordley et al. [10] gave ACTH (5–25 mg) intramuscularly every 6 h for 2–3 weeks to 5 asthmatic subjects. Three had intrinsic asthma and 2 had a combination of intrinsic/extrinsic asthma. Their age range was 26–63 years and the duration of their asthma ranged from 5–23 years. Drugs such as aminophylline and adrenaline provided only partial and very temporary relief. The patients produced sputum which was eosinophilic. The airway mucosa had a significant pathology with a pale edematous appearance covered by a thick purulent mucus. These characteristics were evident from close inspection of the nasal airways [10]. After the beginning of the ACTH therapy, unequivocal clinical benefit was noted in all patients in 4–48 h and the asthmatic symptoms had gone in 1–8 days. The sputum production ceased entirely in 4 of the 5 patients and during treatment their airway mucosa attained a normal pink appearance with a clear mucus. These authors also noted that after stopping therapy, symptoms and signs of mucosal pathology gradually returned over several days (or weeks) in some of the patients who had seemed cured by the drug.

These similarities between the lower and upper airways have lately to some extent been 'rediscovered', and may prove to be of major interest for the advancement of research in the field of airway inflammation and its treatment [46].

Topical Glucocorticoids in Asthma – Pioneering Work by Paul Kallós

The many early attempts made during the 1950s to place glucocorticoids directly in the airways are now generally referred to as failures. This may not be entirely correct. For example, a study carried out by Diamant and Kallós [13] in 1954 has not received due attention. At that time there was great uncertainty as to whether or not an inhaled nebulized drug really arrived at the site of action in the lower airways. Later on inhaled glucocorticoids were also to be criticized for causing no less systemic side effects than when given by the other routes of administration. The study carried out by Diamant and Kallós partly circumvented these problems.

Before Kallós embarked on the study in asthmatic subjects he had convinced himself of the beneficial local glucocorticoid effects in 'asthmatic' guinea pigs, To demonstrate an anti-asthma effect a special form of chronic experimental asthma was required, since glucocorticoids are not effective against ordinary allergen-induced broncho-obstruction in guinea pigs. Over several years beginning in the mid-1930s, Kallós, together whith his wife Liselotte Kallós-Deffner, and the pathologist Walter Pagel, had developed such a chronic allergen-induced asthma in guinea pigs which, essentially from symptoms through pharmacology to histopathology, had several similarities to human asthma [25–27].

Diamant and Kallós [13] reported on eleven effective treatments of 2 patients with severe intractable asthma. To ensure a local application, they gave instillations of cortisone, usually 25 mg, into either main bronchus. The improvement was both prompt and sustained. Since the patients had severe asthma that had not been improved by high intramuscular or oral doses of cortisone (>50 mg), the therapeutic action was particularly marked. The success of this study necessitated the transformation of cortisone into active hydrocortisone in the lung; such a metabolism has now been demonstrated [58]. On the basis of their limited but dose-controlled study Diamant and Kallós [13] proposed that local administration of glucocorticoids had advantages in the treatment of asthma. The combination of critical animal and patient studies gives Kallós a significant role in the development of inhalational glucocorticoids which today are the most effective and important drugs in the treatment of asthma.

Physiological Actions – An Antistress Hormone

Glucocorticoids are naturally-occurring hormones which have important physiological roles in the normal hemostasis of the human being [1, 7]. When glucocorticoids are given in a pharmacological sense, most effects both desired and undesired are exaggerations of the normal physiological functions of the hormones, even if certain effects are never normally seen [39]. It has been proposed that one of the major physiological roles of glucocorticoid actions is to modulate the response to stress. Even at physiologic levels an anti-inflammatory action is evident as demonstrated by the exaggerated response of adrenalectomized animals to inflammatory stimuli [18, 32, 34].

Glucocorticoid effects are characteristically dependent on the concentration of active drugs *at target site* and the *duration of treatment*. The duration aspect refers to two important times, on one side the period when the glucocorticoid needs to be present in active concentration in the biophase of the target cell (contact or *trigger time*) and on the other side the *latency time* (time from application of drug to appearance of biological effects). Unfortunately, these factors have not been sufficiently examined. It is known that the vascular antipermeability action may need a trigger time of some minutes only and the effect is developed after 30–60 min [60]. Other effects, e.g. the block of the mixed leukocytes reaction, requires incubation with glucocorticoids for 24 h [8]. Protracted treatment times, longer than 24–28 h, are required in order to demonstrate certain types of effects. Such long treatment times can only be properly achieved in in vivo studies.

The characteristics of maintenance therapy with glucocorticoids are often stressed in terms of unwanted side effects, such as atrophy and reduction of growth [39], but should be equally stressed for certain desired effects. Whereas short-term glucocorticoid treatment is without either dermal atrophy effects or inhibitory effects on allergen-induced immediate weal and flare responses [54, 64], long-term local dermal therapy with potent preparations is known to induce dermal atrophy [59]. Also, a recent study demonstrated that 4 weeks of topical glucocorticoid treatment induced an 80% reduction of the allergen-induced weal and flare responses [53], which is an effect of potential benefit. A range of observations in asthma and rhinitis which follows later, further supports the view that, for many important aspects of therapeutic actions of glucocorticoids, long-term treatment is mandatory.

Action at a Molecular Level – Through the Synthesis of Specific Messenger Proteins

With few exceptions there is a delay between the administration of glucocorticoids into the systemic circulation and the appearance of effects. This delay is consistent with the view that the effects are dependent on synthesis of new proteins. Progress in recent years has substantiated this possibility. Specific proteins have thus been described as appearing subsequent to administration of glucocorticoids. The most studied induced protein is lipocortin, first described under other names [9]. This 35- to 40-kdalton protein has subsequently been sequenced and cloned in *Escherichia coli*. The semipurified preparations of native lipocortin have demonstrable biological activity, but with access to more purified or recombinant preparations, the proposed mechanisms as well as the in vivo activity of lipocortin have been questioned. Whether studies of the increasing number of glucocorticoid-induced proteins will provide definite clues to the understanding of glucocorticoid effects is still speculative. So far, initial hopes that these proteins would furnish new tools for pharmacological separation of different glucocorticoid effects, and thereby interesting new drugs, have not been fulfilled.

Receptor Activation - Selectivity at This Level - Local Therapy

It has not been possible to demonstrate differences in the glucocorticoid receptors on different cells or tissues. Hence the successful approach in drug research with specific agonists (and antagonists) for receptor subtypes has not been feasible with glucocorticoids.

It would, of course, be of considerable interest if those cells activated in the inflammatory response possess a receptor type different from that involved in the various normal physiological functions. However, no such receptor heterogeneity has so far been described. For improvement of specificity and selectivity of glucocorticoids other solutions have been found. Application of glucocorticoids topically on the target tissue has been extensively and increasingly used.

Dermal Glucocorticoids - A Major Application

Cortisol per se did not produce anti-inflammatory effects when topically applied on the skin. However, changes in the side chains and the glucocorticoid skeleton showed, as a first step, that cortisol-21 acetate was effective in inflammatory dermatosis [21]. Increased penetration of dermal glucocorticoids through the stratum corneum of the skin has been achieved, for example, by omitting either a 17-alpha or a 21-hydroxyl group, by making acetonides or 17- and 21-ester side chains. The anti-inflammatory potency and half-life were concomitantly increased by inserting double bonds and various halogens into the steroid structure. In these new steroids the relationship between the systemic activity and the topical anti-inflammatory activity was largely unchanged. This posed minor problems when small parts of the dermal costume were treated, even with very potent steroids, because the total daily dose of steroids was low, hence the influence on the systemic HPA axis was limited.

Mucosal Applications Increase the Demands on Target Selectivity

A new era of topical glucocorticoids began with the separation of topical anti-inflammatory potency from systemic anti-inflammatory effects. This could be achieved by producing drugs with rapid metabolism in the liver abrogating their survival in the systemic circulation. Biotransformation of glucocorticoids is usually to compounds with no or reduced effect, but bioactivation may also occur. If large dermal areas are to be treated, compounds with low systemic side effects are warranted. This requirement is even more essential when mucosal surfaces are exposed to these drugs, because the rate of absorption is higher from the mucosa than from the skin.

The drug which regained the interest of the clinician to use glucocorticoids topically in the airways, betametasone-17-valerate, was rapidly succeeded by beclomethasone-dipropionate. This drug was first promoted as a systemic-steroid-sparing agent when used at rather low doses. Since then somewhat more airway-selective compounds such as budesonide [11, 41] have been introduced, further increasing the interest in topical airway treatment. The trend has also been to increase the local dose when required. Taking both efficacy and safety into account, topical glucocorticoids must be considered first hand alternatives for the treatment of asthma and allergic rhinitis.

Efficacy in Treating Allergic Manifestations – Use of the Nose in Mucosal Allergy Research

Historically, observations of the nasal mucosa have also led to significant progress in the understanding of asthma, its pathology and treatment [47]. The human upper airways have an accessibility to in vivo repeated samplings of cells, secretions and measurements of various vascular parameters such as blood flow and plasma leakage, not obtainable in the lower airways [22, 40, 50, 61]. Nasal allergen and other challenge experiments can be easily done, specifically, and safely in humans, and so also can natural exposure to allergen be studied in hay fever patients [3, 53]. With a novel device it is also possible to control the exact concentration of allergen, mediators and drugs in contact with a defined mucosal surface area [22]. Using the upper airway approach, several aspects of putative glucocorticoid effects have been studied and the results may also be relevant for bronchial actions [46].

The inhibitory effect of glucocorticoids on the immediate response to allergen could, in theory, be an inhibition on several of the steps leading to mediator release and subsequent symptoms. Reduced release of mediators was suggested from in vitro studies of human nasal tissue treated in vivo for 1 week with topical glucocorticoids. Subsequent in vivo data on histamine release in nasal secretions following allergen challenge [51] have confirmed such an action. One week of topical glucocorticoid treatment induced a 40–50% reduction of allergen-induced histamine release. This result could not be explained by a reduced number of mediator cells in the tissue, a feature which has been shown in the skin after prolonged topical treatment [53]. Interestingly, it has been shown that the density of mast cells close to the epithelial surface is closely related to the reactivity of the airways in symptomatic disease [17, 53].

There is now considerable interest in the inhibitory effects of glucocorticoids on eosinophil activity and migration. Already in the 1930s Kallós maintained that eosinophils might play a major role in asthma. He based his view in part on the abundance of airway eosinophils seen both in his 'asthmatic' guinea pigs and in real asthma [25–27].

Using a lavage approach for repeated harvesting of cells in the nose, it has been demonstrated that glucocorticoids reduce the accumulation of cells, specifically the eosinophils in allergic rhinitis. Systemic glucocorticoid treatment prior to an allergen challenge reduced influx of eosinophils but not neutrophils [6]. Topically applied glucocorticoids blocked the influx of all inflammatory cells [5]. Since the eosinophils have been implicated in the genesis of the late occurring symptoms, as well as in the altered reactivity following the allergen challenge [20], the prevention of the accumulation of eosinophils to the mucosa would hypothetically be a key pharmacological effect. However, IgE-driven late broncho-constriction and plasma exudation after allergen challenge in guinea pigs were inhibited by budesonide without a reduction of the influx of eosinophils [4].

Although there appear to be several animal studies in support of a link between change in reactivity and a local release of eosinophil products [20], human nasal data are not conclusive on this point [3, 6]. Also, human in vivo data on postallergen challenge show a lack of influence of glucocorticoids on the formation of LTB4 [19], a potent chemotactic factor appearing as part of the allergic response.

Further work, where the versatile experimental possibilities of human nasal studies are fully used, will have to reveal further details of mucosal inflammatory processes in allergy, including any involvement of eosinophils. In this work glucocorticoids are important tools.

Effects of Glucocorticoids on Airway Endothelial-Epithelial Barriers

Previous reviews on glucocorticoid actions in airway diseases deal extensively with effects on inflammatory cells. The anti-inflammatory effect of glucocorticoids at the vascular level is at best mentioned but is frequently omitted, or misunderstood, in that it is related to effects on vascular muscle tone. A major sign of inflammation in the airways is exudation of plasma across the venular and mucosal walls [16, 42]. This passage of macromolecules and fluid is an active process, induced by direct effects of mediators and other inflammatory factors on endothelial cells of postcapillary venules. Capillary-venular microvessels form extensive networks just beneath the epithelial lining, probably both in the bronchi [31] and in the nose.

It has only recently been realized that plasma exudation is capable of producing several of the physical (edema, epithelial shedding, mucus thickening and plugging) and physiological (inflammatory, bronchoconstrictor and priming effects of plasma-derived mediators) abnormalities that characterize the asthmatic airway wall and lumen [42, 43]. It was accordingly suggested that the reduction of plasma leakage may be a major anti-asthma effect of glucocorticoids [4, 42, 43].

Already in 1940 Menkin found that extract of adrenal cortex inhibited inflammatory vascular leakage [36]. Leme and Wilhelm [34] demonstrated in rats that corticosterone prevented mediator-induced increases in venular permeability and the enhanced venular responsiveness brought about by adrenalectomy. In guinea pig airways, budesonide reduced leakage of plasma across both endothelial and epithelial barriers [15]. In the late phase reaction to allergen in guinea pigs and in man, there is leakage of plasma which is reduced by glucocorticoids [4].

Ryley and Brogan [56] first observed a relationship between a lowering of the albumin concentration in sputa with steroid therapy together with clinical improvement in 1 asthmatic subject. Moretti et al. [37] studied patients with both reversible airway obstruction and bronchitis and showed that 2 weeks' treatment with methylprednisolone brought about a dramatic reduction in the sputum concentration of plasma proteins but no change in secretory tracers. After a few days of glucocorticoid treatment, a significant reduction was recorded in the ratio of sol-phase sputum concentration to serum concentration of albumin in patients with chronic obstructive bronchitis [68]. An anti-exudative effect would also reduce the entry of plasma proteins with protective functions in the airway, but the overall effect of glucocorticoids on the airway liquid proteinase-antiprotease balance has, in fact, been demonstrated to be beneficial [38].

Further data showing glucocorticoid-induced inhibition of plasma exudation come from studies of the human nasal mucosa [51, 52]. Sörensen et al. [63] treated patients with nasal polyps with a topical glucocorticoid for 1 year. Their main finding was that plasma proteins (albumin, IgG and IgE) were initially high but were reduced to almost normal (low) concentrations in the mucosal surface liquid by the steroid. The tissue and surface eosinophilia in these patients was not reduced. A secretory protein, IgA, was not initially elevated and was not reduced by the treatment.

It has obviously been difficult to demonstrate a significant effect of glucocorticoids on either the actual secretory processes in the airways [29, 35] or on the number of mucus cells [28]. This apparently contrasts the significant inhibitory effects of these drugs on airway 'hypersecretion'. Glucocorticoids reduce that component of 'hypersecretion' which is not a secretory product but a plasma exudate/transudate supporting the view that the latter is of greater significance in the pathogenesis of asthma and rhinitis [48].

Lessons Learned about Diseases from Glucocorticoids

Glucocorticoids have taught us a great deal about asthma. The fact that even by prolonged treatment they do not seem to alter airway contractility [23], has lessened the focus on bronchial smooth muscle. The general and unmatched efficacy of glucocorticoids, also demonstrated in very mild cases and in patients believed to be well controlled with other drugs [11], strongly suggests that an inflammatory process has a general and primary pathogenetic role in asthma.

The present state of knowledge of the complex anti-inflammatory pharmacology of glucocorticoids does not give support to current single cell or single mediator hypotheses of asthma. The importance of inflammatory processes in chronic bronchitis, in particular for the annual decline in lung function, may also be revealed by employment of long-term glucocorticoid treatment in these patients.

With glucocorticoids there is concern that the anti-inflammatory actions would increase the susceptibility to and severity of infections in the airways. This appears logical, since inflammation in general can be defined as exaggerated defense reactions and undue suppression of these should not be desirable. However, inhaled glucocorticoids, used chronically by asthmatic or rhinitic subjects, have not been associated with an increased incidence of airway infections [2, 49, 65, 66]. Indeed, there are data suggesting that glucocorticoid-induced normalization of the mucosa in inflammatory airway disease is an important therapeutic action also when the airways appear infected. Thus, Sykes et al. [62], in a study of 50 patients with chronic mucopurulent rhinosinusitis (normally considered a contraindication for glucocorticoids), demonstrated that topical glucocorticoid and decongestant treatment was effective and that nothing further was gained by adding antibiotics to this regimen. There may be a development in the future where anti-inflammatory glucocorticoid treatment may be increasingly instituted in select diseases which traditionally have only been considered for antibiotic therapy. Recent successful attempts along this line include diseases such as sinusitis [62], chronic bronchitis [55] and bacterial meningitis [33].

References

1 Addison T: On the constitutional and local effects of disease of the suprarenal capsules. London, Samuel Highley, 1855.
2 Agnew RAL, Walker DJ, Phillips LA: A prospective study of respiratory infection in asthmatic patients treated with beclomethasone dipropionate and sodium cromoglycate. Clin Allergy 1977;7:183–188.
3 Andersson M, Andersson P, Venge P, et al: Eosinophils and eosinophil cationic protein (ECP) in nasal lavages during allergen-induced hyperresponsiveness. Effect of topical glucocorticosteroid treatment. Allergy 1989;44:342–348.
4 Andersson PT, Persson CGA: Developments in antiasthma glucocorticoids; in O'Donnell SR, Persson CGA (eds): Directions for New Antiasthma Drugs. Basel, Birkhäuser, 1988, pp 239–260.
5 Bascom R, Wachs M, Naclerio RM, et al: Basophil influx occurs after nasal antigen challenge. Effects of topical corticosteroid pretreatment. J Allergy Clin Immunol 1988;81:580–589.
6 Bascom R, Pipkorn U, Lichtenstein LM, et al: The influx of inflammatory cells into nasal washings during the late response to antigen challenge. Effect of systemic corticosteroids. Am Rev Respir Dis 1989;128:406–412.

7 Baxter JD, Rosseau GG: Glucocorticoid hormone action: an overview; in Baxter JD, Rosseau GG (eds): Glucocorticoid Hormone Action. New York, Springer, 1979.
8 Bertin B, Bure J, Junien JL, et al: New insights into dexamethasone activity in lymphoblast transformation of mouse thymocytes. Agents Actions 1986;19:346–348.
9 Blackwell G, Carnuccio R, Di Rosa M, et al: Macrocortin, a polypeptide causing the antiphospholipase effect of glucocorticoids. Nature 1980;287:147–149.
10 Bordley JE, Carey RA, Harvey AM, et al: Preliminary observations on the effect of adrenocorticotropic hormone (ACTH) in allergic disease. Bull Johns Hopkins Hosp 1949;85:396–398.
11 Hargreave FE, Hogg JC, Malo JL, Toogood JH (eds): Glucocorticoids and mechanisms of asthma. Amsterdam, Excerpta Medica 1989.
12 Carryer HM, Koelsche GA, Prickman LE, et al: Effects of cortisone on bronchial asthma and hay fever occurring in subjects sensitive to ragweed pollen. Proc Mayo Clin 1950;25:482–486.
13 Diamant M, Kallós P: Intrabronchiale Cortisonbehandlung von schweren Asthmafällen. Int Arch Allergy Appl Immunol 1954;5:283–288.
14 Duke WW: Details in the treatment of hay-fever, asthma and other manifestations of allergy. Am J Med Sci 1923;166:645–663.
15 Erjefält I, Persson CGA: Antiasthma drugs attenuate inflammatory leakage of plasma into airway lumen. Acta Physiol Scand 1986;128:653–654.
16 Erjefält I, Persson CGA: Inflammatory passage of plasma macromolecules into airway wall and lumen. Pulm Pharmacol 1989;2:93–102.
17 Flint KC, Leung KBP, Hudspith BN, et al: Bronchoalveolar mast cells in extrinsic asthma: a mechanism for the initiation of antigen specific bronchoconstriction. Br Med J 1985;291:923–926.
18 Flower RJ, Parente L, Persico P, et al: A comparison of the acute inflammatory response in adrenalectomised and shamoperated rats. Br J Pharmacol 1986; 87:57–62.
19 Freeland H, Pipkorn U, Proud D, et al: Leukotriene B4 as a mediator of early and late reactions to antigen in humans. The effect of systemic glucocorticoid treatment in vivo. J Allergy Clin Immunol 1989;83:634–642.
20 Frigas E, Gleich GJ: The eosinophil and the pathophysiology of asthma. J Allergy Clin Immunol 1986;77:527–537.
21 Goldman L, Preston R, Rockwell E: The local effects of 17-hydrocorticosterone-21-acetate (compound F) on the diagnostic patch test reaction. J Invest Dermatol 1952;19:89–93.
22 Greiff L, Pipkorn U, Alkner U, et al: The 'nasal pool-device' applies controlled concentrations of solutes on human nasal airway mucosa and samples its surface exudations/secretions. Clin Exp Allergy 1989; submitted.
23 Gustafsson B, Persson CGA: Effect of three weeks' treatment with budesonide on in vitro contractile and relaxant airway effects in the rat. Thorax 1989;44:24–27.
24 Hench PS, Kendall EC, Slocumb CH, et al: The effect of a hormone of the adrenal cortex, cortisone (17-hydroxy-11-dehydrocorticosterone: compound E) and of pituitary adrenocorticotropic hormone on rheumatoid arthritis and acute rheumatic fever: preliminary report. T A Am Physicians 1949;62:64–80.
25 Kallós P (ed): Introduction. Prog Allergy. Basel, Karger 1952, vol III, pp 1–20.

26 Kallós P, Pagel W: Experimentelle Untersuchungen über Asthma bronchiale. Acta Med Scand 1937;91:292–305.
27 Kallós P, Kallós-Deffner L: Die experimentellen Grundlagen der Allergielehre. Prog Allergy. Basel, Karger 1939, vol I, pp 5–18.
28 Karlsson G, Pipkorn U: Natural allergen exposure does not influence the density of goblet cells in the nasal mucosal of patients with seasonal allergic rhinitis. ORL 1989;51:171–174.
29 Keal EE: Physiological and pharmacological control of airways secretions; in Brain J, Proctor DF, Reid LM (eds): Respiratory Defense Mechanisms. Dekker, New York, 1977, pp 357–402.
30 Kendall EC: Hormones of the adrenal cortex. Endocrinology 1942;30:853–860.
31 Laitinen LA, Laitinen A, Widdicombe: Effects of inflammatory and other mediators on airway vascular beds. Am Rev Respir Dis 1987;135:S67–S70.
32 Laue L, Kawai S, Brandon DD, et al: Receptor-mediated effects of glucocorticoids on inflammation: enhancement of the inflammatory response with a glucocorticoid antagonist. J Steroid Biochem 1988;29:591–598.
33 Lebel MH, Freij BJ, Syrogiannopoulos GA, et al: Dexamethasone therapy for bacterial meningitis. Results of two double-blind, placebo-controlled trials. N Engl J Med 1988;15:964–971.
34 Leme GJ, Wilhelm DL: The effects of adrenalectomy and corticosterone on vascular permeability responses in the skin of the rat. Br J Exp Pathol 1975;56:402–407.
35 Malm L, Wihl J-Å, Lamm CJ, et al: Reduction of metacholine induced nasal secretions by the treatment with a new topical steroid in perennial non-allergic rhinitis. Allergy 1982;36:209–214.
36 Menkin V: Effect of adrenal cortex extract on capillary permeability. Am J Physiol 1981;129:1592–1595.
37 Moretti M, Giannico G, Marchioni CF, et al: Effects of methylprednisolone on sputum biochemical components in asthmatic bronchitis. Eur J Respir Dis 1984;65: 365–370.
38 Morrison HM, Afford SC, Stockley RA: Inhibitory capacity of $alpha_1$ antitrypsin in lung secretions: variability and the effect of drugs. Thorax 1984;39:510–516.
39 Munch A, Guyre PM, Holbrook NJ: Physiological functions of glucocorticoids in stress and their relation to pharmacological actions. Endocr Rev 1984;5:25–44.
40 Naclerio RM, Meier HL, Kagey-Sobotka A, et al: Mediator release after nasal airway challenge with allergen. Am Rev Respir Dis 1983;128:597–602.
41 Pedersen S, Fuglsang G: Urine cortisol excretion in children treated with high doses of inhaled corticosteroids. A comparison of budesonide and beclomethasone. Eur Respir J 1988;1:433–435.
42 Persson CGA: Role of plasma exudate in asthmatic airways. Lancet 1986;ii:1126–1129.
43 Persson CGA: Plasma exudation and asthma. Lung 1988;166:1–23.
44 Persson CGA: Glucocorticoids for asthma. Early contributions. Pulm Pharmacol 1989;2:163–166.
45 Persson CGA: Antiinflammatory therapy with glucocorticoids in intrinsic asthma. Agent Actions Suppl 1989; in press.
46 Persson CGA, Pipkorn U: Pathogenesis and pharmacology of asthma and rhinitis; in Mygind N, Dahl R, Pipkorn U (eds): Rhinitis and asthma: similarities and differences. Copenhagen, Muncksgaard 1989; in press.

47 Persson CGA: On the medical history of asthma and rhinitis; in Mygind N, Dahl R, Pipkorn U (eds): Rhinitis and asthma: similarities and differences. Copenhagen, Muncksgaard, 1989; in press.
48 Persson CGA: Permeability changes in obstructive airway diseases; in Sluiter HJ, Van Der Lende R (eds): Bronchitis IV. Assen, Van Gorcum, 1989, pp 236–28.
49 Pipkorn U, Pukander J, Suonpää J, et al: Long-term safety of budesonide nasal aerosol 5 1/5 year follow up study. Clin Allergy 1988;18:253–259.
50 Pipkorn U, Karlsson G: Methods for obtaining specimens from the nasal mucosa for morphological and biochemical analysis. Eur Respir J 1988;1:856–862.
51 Pipkorn U, Proud D, Schleimer RP, et al: Effects of short term systemic glucocorticoid treatment on human nasal mediator release after antigen challenge. J Clin Invest 1987;80:957–961.
52 Pipkorn U, Proud D, Lichtenstein LM, et al: Inhibition of mediator release in allergic rhinitis by pretreatment with topical glucocorticosteroids. N Engl J Med 1988;316:1506–1510.
53 Pipkorn U, Hammarlund A, Enerbäck L: Prolonged treatment with topical glucocorticoids results in an inhibition of the allergen-induced weal and flare response and a reduction in skin mast cell numbers and histamine content. Clin Exp Allergy 1989;19:19–25.
54 Poothullil J, Umemoto L, Dolovich J, et al: Inhibition by prednisone of late cutaneous responses induced by antiserum to human IgE. J Allergy Clin Immunol 1976;57:164–167.
55 Postma DS, Peters I, Steenhuis EJ, et al: Severe chronic airflow obstruction, can corticosteroids slow down progression? Eur Respir J 1988;1:22–26.
56 Ryley HC, Brogan TD: Variation in the composition of sputum in chronic chest diseases. Br J Exp Pathol 1968;49:625–633.
57 Solis-Cohen S: The use of adrenal substance in the treatment of asthma. JAMA 1900;34:1164–1166.
58 Sowell JG, Kagen AA, Troop RC: Metabolism of cortisone-4-^{14}C by rat lung tissue. Steroids 1971;18:289–301.
59 Stefanovic DS: Corticosteroid induced atrophy of the skin with teleangiectasia. Br J Dermatol 1972;87:548–553.
60 Svensjö E, Roempke K: Time-dependent inhibition of bradykinin and histamine-induced increase in microvascular permeability by local glucocorticosteroid treatment; in Hogg JC, Ellul-Micallef R, Brattsand R (eds): Glucocorticoids, Inflammation and Bronchial Hyperreactivity. Amsterdam, Excerpta Medica 1985, pp 136–144.
61 Svensson C, Pipkorn U, Baumgarten CR, et al: Reversibility and reproducibility of histamine-induced plasma leakage of the human nasal airways. Thorax 1989;44: 13–18.
62 Sykes DA, Wilson R, Chan KL, et al: Relative importance of antibiotic and improved clearance in topical treatment of chronic mucopurulent rhinosinusitis. Lancet 1986;ii:359–360.
63 Sörensen H, Mygind N, Pedersen CB, et al: Long term treatment of nasal polyps with beclomethasone dipropionate aerosol. III. Morphological studies and conclusions. Acta Otolaryngol 1976;82:260–262.
64 Talbot S, Atkins PC, Zweiman B: In vivo effects of corticosteroids on human allergic responses. I. Effects of systemic administration of steroids. Ann Allergy 1987;58: 363–365.

65 Tarlo S, Broder I, Spence L: A prospective study of respiratory infection in adult asthmatics and their normal spouses. Clin Allergy 1979;9:293–301.
66 Toogood JH: Corticosteroids; in Jenne JW, Murphy S (eds): Drug Therapy for Asthma. New York, Dekker, 1987, vol 31, pp 463–516.
67 Thorn GW, Forsham PH, Frawley TF, et al: Clinical usefulness of ACTH and cortisone. N Engl J Med 1950;242:783–793.
68 Wiggins J, Elliot JA, Stevenson RD, et al: Effect of corticosteroids on sputum sol-phase protease inhibitors in chronic obstructive pulmonary disease. Thorax 1982;37:652–656.

Carl Persson, MD, AB Draco, Box 34, S–221 00 Lund (Sweden)

Subject Index